Applied Mathematics for Engineers

Applied Mathematics for Engineers

Edited by **Gregory Rago**

Willford Press

New York

Published by Willford Press,
118-35 Queens Blvd., Suite 400,
Forest Hills, NY 11375, USA
www.willfordpress.com

Applied Mathematics for Engineers
Edited by Gregory Rago

International Standard Book Number: 978-1-68285-205-7 (Hardback)

Printed in China.

Contents

Preface

Over the recent decade, advancements and applications have progressed exponentially. This has led to the increased interest in this field and projects are being conducted to enhance knowledge. The main objective of this book is to present some of the critical challenges and provide insights into possible solutions. This book will answer the varied questions that arise in the field and also provide an increased scope for furthering studies.

Application of mathematical principles can help solve a variety of problems in various industries, especially those concerned with cost and economics. This book provides comprehensive insights into the field of applied mathematics by discussing a host of different topics of high significance, such as numerical analysis, computational algebra, statistics, etc. It aims to bring forth case studies of problem solving and mathematical developments to solve the challenges faced by modern industry. It will provide in-depth knowledge to the readers and will serve as a valuable source of reference for graduate and post graduate students.

I hope that this book, with its visionary approach, will be a valuable addition and will promote interest among readers. Each of the authors has provided their extraordinary competence in their specific fields by providing different perspectives as they come from diverse nations and regions. I thank them for their contributions.

Editor

Numerical simulation of thin paint film flow

Bruno Figliuzzi[1*], Dominique Jeulin[1], Anaël Lemaître[2], Gabriel Fricout[3], Jean-Jacques Piezanowski[3] and Paul Manneville[4]

*Correspondence: figliuzzi.bruno@gmail.com
[1]Centre de Morphologie Mathématique, École des Mines ParisTech, 77300, Fontainebleau, France
Full list of author information is available at the end of the article

Abstract

Purpose: Being able to predict the visual appearance of a painted steel sheet, given its topography before paint application, is of crucial importance for car makers. Accurate modeling of the industrial painting process is required.

Results: The equations describing the leveling of the paint film are complex and their numerical simulation requires advanced mathematical tools, which are described in detail in this paper. Simulations are validated using a large experimental data base obtained with a wavefront sensor developed by Phasics™.

Conclusions: The conducted simulations are complex and require the development of advanced numerical tools, like those presented in this paper.

Keywords: thin films; numerical simulation; industrial painting process; roughness; lubrication approximation.

1 Introduction

The visual appearance of painted steel sheets forming the body of a car is a prominent factor in appreciating its quality. Being able to predict it is thus of crucial importance to car makers, while remaining a serious mathematical challenge requiring accurate modeling of the industrial painting process.

The deposition of the successive coating layers on a car body involves complex physical and chemical processes, with many variants. Here, we consider the sheet in its initial surface state (galvanized and phosphated) and summarize the painting process as follows: once assembled, the car body is immersed in an electrophoresis bath, where a layer of corrosion-protecting paint is deposited. The vehicle body is then baked in an oven. A second paint layer, the *sealer*, is applied and the vehicle baked again. Finally, a layer of *lacquer* is applied before a last baking. The steel sheet is thus covered with three layers of coating as shown in Figure 1.

The last two paint coatings are mainly designed to provide an aesthetically pleasing appearance to the car. During the painting process, the final topography of each layer results from two main processes:

- the leveling of the film (flow and evaporation) which occurs during the flash time, i.e. the time period just following the end of the deposit,
- baking in an oven, which favors evaporation.

The leveling process has received considerable attention in the literature, although not in the context of the industrial paints used in the automotive industry. In 1961, Orchard [1] was the first to note that the leveling dynamics is controlled by an interplay between

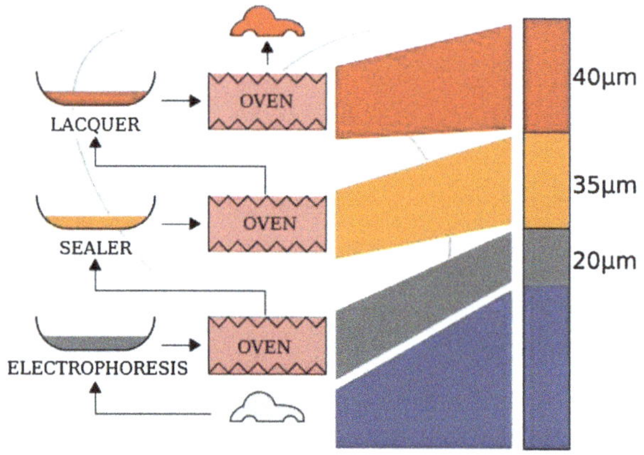

Figure 1 **Successive paint layers on a steel sheet, with approximate thicknesses as indicated.**

surface tension, with capillary forces tending to reduce surface irregularities, and the fluid viscosity limiting the flow induced by that leveling. Orchard's model is mainly based on two assumptions: the paint exhibits a Newtonian behavior and evaporation effects are negligible. To take into account the effects of evaporation, Overdiep [2] considered a fluid made of a resin and a solvent, where only the solvent can evaporate, demonstrating the potential importance of the surface tension spatial variations. Surface tension indeed depends on the paint composition, in particular on the respective proportions of resin and solvent. In the presence of evaporation, thinner regions tend to dry faster, and therefore to have lower solvent concentrations, which causes surface tension gradients, a physical phenomenon known as Marangoni effect, hence a shearing effect at the film surface, understood as the main physical effect involved in the leveling of the paint film by Overdiep. This approach was taken up and developed in several subsequent articles. Wilson [3] and later Howison *et al.* [4] analyzed and generalized Overdiep's model, performing numerical simulations that showed good agreement with experimental data collected for simple deposit geometries.

The topography of the substrate on which the coating is deposited plays an important role in the flow dynamics. In 1995, Weidner *et al.* [5] studied the effect of substrate curvature on the film flow in a two-dimensional context. Subsequently Eres *et al.* [6] and later Schwartz *et al.* [7] generalized the work to the three-dimensional case. In these papers, numerical models have been implemented for specific topographies, showing good agreement with experimental measurements. Gaskell *et al.* [8, 9] finally considered the generalization of the different models to the case of inclined substrates, where gravity plays a significant physical role in the flow dynamics.

Industrial paints used in the context of the automotive industry are complex media that have not been extensively studied. Their detailed rheology is not well known, though its effects on the leveling are a key issue. In view of the complexity of the phenomena, experiments aiming at the identification of the physical effects within the film and the evaluation of their relative importance appear to be a prerequisite to film flow modeling. Using a wavefront sensor developed by Phasics™ [10], we could determine the evolution of rough surfaces accurately and with a high temporal resolution throughout the whole painting process [11]. In Section 2, we describe the mathematical model used to model the evolution of the painted film topography and its numerical simulation. Section 3 is devoted to

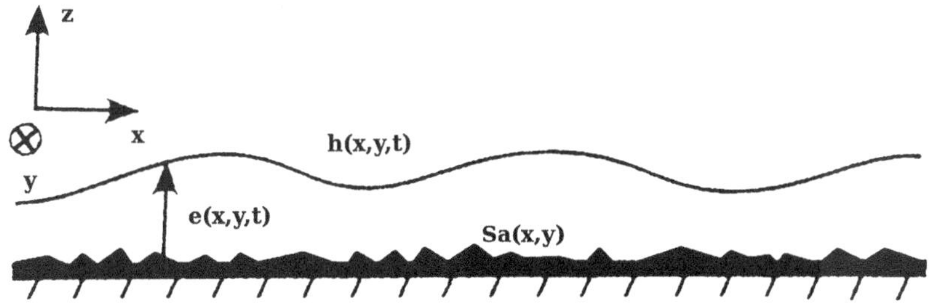

Figure 2 Thin paint film flow.

the presentation of the experimental data obtained with the wavefront sensor. Rheological parameters extracted from the experimental data are used in Section 4 to perform a simulation of the topography evolution during the painting process. Conclusions are drawn in Section 5.

2 The mathematical model and its implementation

Following the accepted practice, we study the leveling process within the framework of a lubrication approximation, but more elaborate theories can be developed from the Navier-Stokes equations [12–17]. The lubrication approximation builds on two observations: firstly, the thin film flow is very slow, so that it becomes possible to neglect the inertia terms in the Navier-Stokes equation; secondly, the thickness of the film is much smaller than the wavelength of the modulations along the surface, which also implies that the fluid velocity is essentially directed parallel the surface. All this allows a substantial simplification of the equations describing the flow of the thin paint film.

2.1 Physical model

Here, we consider the leveling of a thin incompressible film deposited on an horizontal steel sheet, as represented on Figure 2. The topography of the bare sheet is denoted as $S_a(x,y)$, the film thickness as $e(x,y,t)$, and the height of film free surface as $h(x,y,t)$. The paint film is deposited at $t = 0$, and evolves until solidification due to polymer curing, which happens at t_{ret} during the baking. The final film height is then $S_a(x) + e(x, t_{ret})$. The film thickness at the beginning of the leveling is approximately $H = 70\ \mu$m. A typical value of the paint velocity is $U = 10\ \mu$m/s. The Reynolds number $Re = \rho UH/\eta$ is approximately $Re \cong 7.8 \times 10^{-7} \ll 1$. It is also of interest to compute the Ohnesorge number of the film flow, which relates the viscous forces to inertial and surface tension forces:

$$Oh = \frac{\eta}{\rho\gamma L},$$

where L denotes a characteristic length in the horizontal direction. With $\eta = 0.9$ Pa.s, $\rho = 1000$ kg/m^3, $\gamma = 2.71 \times 10^{-2}$ N/m and $L = 150\ \mu$m, we find $Oh \cong 221$, which indicates a preponderant influence of the viscosity in the leveling phenomenon.

2.1.1 Lubrication approximation

Without making any assumption about the paint rheology, neglecting gravity, the mechanical equilibrium equation reads

$$-\nabla p + \nabla \cdot \bar{\bar{\sigma}} = 0, \tag{1}$$

where $\bar{\bar{\sigma}}$ denotes the deviator stress tensor and p the local pressure within the film. Letting u and v be the velocity components along x and y, the z-component being neglected in the lubrication approximation, the strain rate tensor reads:

$$\frac{1}{2}(\nabla \mathbf{u} + {}^{T}\nabla \mathbf{u}) = \begin{pmatrix} \frac{\partial u}{\partial x} & \frac{1}{2}\left(\frac{\partial u}{\partial y} + \frac{\partial v}{\partial x}\right) & \frac{1}{2}\frac{\partial u}{\partial z} \\ \frac{1}{2}\left(\frac{\partial u}{\partial y} + \frac{\partial v}{\partial x}\right) & \frac{\partial v}{\partial y} & \frac{1}{2}\frac{\partial v}{\partial z} \\ \frac{1}{2}\frac{\partial u}{\partial z} & \frac{1}{2}\frac{\partial v}{\partial z} & 0 \end{pmatrix}. \tag{2}$$

Within the lubrication approximation, the gradients of u and v along x and y can be neglected. The strain rate tensor is then reduced to:

$$\frac{1}{2}(\nabla \mathbf{u} + {}^{T}\nabla \mathbf{u}) = \begin{pmatrix} 0 & 0 & \frac{1}{2}\frac{\partial u}{\partial z} \\ 0 & 0 & \frac{1}{2}\frac{\partial v}{\partial z} \\ \frac{1}{2}\frac{\partial u}{\partial z} & \frac{1}{2}\frac{\partial v}{\partial z} & 0 \end{pmatrix}. \tag{3}$$

One can expect the deviator stress tensor to be parallel to the strain rate tensor. Tensor $\bar{\bar{\sigma}}$ then reads

$$\bar{\bar{\sigma}} = \begin{pmatrix} 0 & 0 & \sigma_{xz} \\ 0 & 0 & \sigma_{yz} \\ \sigma_{xz} & \sigma_{yz} & 0 \end{pmatrix} \tag{4}$$

so that Equation 1 becomes:

$$\begin{cases} -\dfrac{\partial p}{\partial x} + \dfrac{\partial \sigma_{xz}}{\partial z} = 0, \\ -\dfrac{\partial p}{\partial y} + \dfrac{\partial \sigma_{yz}}{\partial z} = 0, \\ -\dfrac{\partial p}{\partial z} = 0. \end{cases} \tag{5}$$

Boundary conditions are given by a no slip kinematic condition at the substrate surface, $u(z = 0) = v(z = 0) = 0$, and by a mechanical condition expressing that the constraint is zero at the free surface, $\sigma_{xz}(z = h) = \sigma_{yz}(z = h) = 0$. In what follows, we will set the origin of the altitudes at the mean substrate level. Equation 5 can consequently be integrated to yield:

$$\begin{cases} \sigma_{xz} = -\dfrac{\partial p}{\partial x}(h - z), \\ \sigma_{yz} = -\dfrac{\partial p}{\partial y}(h - z). \end{cases} \tag{6}$$

The pressure is given as the product of the surface tension and the free surface curvature C which at lowest order reads:

$$C = -\frac{\partial^2 h}{\partial x^2} - \frac{\partial^2 h}{\partial y^2}, \tag{7}$$

where $h(x, y, t) = e(x, y, t) + S_a(x, y)$ is the altitude of the fluid surface. Finally, the local altitude is linked to the evaporation rate E and the local flow rate $\mathbf{q}$ by the mass conservation equation

$$\frac{\partial h}{\partial t}(x,y,t) = -\nabla_{\mathrm{h}} \cdot \mathbf{q}(x,y,t) - E(x,y,t), \tag{8}$$

where ∇_{h} is the gradient along the plane (x, y).

2.1.2 Paint rheology

Equations 6-8 have been derived without making any assumptions about the paint rheology. To close these equations, we have to prescribe how the mass flux $\mathbf{q}$ depends on the local pressure gradient.

In [11], $\mathbf{q}$ was computed from the data obtained with the wavefront sensor by solving the following problem:

$$-\nabla_{\mathrm{h}}\left(\frac{\partial h}{\partial t}(x,y,t) + E(x,y,t)\right) = \Delta_{\mathrm{h}}\mathbf{q}(x,y,t), \tag{9}$$

which comes after noting that within the lubrication approximation:

$$\mathbf{curl}\,\mathbf{curl}(\mathbf{q}) = \mathbf{0}. \tag{10}$$

Estimating the left hand side of Equation 9 indeed allows the access to the local values of the mass flux by solving the Poisson equation, and hence permits us to test the rheological model. The so-obtained data showed that for the space and time scales involved in the problem, the film can be considered as Newtonian.

Assuming a Newtonian rheology, the deviator stress tensor can then easily be expressed as a function of the strain rate tensor:

$$\bar{\bar{\sigma}} = \frac{\eta}{2}(\nabla\mathbf{u} + {}^{T}\nabla\mathbf{u}), \tag{11}$$

so that Equation 5 can be rewritten as:

$$\begin{cases} -\dfrac{\partial p}{\partial x}(x,y,t) + \eta\dfrac{\partial^2 u}{\partial z^2}(x,y,t) = 0, \\ -\dfrac{\partial p}{\partial y}(x,y,t) + \eta\dfrac{\partial^2 v}{\partial z^2}(x,y,t) = 0. \end{cases} \tag{12}$$

2.1.3 Newtonian model equation

Since the pressure p is independent of z, Equations 12 can easily be integrated. Boundary conditions were indeed given by a no slip kinematic condition at the substrate surface, $u(z = S_{\mathrm{a}}) = 0$, $v(z = S_{\mathrm{a}}) = 0$, and by a mechanical condition expressing that the constraint is zero at the free surface, $\partial u/\partial z(z = h) = 0$, $\partial v/\partial z(z = h) = 0$.

$$\begin{cases} u(x,y,z,t) = \dfrac{1}{\eta}\dfrac{\partial p}{\partial x}(x,y,t)\left(\dfrac{1}{2}z^2 - h(z - S_{\mathrm{a}}) - \dfrac{1}{2}S_{\mathrm{a}}^2\right), \\ v(x,y,z,t) = \dfrac{1}{\eta}\dfrac{\partial p}{\partial y}(x,y,t)\left(\dfrac{1}{2}z^2 - h(z - S_{\mathrm{a}}) - \dfrac{1}{2}S_{\mathrm{a}}^2\right). \end{cases} \tag{13}$$

Consequently, the local flow components on the film thickness along the horizontal directions read

$$\begin{cases} q_x = \int_{S_a}^{h} u(x,y,z,t)\,dz = \dfrac{\gamma}{3\eta}(h - S_a)^3\left(\dfrac{\partial^3 h}{\partial x^3} + \dfrac{\partial^3 h}{\partial x\,\partial y^2}\right), \\ q_y = \int_{S_a}^{h} v(x,y,z,t)\,dz = \dfrac{\gamma}{3\eta}(h - S_a)^3\left(\dfrac{\partial^3 h}{\partial y^3} + \dfrac{\partial^3 h}{\partial y\,\partial x^2}\right). \end{cases} \tag{14}$$

Using the mass conservation Equation 8, the complete model equation is

$$\begin{aligned} \frac{\partial h}{\partial t} = &-\frac{\gamma}{3\eta}\frac{\partial}{\partial x}\left[(h - S_a)^3\left(\frac{\partial^3 h}{\partial x^3} + \frac{\partial^3 h}{\partial x\,\partial y^2}\right)\right] \\ &-\frac{\gamma}{3\eta}\frac{\partial}{\partial y}\left[(h - S_a)^3\left(\frac{\partial^3 h}{\partial y^3} + \frac{\partial^3 h}{\partial y\,\partial x^2}\right)\right] - E. \end{aligned} \tag{15}$$

We will assume that the paint is composed of a resin in concentration $1 - c$ and a solvent in concentration c. Only the solvent can evaporate, while the evaporation rate will essentially depend on the solvent concentration. Accordingly, we shall assume that the largest scales patterns attenuation is mainly caused by evaporation, for a leveling caused by surface tension would suppose a huge mass transport which would be unrealistic considering the geometric characteristics of the painted film. A method based on this idea is presented in [11], which allows a determination of the evaporation rate as a function of c. If we neglect the local variations of the solvent concentration, the evaporation rate will consequently be spatially constant, and will only vary with time.

2.1.4 Marangoni effect

The local variations in the solvent concentration may generate a surface tension gradient. This surface tension gradient modifies the mechanical equilibrium conditions on the free film surface which become

$$\eta\frac{\partial u}{\partial z} = \frac{\partial \gamma}{\partial x}, \qquad \eta\frac{\partial v}{\partial z} = \frac{\partial \gamma}{\partial y}. \tag{16}$$

Expressions 14 become then

$$\begin{cases} q_x = \dfrac{1}{3\eta}(h - S_a)^3\dfrac{\partial p}{\partial x} + \dfrac{1}{2\eta}\dfrac{\partial \gamma}{\partial x}(h - S_a)^2, \\ q_y = \dfrac{1}{3\eta}(h - S_a)^3\dfrac{\partial p}{\partial y} + \dfrac{1}{2\eta}\dfrac{\partial \gamma}{\partial y}(h - S_a)^2. \end{cases} \tag{17}$$

The Laplace pressure is given as a function of the surface derivatives by Equation 7. Using the mass conservation equation, one gets:

$$\frac{\partial h}{\partial t} + \frac{\partial q_x}{\partial x} + \frac{\partial q_y}{\partial y} + E = 0. \tag{18}$$

In Equation 18, as the concentration locally vary, the evaporation rate varies both in time and in space. The equation governing the concentration c is obtained by using the solvent mass conservation equation:

$$\frac{\partial(ce)}{\partial t} = -E - \frac{\partial(cq_x)}{\partial x} - \frac{\partial(cq_y)}{\partial y}, \tag{19}$$

hence using Equation 18:

$$\frac{\partial c}{\partial t} = -\left(\frac{1-c}{e}\right)E - \left(\frac{\partial c}{\partial x}\right)\frac{q_x}{e} - \left(\frac{\partial c}{\partial y}\right)\frac{q_y}{e}. \tag{20}$$

The combination of Equations 18 and 20 completely describes the evolution of the film topography. The physical parameter γ is related to the solvent concentration by the law presented later on Figure 7.

2.2 Numerical implementation

The leveling of the paint layer is described by high order non-linear partial differential equations. The numerical handling of these equations is therefore a delicate problem. The model Equations 15, 18 and 20 can be written in the form:

$$\frac{\partial \psi}{\partial t} = F\left(\psi, \frac{\partial \psi}{\partial x}, \frac{\partial \psi}{\partial y}, \ldots, \frac{\partial^n \psi}{\partial x^n}, \frac{\partial^n \psi}{\partial y^n}, \ldots\right), \tag{21}$$

where F is a non-linear function of the spatial derivatives. The method of lines [18] is used to solve Equation 21, in combination with a pseudo-spectral method: Function F is evaluated in the Fourier space and Equation 21 is integrated using an adaptative step size Runge-Kutta scheme.

2.2.1 Evaluation of spatial gradients

We assume that Equation 21 is submitted to periodic spatial boundary conditions. Using the Fourier transform helps us computing high-order space derivatives present in Equations 15, 18, and 20 in a simple way. However the Fourier transform of a product of functions in physical space is the convolution of the Fourier transforms of the functions. Numerically, care has to be taken when the Fourier transform of the product is calculated, since sampling implies aliasing. Let f and g be two functions which are sampled with a step equal to one. The Fourier series expansion of these functions are

$$f[n] = \sum_{k=-N/2}^{N/2} \hat{f}[k]e^{i\frac{2\pi}{N}kn}, \qquad g[n] = \sum_{k=-N/2}^{N/2} \hat{g}[k]e^{i\frac{2\pi}{N}kn}. \tag{22}$$

A consequence of the function sampling is that its Fourier transform is artificially periodized. Considering Equation 22, the Fourier series expansion of the product function fg is

$$\hat{fg}[k] = \sum_{n=0}^{N} fg[n]e^{-i\frac{2\pi}{N}kn} = \sum_{n=0}^{N} \sum_{k_1,k_2=-N/2}^{N/2} \hat{f}[k_1]\hat{g}[k_2]e^{(-i\frac{2\pi}{N}(k-k_1-k_2)n)}. \tag{23}$$

The quantity $\sum_{n=0}^{N} e^{(-i\frac{2\pi}{N}(k-k_1-k_2)n)}$ cancels for all values of k, k_1 and k_2, except when $k = k_1 + k_2 + mN$, with $m \in \mathbb{Z}$. Considering the values taken by k, k_1 and k_2, we verify that

$$\begin{aligned}\hat{fg}[k] = & \sum_{k_1=-N/2}^{N/2} \hat{f}[k_1]\hat{g}[k-k_1] + \sum_{k_1=-N/2}^{N/2} \hat{f}[k_1]\hat{g}[k+N-k_1] \\ & + \sum_{k_1=-N/2}^{N/2} \hat{f}[k_1]\hat{g}[k-N-k_1].\end{aligned} \tag{24}$$

The first term on the right hand side of Equation 24 corresponds to the convolution product of the Fourier transforms of f and g. The two other terms arise from aliasing and have to be removed. To do this, a simple method is to consider M frequencies instead of N, with $N < M$, where all terms whose frequencies belong to the intervals $]-\frac{M}{2}, -\frac{N}{2}[$ and $]\frac{N}{2}, \frac{M}{2}[$, are cancelled [19]. This method simply consists in oversampling the projection of our function on the basis constituted by the N initial harmonics, from a spatial sampling step of size $\frac{L}{N}$ to a spatial sampling step of size $\frac{L}{M}$:

$$f[n] = \sum_{k=-M/2}^{M/2} \hat{f}[k]e^{i\frac{2\pi}{M}kn}, \qquad g[n] = \sum_{k=-M/2}^{M/2} \hat{g}[k]e^{i\frac{2\pi}{M}kn}. \tag{25}$$

As the frequencies between $N/2$ and $M/2$ are equal to zero, Equation 24 reads

$$\begin{aligned}\hat{fg}[k] = & \sum_{k_1=-N/2}^{N/2} \hat{f}[k_1]\hat{g}[k-k_1] + \sum_{k_1=-N/2}^{N/2} \hat{f}[k_1]\hat{g}[k+M-k_1] \\ & + \sum_{k_1=-N/2}^{N/2} \hat{f}[k_1]\hat{g}[k-M-k_1].\end{aligned} \tag{26}$$

The most dangerous term considering aliasing is obtained for $k = -N/2$ and $k_1 = -N/2$ (respectively $k = N/2$ and $k_1 = N/2$) in the second (respectively third) sum of Equation 26. The corresponding value of $\hat{g}$ will be equal to zero if

$$M > \frac{3N}{2}. \tag{27}$$

This inequality ensure that the quantities $k - M - k_1$ and $k + M - k_1$ fall into the intervals $]-\frac{M}{2}, -\frac{N}{2}[$ and $]\frac{N}{2}, \frac{M}{2}[$. The argument is easily extended to higher degree nonlinearities. Since Equation 15 involves fourth-degree monomials, full desaliasing requires $M = \frac{5N}{2}$.

2.2.2 Integration of the equation

Equation 21 is integrated using a Runge-Kutta scheme. This scheme uses evaluations of the time derivative at intermediate points to achieve the integration, given by the formula:

$$\psi_{n+1} = \psi_n + \Delta t \sum_{i=1}^{s} b_i k_i, \tag{28}$$

Table 1 General Butcher table of a Runge-Kutta scheme.

0					
c_2	a_{21}				
c_3	a_{31}	a_{32}			
$\vdots$	$\vdots$		$\ddots$		
c_s	a_{s1}	a_{s2}	$\dots$	a_{ss-1}	
	b_1	b_2	$\dots$	b_{s-1}	b_s
	b_1^*	b_2^*	$\dots$	b_{s-1}^*	b_s^*

Table 2 Butcher table of the Heun scheme.

0		
1	1	
	1/2	1/2
	1	0

with:

$$\begin{cases} k_1 = F(t_n, \psi_n), \\ k_2 = F(t_n + c_2\Delta t, \psi_n + a_{21}\Delta t k_1), \\ k_3 = F(t_n + c_3\Delta t, \psi_n + a_{31}\Delta t k_1 + a_{32}\Delta t k_2), \\ \dots \\ k_N = F(t_n + c_s\Delta t, \psi_n + a_{N1}\Delta t k_1 + a_{N2}\Delta t k_2 + \dots + a_{N,N-1}\Delta t k_{N-1}). \end{cases} \tag{29}$$

To specify a particular method, one simply has to set the coefficients a_{ij}, b_i and c_i which characterize the discretization of the equation for $i = 1, 2, \dots, N$ and $j = 1, 2, \dots, i$. The selected coefficients can be represented in a table called the Butcher table.

Consistency of the scheme is ensured if $\sum_{j=0}^{i} a_{ij} = c_i$. A Runge-Kutta scheme of order N is accurate at order N in Δt. It is possible to control the approximation error at each step by estimating the difference between approximations at order $N - 1$ and N. By wisely choosing the coefficients a_{ij} and c_i, intermediate points calculated in the method can be used to calculate two separate evaluations of the solution:

- A first evaluation $\psi_{n+1} = \psi_n + \Delta t \sum_{i=1}^{N} b_i k_i$ accurate at order N.
- A second evaluation $\psi_{n+1} = \psi_n + \Delta t \sum_{i=1}^{N-1} b^*{}_i k_i$ accurate at order $N - 1$, which uses an other ponderation $\{b^*{}_i\}$, $i = 1, 2, \dots, N$.

The difference between these two evaluations gives an estimate of the approximation error of the scheme:

$$\epsilon = \Delta t \sum_{i=1}^{N-1} (b_i - b^*{}_i) k_i. \tag{30}$$

The corresponding Butcher table is given in Table 1.

The Heun scheme (order 2), the Bogacki-Shampine scheme [20] (order 3) and the Cash-Karp scheme [21] (order 5) were implemented. All these methods realize an explicit integration of Equation 21, and the schemes are conditionnaly stable. Tables 2, 3, and 4 show the Butcher tables of the schemes.

The dynamics of the paint levelling varies considerably during the painting process, and it is then of interest to use an adaptive stepsize integration scheme. A method described

Table 3 Butcher table of the Bogacki-Shampine scheme.

0				
1/2	1/2			
3/4	0	3/4		
1	2/9	1/3	4/9	
	2/9	1/3	4/9	0
	7/24	1/4	1/3	1/8

Table 4 Butcher table of the Cash-Karp scheme.

0						
1/5	1/5					
3/10	3/40	9/40				
3/5	3/10	−9/10	6/5			
1	$-\frac{11}{54}$	5/2	$-\frac{70}{27}$	$\frac{35}{27}$		
7/8	$\frac{1631}{55296}$	$\frac{175}{512}$	$\frac{44275}{110592}$	$\frac{253}{4096}$		
	$\frac{37}{378}$	0	$\frac{250}{621}$	$\frac{125}{594}$	0	$\frac{512}{1771}$
	$\frac{2825}{27648}$	0	$\frac{18575}{48384}$	$\frac{13525}{55296}$	$\frac{277}{14336}$	1/4

in [22] is used to adjust the time step, which uses the error estimate returned by the integration scheme.

2.3 Validation of the numerical scheme

Assuming that the amplitude of surface modulation is small, Equation 15 can be linearized by setting $h = h_0 + \delta h$, expanding it in powers of δh, and keeping lowest order terms. Denoting the mean paint thickness as e_0, Equation 15 reads:

$$\begin{aligned}\frac{\partial \delta h}{\partial t}(x,y,t) \\ = -\frac{\gamma}{3\eta}e_0^3\left(\frac{\partial^4 \delta h}{\partial x^4}(x,y,t) + 2\frac{\partial^4 \delta h}{\partial x^2\,\partial y^2}(x,y,t) + \frac{\partial^4 \delta h}{\partial y^4}(x,y,t)\right),\end{aligned} \tag{31}$$

which can be solved analytically using Fourier transforms. If Equation 31 has a solution in $L^2(\mathbb{R})$, in the Fourier space it fulfills:

$$\frac{\partial \widehat{\delta h}}{\partial t} = -\frac{\gamma}{3\eta}e_0^3(\xi_x^4 + \xi_x^2\xi_y^2 + \xi_y^4)\widehat{\delta h}, \tag{32}$$

in which $\boldsymbol{\xi} = (\xi_x, \xi_y)$ is the Fourier wavevector. Consequently,

$$\widehat{\delta h}(t) = \widehat{\delta h}(0)\exp\left(-\frac{\gamma}{3\eta}e_0^3(\xi_x^4 + \xi_x^2\xi_y^2 + \xi_y^4)t\right) \tag{33}$$

and

$$\delta h(t) = \frac{1}{4\pi^2}\int_{\mathbb{R}^2}\widehat{\delta h}(0)\exp\left(-\frac{\gamma}{3\eta}e_0^3(\xi_x^4 + \xi_x^2\xi_y^2 + \xi_y^4)t\right)\exp(i(\xi_x x + \xi_y y))\,\mathrm{d}\boldsymbol{\xi}. \tag{34}$$

Figure 3 compares the results obtained for a cataphoresis layer at $t = 600$, from the initial condition displayed in the top panel, derived from the analytic solution Equation 34 (bottom-left) or obtained with the numerical scheme (bottom-right). It can be noticed that the results are very close, which validates the numerical scheme used to solve Equation 31. The parameters chosen for the comparison are given in Table 5, taken from the

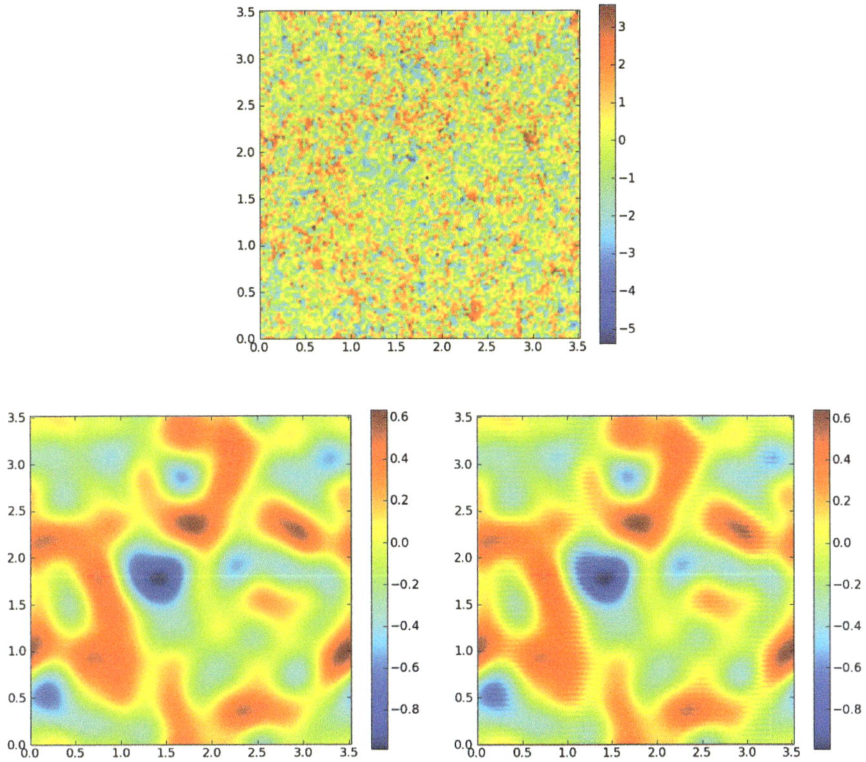

Figure 3 Validation of the numerical scheme. Top: Initial topography as measured using the Phasics™ interferometric microscope. Bottom: Solution at $t = 600$, analytical (left) and numerical (right). Each image corresponds to the same 3.52 × 3.52 mm^2 domain decomposed into 27.5 × 27.5 μm^2 subdomains or pixels, over which the altitude is averaged (color coding).

Table 5 Physical parameters used for the numerical scheme validation.

Parameter	Symbol	Value	Unit
Surface tension	γ	3.0×10^{-2}	N/m
Paint viscosity	η	1.0	Pa.s
Initial thickness	e_0	20.0	μm

relevant literature [2]. It is also assumed that the surface tension is constant and that the substrate is perfectly flat ($S_a(x, y) = 0$ everywhere).

3 Experimental measurements

We now present the experiments performed with the high resolution wavefront sensor developed by Phasics™ [10], using a technology based on a modified Hartmann test to measure wavefront distortions: by means of 2D diffraction grating, a beam is replicated into four identical waves which are propagated along slightly different directions. The direction differences create interference patterns and the interference fringes are used to reconstruct the measured surface topography. We use it to map painted steel sheets at regular time intervals during the whole painting process, for a sealer or a lacquer layer. The experimental procedure is as follows:

- Paint is deposited over a sample of metal sheet (polished or already covered with an electrophoresis layer) in a painting cabin using a paint gun.

- The sheet is then placed on a baking plate. During the first few minutes, complete samplings of the surface are performed at regular time intervals (typically 2.5 Hz), in order to record the evolution of the painted layer topography at the beginning of the flash time in detail.
- After two minutes the sampling rate is decreased to 0.1 Hz, for the flow dynamic next slows down considerably.
- The baking cycle starts after 10 minutes, with the sampling frequency reincreased to 1.25 Hz.
- Chemical bonds begin to form within the paint 5 minutes after the beginning of the baking. Cross-linking then stops the evolution so that the sampling frequency can be decreased to 0.1 Hz.

The wavefront sensor collects information over a surface of 18×18 mm^2. The topography is analyzed as a 128×128 square image. Each pixel represents the mean altitude over a 60×60 μm^2 surface. The precision of vertical measurements is up to 10^{-2} μm.

During the flash time, the outside temperature is 25°C. The baking cycle is divided into two stages: a linear temperature rise during 300 sec until reaching 150°C, followed by a 15 minutes plateau at this temperature.

3.1 Surface evolution

The following figures show the evolution of the topography of a lacquer layer during the whole painting process. The lacquer is deposited on a smooth substrate. Altitudes are given in μm. On each surface, during measurements, the minimum is arbitrarily set to zero since only relative but not absolute altitudes can be obtained from the device.

Figure 4 displays the beginning of the flash time. A rapid leveling of the paint is observed, due to the combined effects of the rapid evaporation of the light sealer solvent and the flow caused by surface tension. The phenomenon is specially important at the beginning of the flash time when the viscosity of the paint is still relatively low. At the end of the flash time, the leveling slows down until the topography of the layer stops evolving.

Figure 5 shows the evolution of the lacquer layer during the baking. The same altitude scale has been kept, which allows a comparison with the previous sequence. A second stage of leveling and evaporation takes place during the baking of the lacquer. Temperature increase promotes the evaporation of heavier solvents contained in paint and the subsequent cross-linking of the molecules.

3.2 Evolution of the roughness

Roughness evolution during the painting process helps us quantifying the paint leveling capability. Since the physical effects involved develop at different scales, it is of interest to play with tools able to separate the different roughness scales. The surface is sampled with a 60 μm horizontal step, yielding a 128×128 image $S = S[n_1, n_2]$. An algorithm based on the wavelet packet transform [23] and the reconstruction formula is used, that allows a decomposition of the roughness into a sum of contributions [24–26].

A wavelet $\psi(t)$ is a function in $L^2(\mathbb{R})$ which has zero average, such that $\|\psi\| = 1$ using the L^2 norm, and centered around $t = 0$ [23]. We obtain a wavelet family by dilating this function with a scale parameter $s > 0$ and translating it by $u \in \mathbb{R}$:

$$\psi_{u,s}(t) = \frac{1}{\sqrt{s}}\psi\left(\frac{t-u}{s}\right). \tag{35}$$

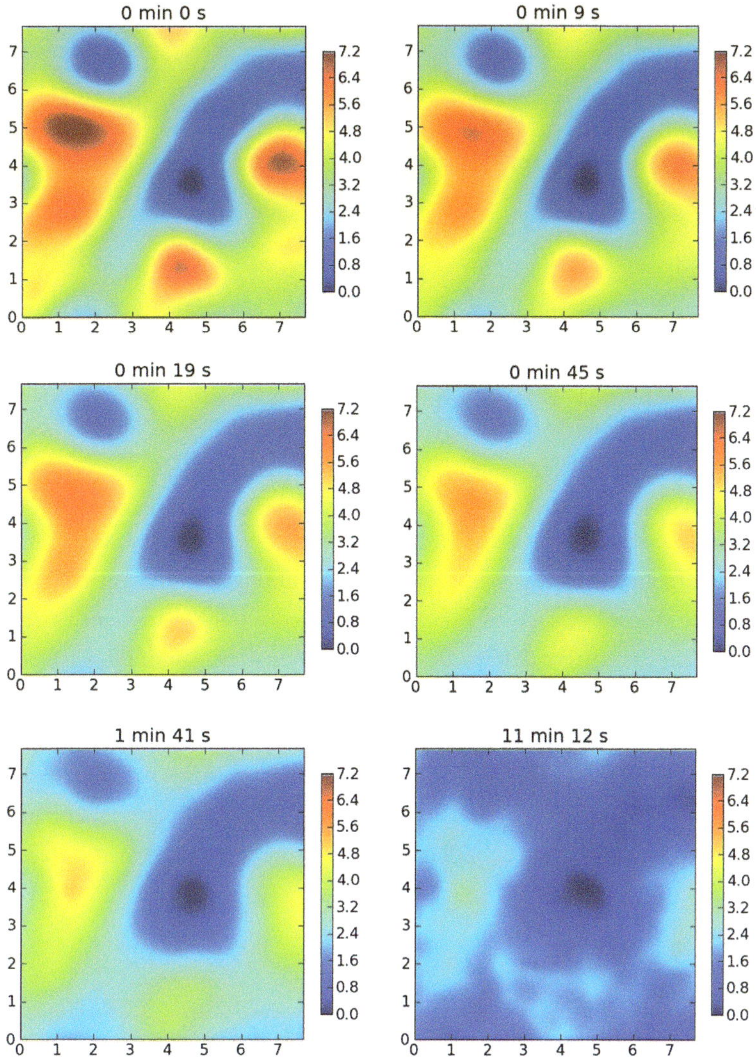

Figure 4 Evolution of the topography of the lacquer film during the flash time. The topography of a $7.68 \times 7.68\ \text{mm}^2$ surface is represented in a 128×128 square image. Color coding at each pixel gives the mean altitude over a $60 \times 60\ \mu\text{m}^2$ surface.

Let f be a function in $L^2(\mathbb{R})$. The continuous wavelet transform of f is the function:

$$Wf(u,s) = \int_{-\infty}^{\infty} f(t) \frac{1}{\sqrt{s}} \psi\left(\frac{t-u}{s}\right) \mathrm{d}t. \tag{36}$$

The original signal can be reconstructed from its wavelet coefficients:

$$f(t) = \frac{1}{C_\psi} \int_0^{\infty} \int_{-\infty}^{\infty} Wf(u,s) \frac{1}{\sqrt{s}} \psi\left(\frac{t-u}{s}\right) \mathrm{d}u \frac{\mathrm{d}s}{s^2}, \tag{37}$$

where

$$C_\psi = \int_0^{\infty} \frac{|\hat{\psi}(\omega)|^2}{\omega} \mathrm{d}\omega < +\infty. \tag{38}$$

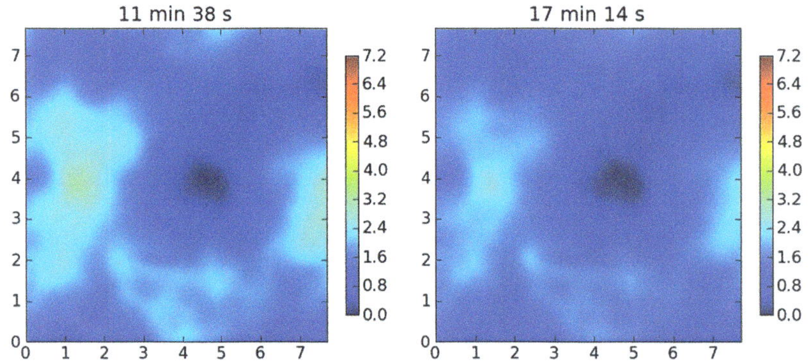

Figure 5 Lacquer layer topography evolution during the baking. The topography of a 7.68 × 7.68 mm^2 domain is represented as a square image of 128 × 128 pixels. Each pixel accounts for the mean altitude on a 60 × 60 μm^2 sub-domain.

A wavelet is a function which is well localized both in physical and Fourier spaces. The projection of the studied surface on a family of wavelets allows us to identify the scale hierarchy present in the pattern. The scale-by-scale reconstruction of the surface can next isolate the different contributions to the roughness of the given topography. The continuous wavelet transform was adapted to the discrete case, yielding the orthonormal wavelet and wavelet packet transforms [23]. The roughness characterization algorithm therefore decomposes the studied surface $S[n_1, n_2]$ onto a family of wavelet packets, and next reconstructs it by keeping only wavelet coefficients corresponding to successive scale:

$$S[n_1, n_2] = \sum_{j=0}^{J} S_j[n_1, n_2], \tag{39}$$

J denoting the number of scales and S_j the reconstructed surface at the scale j. For each reconstructed surface, it is possible to define a parameter Mq characterizing the mean deviation from the average surface at scale j. The resulting curve $Mq[j]$ then accounts for the frequency content of the studied roughness:

$$Mq[j] = \frac{1}{N^2}\sqrt{\sum_{n_1=1}^{N}\sum_{n_2=1}^{N}(S_j[n_1, n_2] - E[S_j])^2}. \tag{40}$$

Figures 6 show how Mq changes from one to another step of the painting process. It is interesting to note that the small scales are not completely attenuated during the painting process, due to resurgence of the underlying substrate residual roughness. On the other hand, the baking has little impact on the paint leveling for the lacquer.

The scale-by-scale study of the surface roughness provides valuable information on the dynamics of the leveling. The curves in Figure 6 show little leveling during the baking, the difference being mainly due to evaporation since the two curves are quite similar. In the next section we therefore focus only on the simulation of the surface dynamics during flash time, when both flow and evaporation are involved.

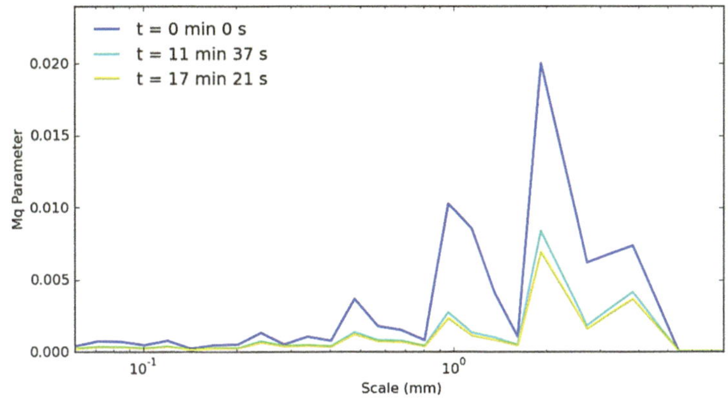

Figure 6 Evolution of Mq as a function of the scale during the whole painting process.

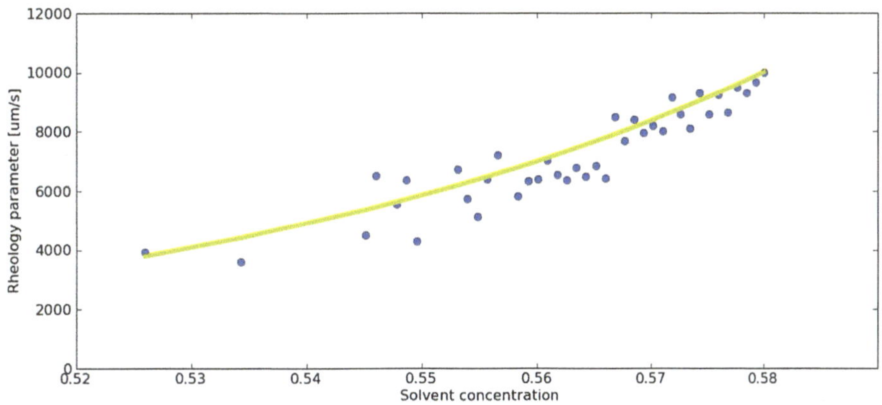

Figure 7 Variation of $\frac{\gamma}{3\eta}$ as a function of the solvent concentration.

Table 6 Simulation parameters.

Parameter	Symbol	Value	Value in [2]	Unit
Resin surface tension	γ_r	-	3.0×10^{-2}	N/m
Solvent surface tension	γ_s	-	2.5×10^{-2}	N/m
Initial paint viscosity	η_0	-	0.55-1.59	Pa.s
Initial solvent concentration	c_0	0.58	0.5	-
Viscosity exponent	a	18	15 [5]	-
Evaporation parameter	λ	4.0×10^{-9}	2.0×10^{-9}	m/s
Initial rheological parameter	$\gamma_0/3\eta_0$	1.0×10^4	1.0×10^4	μm/s

4 Direct simulation

We now compare simulation results to experimental data obtained from the wavefront sensor. In [11], the value of the rheological parameter $\gamma/3\eta$ was determined as a function of the solvent concentration c from mass fluxes computed by solving an inverse problem relying on the Newtonian model. The surface tension is assumed to be related to the solvent concentration via:

$$\gamma = \gamma_r + c(\gamma_s - \gamma_r) = \gamma_r + c\Delta\gamma, \tag{41}$$

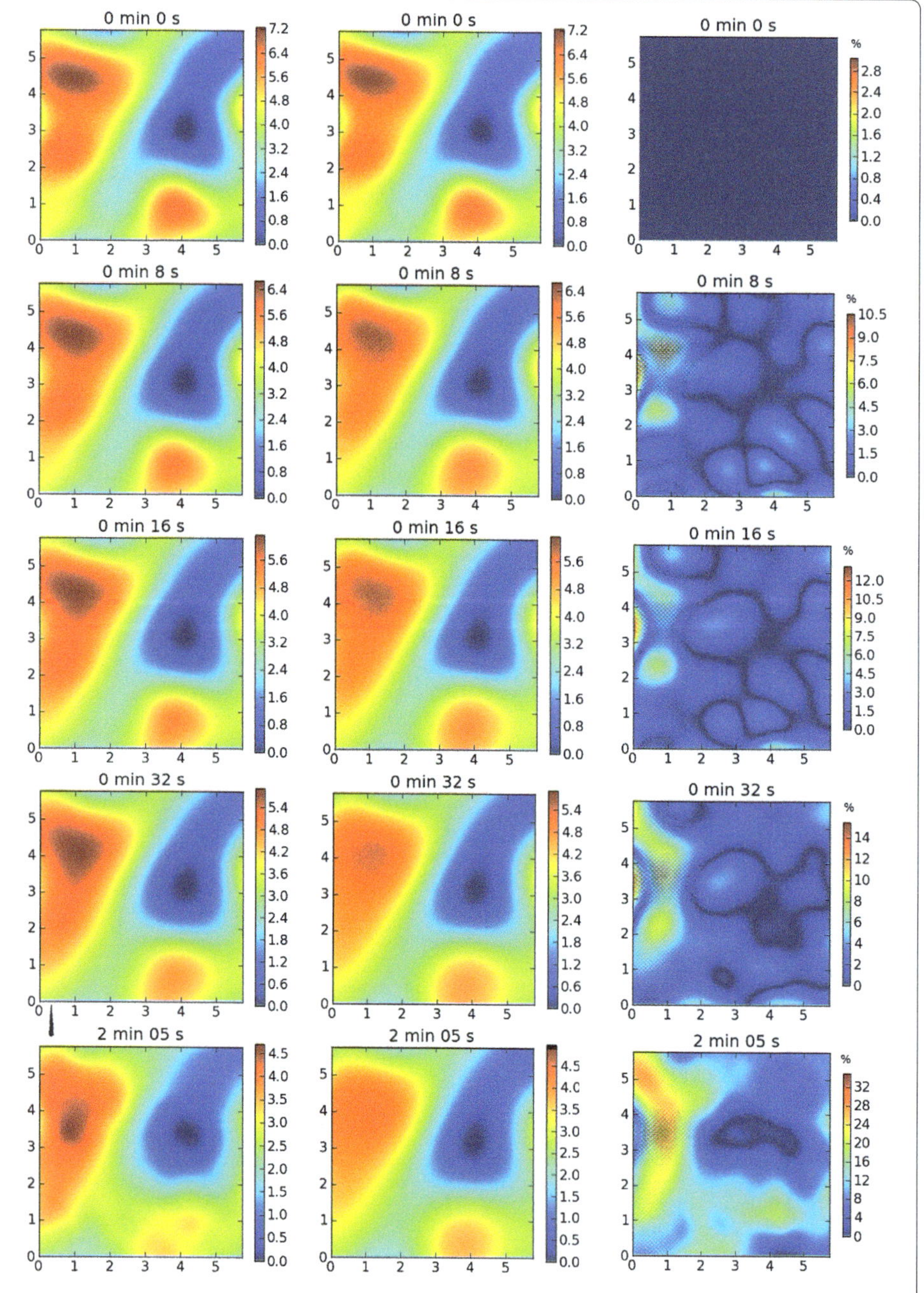

Figure 8 Evolution of the film topography: Experiments (center) compared to simulation results from the model (left) without Marangoni effect included. Relative local error ϵ is represented (right). The vertical scale varies from a topography to the other.

where γ_r (γ_s) denotes the resin (solvent) surface tension. On the other hand, viscosity depends upon the solvent concentration according to:

$$\eta(c) = \eta_0 e^{-ac}. \tag{42}$$

Using these relations to fit the experimental data during the flash time (Figure 7), we find:

$$\frac{\gamma}{3\eta}(c) = \frac{\gamma_0}{3\eta_0} e^{a(c-c_0)}, \tag{43}$$

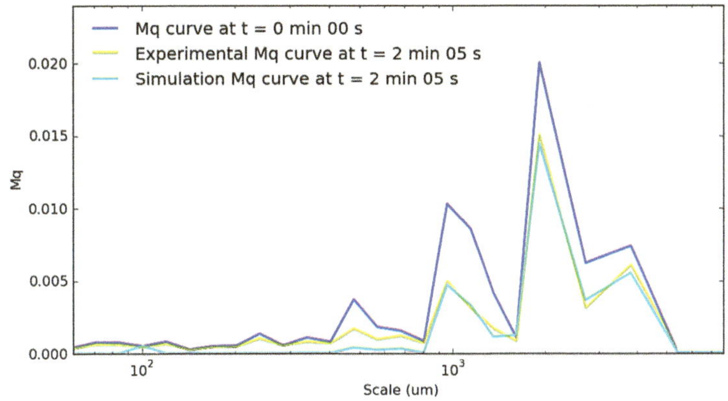

Figure 9 Evolution of the *Mq* curves: Experiments are compared to simulation results.

with $\frac{\gamma_0}{3\eta_0} = 0.1$ m/s, $c_0 = 0.58$ and $a = 18.0$. Table 6 shows that the rheological parameters close to those reported in [2, 5].

Using the experimental data obtained with the wavefront analyzer and the evaporation law deduced from these measurements, simulations were performed with the two models described in Section 2. These simulations start from the first reconstructed topography and aim at reproducing the entire evolution of the film during the flash time. Parameters used are given in Table 6 obtained as explained above. We consider that the substrate is completely smooth. The numerical resolution code was described at the end of Section 2. The simulations are performed using a 3.40 GHz Intel(R) Xeon(TM) processor, and last about five hours.

The local relative error defined as

$$\epsilon(x,y) = \frac{|h^s(x,y) - h^e(x,y)|}{\max_{(x,y)} h^e}, \tag{44}$$

where h^s denotes the simulated topography and h^e the experimental one is displayed in Figure 8 (right). The main error is caused by side effects, as the surface is artificially periodized. The local simulation error remains small, and the model seems to be able to reproduce the patterns evolution, at least at the millimeter scale. It remains interesting to compare the *Mq* curves of the simulated and experimental topographies. The curves of Figure 9 show that the simulation model is able to reproduce the leveling dynamics for the millimetric scales. It was supposed that the substrate of the paint layer was completely smooth. Nevertheless, the substrate exhibits some roughness at the smallest scales, and evaporation causes the resurgence of the underlying substrate roughness, which explains the differences between the simulation and the experimental *Mq* curves. For the last represented topography, the resurgence of the underlying topography also introduces an error.

Figure 10 compares simulations realized with the model including Marangoni effect (right) and not (left), showing that the model developed by Orchard [1] seems sufficiently accurate to explain the physics of the leveling of the coating. In fact, the geometric situation considered here is somewhat different from those studied by Weidner *et al.* [5] or Gaskell *et al.* [9]. For automotive paints, fluctuations in the thickness of the coating are indeed relatively small (<7 μm) compared to the thickness itself ($\cong 70$ μm), so that the

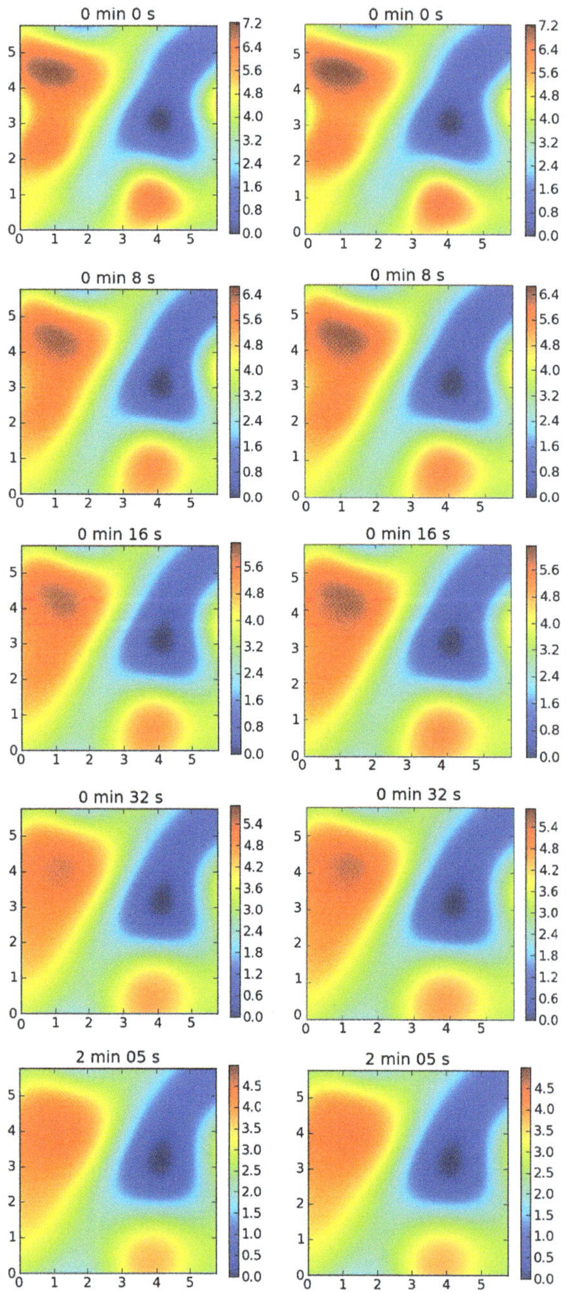

Figure 10 Comparison between simulation results from the model with (right) or without (left) Marangoni effect included.

rate of evaporation and consequently the solvent concentration remain rather uniform in the paint layer.

5 Conclusion

Painting of steel sheets is a complex phenomenon that depends on many physical processes. With the wavefront sensor developed by Phasics™, it was possible to perform experiments allowing an accurate monitoring of the topography of a film during its deposition. The fast response time of the wavefront sensor allowed us to access the rheological parameters of the paint in an original way by solving an inverse problem. The obtained parameters were used to perform a complete simulation of the film evolution during the painting process, which demonstrated that the Newtonian model was able to reproduce the leveling of the paint layer accurately and that Marangoni effect could be neglected at the beginning of the flash time, when significant flow occurs. At the end of the flash time, the flow rates decreases and it is clear that the film then exhibits a more complex rheology due to the solvent evaporation, but the leveling dynamic is then considerably attenuated, and the influence on the surface topography is negligible. The conducted simulations are however complex and require the development of advanced numerical tools, like those presented in this paper.

Competing interests
Our research was supported by ArcelorMittal Global R&D, F-57283 Maizières-lès-Metz Cedex, France.

Author's contribution
BF, DJ and GF designed and implemented the numerical scheme which is presented in this paper. The experiments were realized by BF, GF and J-JP. The physical models and the resolution of the inverse problem were tackled by BF, DJ, AL and PM. All authors participated in writing the manuscript.

Author details
[1]Centre de Morphologie Mathématique, École des Mines ParisTech, 77300, Fontainebleau, France. [2]UMR Navier, 2 allée Kepler, 77420, Champs-sur-Marne, France. [3]ArcelorMittal Global R&D, 57283, Maizières-lès-Metz Cedex, France. [4]LadHyX, École Polytechnique, 91128, Palaiseau, France.

References

1. Orchard SE: **On surface levelling in viscous liquids and gels.** *Appl Sci Res* 1961, **11**:451-464.
2. Overdiep WS: **The levelling of paints.** *Prog Org Coat* 1986, **14**:159-175.
3. Wilson SK: **The levelling of paint films.** *IMA J Appl Math* 1993, **50**:149-166.
4. Howison SD, Moriarty JA, Ockendon JR, Terrill EL: **A mathematical model for drying paint layers.** *J Eng Math* 1997, **32**:377-394.
5. Weidner DE, Schwartz LW, Eley RR: **Role of surface tension gradients in correcting coating defects in corners.** *J Colloid Interface Sci* 1996, **179**:66-75.
6. Eres MH, Weidner DE, Schwartz LW: **Three-dimensional direct numerical simulation of surface-tension-gradient effects on the leveling of an evaporating multicomponent fluid.** *Langmuir* 1999, **15**:1859-1871.
7. Schwartz LW, Roy RV, Eley R, Petrash S: **Dewetting patterns in a drying liquid film.** *J Colloid Interface Sci* 2001, **234**:363-374.
8. Gaskell PH, Jimack PK, Sellier M, Thompson HM, Wilson MCT: **Gravity-driven flow of continuous thin liquid films on non-porous substrates with topography.** *J Fluid Mech* 2004, **509**:253-280.
9. Gaskell PH, Jimack PK, Sellier M, Thompson HM: **Flow of evaporating gravity-driven thin liquid films over topography.** *Phys Fluids* 2006, **18**:031601.
10. **Phasics:** [http://www.phasicscorp.com/]
11. Figliuzzi B, Jeulin D, Lemaitre A, Manneville P, Fricout G, Piezanowski JJ: **Rheology of thin films from flow observations.** *Exp. Fluids*, submitted.
12. Benney DJ: **Long waves on liquid films.** *J Math Phys* 1966, **45**:150-155.
13. Shkadov VY: **Wave flow regimes of a thin layer of viscous fluid subject to gravity.** *Izv AN SSSR Mekhanika Zhidkosti i Gaza* 1967, **2**(1):43-51.
14. Shkadov VY: **Solitary waves in a layer of viscous liquid.** *Izv AN SSSR Mekhanika Zhidkosti i Gaza* 1977, **1**:63-66.
15. Ruyer-Quil C, Manneville P: **Improved modeling of flows down inclined planes.** *Eur Phys J B* 2000, **15**:357-369.
16. Ruyer-Quil C, Manneville P: **Modeling film flows down inclined planes.** *Eur Phys J B* 1998, **6**:277-292.
17. Oron A, Davis SH, Bankoff SG: **Long scale evolution of thin liquid films.** *Rev Mod Phys* 1997, **69**(3):931-980.
18. Fletcher CAJ: *Computational Techniques for Fluid Dynamics.* Berlin: Springer; 1991.
19. Manneville P: *Instabilities, Chaos and Turbulence.* London: Imperial College Press; 2010.
20. Bogacki P, Shampine L: **A 3(2) pair of Runge-Kutta formulas.** *Appl Math Lett* 1989, **2**:321-325.
21. Cash JR, Karp AH: **A variable order Runge-Kutta method for initial value problems with rapidly varying right-hand sides.** *ACM Trans Math Softw* 1990, **16**:201-222.
22. Press WH, Teukolsky SA, Vetterling WT, Flannery BP: *Numerical Recipes: The Art of Scientific Computing.* 3rd edition. Cambridge: Cambridge University Press; 2007.
23. Mallat S: *A Wavelet Tour of Signal Processing.* San Diego: Academic Press; 2008.
24. Lee SH, Zahouani H, Caterini R, Mathia TG: **Morphological characterization of engineered surfaces by wavelet transform.** *Int J Mach Tools Manuf* 1998, **38**:581-589.
25. Chen Q, Yang S, Li Z: **Surface roughness evaluation by using wavelets analysis.** *Precis Eng* 1999, **23**:209-212.

26. Liu JJ, Kim D, Han C: **Use of wavelet packet transform in characterization of surface quality.** *Ind Eng Chem Res* 2007, **46**:5152-5158.

2

Simulation of electromagnetic descriptor models using projectors

Nicodemus Banagaaya* and Wil Schilders

*Correspondence:
n.banagaaya@tue.nl
Department of Mathematics and Computer Science, Eindhoven University of Technology, Eindhoven, The Netherlands

Abstract

Electromagnetic descriptor models are models which lead to differential algebraic equations (DAEs). Some of these models mostly arise from electric circuit and power networks. The most frequently used modeling technique in the electric network design is the modified nodal analysis (MNA) which leads to differential algebraic equations in descriptor form. DAEs are known to be very difficult to solve numerically due to the sensitivity of their solutions to perturbations. We use the tractability index to measure this sensitivity since it can be computed numerically. Simulation of DAEs is a very difficult task especially for those with index greater than one. To solve higher-index DAEs, one needs to use multistep methods such as Backward difference formulas (BDFs). In this paper, we present an easier method of solving DAEs numerically using special projectors. This is done by first splitting the DAE system into differential and algebraic parts. We then use the existing numerical integration methods to approximate the solutions of the differential part and the solutions of the algebraic parts are computed explicitly. The desired solution of the DAE system is obtained by taking the linear combination of the solutions of the differential and algebraic parts. Our method is robust and efficient, and can be used on both small and very large systems.

1 Introduction

Consider a linear Resistor-Inductor-Capacitor (RLC) electric network which connects linear capacitors, inductors and resistors, and independent voltage $v(t) \in \mathbb{R}^{n_V}$ and current sources $\imath(t) \in \mathbb{R}^{n_I}$. Circuit designers are interested to know the unknowns which describe the network which are: the node potentials $e(t) \in \mathbb{R}^{n_e}$, and the currents $\imath_L(t) \in \mathbb{R}^{n_L}$ and $\imath_V(t) \in \mathbb{R}^{n_V}$ through inductors and voltage sources, respectively. Using the commonly used Modified Nodal Analysis (MNA) [1], we introduce the incidence matrices $\mathbf{A}_C \in \mathbb{R}^{n_e,n_C}$, $\mathbf{A}_L \in \mathbb{R}^{n_e,n_L}$ and $\mathbf{A}_R \in \mathbb{R}^{n_e,n_G}$, which describe the branch-node relationships for capacitors, inductors and resistors. Furthermore, we introduce the incidence matrices $\mathbf{A}_V \in \mathbb{R}^{n,n_V}$ and $\mathbf{A}_I \in \mathbb{R}^{n_e,n_I}$, which describe this relationship for voltage and current sources, respectively. This leads to a linear time invariant (LTI) DAE system in descriptor form:

$$\underbrace{\begin{bmatrix} \mathbf{A}_C\mathbf{C}\mathbf{A}_C^T & \mathbf{0} & \mathbf{0} \\ \mathbf{0} & \mathbf{L} & \mathbf{0} \\ \mathbf{0} & \mathbf{0} & \mathbf{0} \end{bmatrix}}_{\mathbf{E}} x' = \underbrace{\begin{bmatrix} -\mathbf{A}_R\mathbf{G}\mathbf{A}_R^T & -\mathbf{A}_L & -\mathbf{A}_V \\ \mathbf{A}_L^T & \mathbf{0} & \mathbf{0} \\ \mathbf{A}_V^T & \mathbf{0} & \mathbf{0} \end{bmatrix}}_{\mathbf{A}} x + \underbrace{\begin{bmatrix} -\mathbf{A}_I & \mathbf{0} \\ \mathbf{0} & \mathbf{0} \\ \mathbf{0} & -\mathbf{I} \end{bmatrix}}_{\mathbf{B}} \underbrace{\begin{bmatrix} \imath \\ v \end{bmatrix}}_{u}, \qquad (1)$$

where $\mathbf{C} \in \mathbb{R}^{n_C,n_C}$, $\mathbf{L} \in \mathbb{R}^{n_L,n_L}$ and $\mathbf{G} \in \mathbb{R}^{n_G,n_G}$ are the capacitance, inductance and conductance matrices, respectively which are usually assumed to be symmetric and positive-definite. We need to find the unknowns $x = (e, \imath_L, \imath_V)^T$ of the system (1). If we let $n = n_e + n_L + n_V$ and $m = n_I + n_V$, then $\mathbf{E}, \mathbf{A} \in \mathbb{R}^{n,n}$, $\mathbf{B} \in \mathbb{R}^{n,m}$ and $u \in \mathbb{R}^m$. Equation (1) can be solved depending to the elements in the electric networks. For example if we consider a resistor network, (1) simplifies to a system of algebraic equations which implies that $\mathbf{E} = \mathbf{0}$. Thus we need to solve a linear system of the form $-\mathbf{A}x = \mathbf{B}u$. But if the network contains a combination of resistors and other elements, this leads to differential algebraic equations (DAEs), *i.e.*, $\det(\mathbf{E}) = 0$. In this paper, we consider electric networks which leads to DAEs.

DAEs are very difficult to solve numerically due to the sensitivity of their solutions to perturbations. This sensitivity is measured by the *index concepts* such as differentiation index, perturbation index and tractability index. We use the tractability index because it can be obtained numerically. According to [1], a DAE-index from electric networks cannot be greater than 2. Thus we assume (1) has maximum index of 2. In order to solve DAEs we assume the following: (i) the matrix pencil $\lambda\mathbf{E} - \mathbf{A}$ must be nonsingular for some $\lambda \in \mathbb{C}$. (ii) the input vector u must be smooth enough. Further more consistent initial values $x(0) = x_0$ must be applied.

A lot of work has been done to solve DAE system using multistep methods such BDFs, NDFs and Runge–Kutta methods [2]. Although these methods are accurate they need a lot of computational effort depending on the index of DAE system, for example index-2 systems, BDF is convergent and globally accurate to $\mathcal{O}(h^k)$ but these require tight Newton solutions accurate to $\mathcal{O}(h^{k+1})$ [3]. The implicit Euler method loses order of accuracy with each increase in index, thus cannot be used in practice to solve DAE systems numerically [3].

In this paper, we propose a new way of solving DAEs using special projectors. This method enables one to use any ODE numerical methods to solve DAE systems. This idea was first introduced by März [4], and it involves splitting the DAE system into differential and algebraic parts using projector and matrix chains. Then one can use existing numerical integrations methods to solve the differential part and the some components of the algebraic part are approximated using numerical differential methods for DAEs with index greater than one. The desired solution is obtained by the linear combination of the solutions of the differential and algebraic parts. This approach leads to a decoupled system of dimension $(\mu + 1)n$, where μ is the tractability index. Moreover, the spectrum of the decoupled system consists not only of the spectrum of the matrix pencil (E, A) of the original system but also of additional infinite eigenvalues [5]. In practice a system should be stable otherwise it may not work properly or may even be destroyed in practical use. There is a direct criterion for the stability of regular descriptor linear systems which indicates that the stability of the system (1) is totally determined by spectrum of the matrix pencil $\sigma(\mathbf{E}, \mathbf{A})$ [6]. The set of all finite eigenvalues of the matrix pencil $(\mathbf{E}, \mathbf{A})$ is denoted by $\sigma_f(\mathbf{E}, \mathbf{A})$ which is called the finite spectrum while the infinite spectrum of the matrix pencil $(\mathbf{E}, \mathbf{A})$ is denoted by $\sigma_\infty(\mathbf{E}, \mathbf{A})$. Thus the spectrum of $(\mathbf{E}, \mathbf{A})$ is donated by $\sigma(\mathbf{E}, \mathbf{A}) = \sigma_f(\mathbf{E}, \mathbf{A}) \cup \sigma_\infty(\mathbf{E}, \mathbf{A})$. System (1) is stable if and only if

$$\sigma_f(\mathbf{E}, \mathbf{A}) \subset \mathbb{C}^- = \{s \in \mathbb{C}, \mathrm{Re}(s) \leq 0, \text{ and any } \mathrm{Re}(s) = 0 \text{ is simple}\}.$$

Thus, we need to make sure the physical properties of the DAE system (1) such as stability to be inherited in its decoupled system. This motivated us to do some modifications in the März decomposition, using special basis vectors, which leads to a modified decoupled system of dimension n. Moreover, this decoupling preserves the spectrum of the matrix pencil (E, A) of the DAE.

This paper is organized as follows: In Section 2, we discuss the decoupling of LTI DAE system using März decomposition and its limitation. In Section 3, we propose the modification of the März decomposition using bases of special projectors. In this section we further discuss the modified decoupling of index-1 and -2 electric systems. We observed that higher index DAEs can decoupled into two ways depending on the spectrum of the matrix pencil $(\mathbf{E}, \mathbf{A})$. In Section 4, we check whether the decoupled system preserves the physical properties of the DAE system. In Section 4.1, we compare the numerical accuracy of the decoupled and undecoupled system. In Section 5, we test the proposed method on both simple and industrial problems. The industrial examples show the feasibility of this method on real-life applications. This paper is concluded by some final remarks in Section 6.

2 Decoupling of LTI DAE systems using projectors

Assume (1) is of tractability index μ. Setting $\mathbf{E}_0 := \mathbf{E}$, $\mathbf{A}_0 := \mathbf{A}$, then the projector and matrix chains of the matrix pair $(\mathbf{E}, \mathbf{A})$ can be written as: $\mathbf{E}_{j+1} = \mathbf{E}_j - \mathbf{A}_j\mathbf{Q}_j$, $\mathbf{A}_{j+1} := \mathbf{A}_j\mathbf{P}_j, j \geq 0$, where the image space of the projector $\mathbf{Q}_j$ is equal to the nullspace of singular matrix $\mathbf{E}_j$, *i.e.*, $\operatorname{Im}\mathbf{Q}_j = \operatorname{Ker}\mathbf{E}_j$ and $\mathbf{P}_j = \mathbf{I} - \mathbf{Q}_j$ is the complementary projector. There exists μ such that $\mathbf{E}_\mu$ is nonsingular while $\mathbf{E}_j$ are singular for all $0 \leq j < \mu - 1$. In order to make sure projector products are also projectors, März [4] suggested an additional constraint $\mathbf{Q}_j\mathbf{Q}_i = 0, j > i$ on the projector construction for higher index DAEs. Using these chains we can rewrite (1) as projected system of index-μ:

$$\mathbf{P}_{\mu-1} \cdots \mathbf{P}_0 x' + \mathbf{Q}_0 x + \cdots + \mathbf{Q}_{\mu-1} x = \mathbf{E}_\mu^{-1}(\mathbf{A}_\mu x + \mathbf{B}u). \tag{2}$$

As a consequence, (2) can be decoupled into 1 differential part and μ algebraic parts. Unfortunately, März decomposition leads to decoupled system of larger dimension $(1 + \mu)n$ than the dimension n of (1) and also it doesn't preserve the spectrum of the matrix pencil. This motivated us to modify her decoupled system using special basis vectors introduced in [7] and [5] for index-1 and -2, respectively. This modification leads to a decoupled system which preserves the dimension and the stability of the DAE system. In fact, the dimension of the differential part is equal to the dimension of the finite spectrum of the matrix pencil $(\mathbf{E}, \mathbf{A})$. Thus, the stability of the solutions of the decoupled system is guaranteed. We discuss this modification for index-1 and -2 systems in the next section.

3 Modification of the März decomposition

In this section, we modify the März decomposition using special basis vectors. This decomposition leads to a decouple system that preserves the dimension and the spectrum of the matrix pencil of the DAE system.

3.1 Decoupling of index-1 electric networks

Here we assume Equation (1) is of index-1 which implies $\mu = 1$ in Equation (2). We then construct basis vectors $(\mathbf{p}_0, \mathbf{q}_0)$ in $\mathbb{R}^n$ for the projectors $\mathbf{P}_0$ and $\mathbf{Q}_0$ with their inversion

$(\mathbf{p}_0^*, \mathbf{q}_0^*)^T$, where $\mathbf{p}_0^T \in \mathbb{R}^{n,n_0}$, $\mathbf{q}_0^T \in \mathbb{R}^{n,k_0}$ and $n = n_0 + k_0$. Thus an index-1 electric network can be decoupled as:

$$\xi_p' = \mathbf{A}_p\xi_p + \mathbf{B}_p u, \qquad \xi_p(0) = \mathbf{p}_0^{*T} x_0, \tag{3a}$$

$$\xi_q = \mathbf{A}_q\xi_p + \mathbf{B}_q u. \tag{3b}$$

Then, the solution of (1) can be computed using the formula:

$$x = \mathbf{p}_0\xi_p + \mathbf{q}_0\xi_q, \tag{4}$$

where $\mathbf{A}_p := \mathbf{p}_0^{*T}\mathbf{E}_1^{-1}\mathbf{A}_0\mathbf{p}_0 \in \mathbb{R}^{n_0,n_0}$, $\mathbf{B}_p := \mathbf{p}_0^{*T}\mathbf{E}_1^{-1}\mathbf{B} \in \mathbb{R}^{n_0,m}$, $\mathbf{A}_q := \mathbf{q}_0^{*T}\mathbf{E}_1^{-1}\mathbf{A}_0\mathbf{p}_0 \in \mathbb{R}^{k_0,n_0}$ and $\mathbf{B}_q := \mathbf{q}_0^{*T}\mathbf{E}_1^{-1}\mathbf{B} \in \mathbb{R}^{k_0,m}$. Observe that (3a) and (3b) can be solved in a hierarchical way and the desired solutions of the DAE system can be obtained using (4). We observe that the differential part (3a) can be solved using ODE numerical methods and algebraic part (3b) can solved explicitly using the solutions from the differential part. This decoupled system always preserves the dimension since $n = n_0 + k_0$ and the stability of (1) since it can easily be proved that $\sigma(\mathbf{A}_p) = \sigma_f(\mathbf{E}, \mathbf{A})$. The proof can be found in [7]. n_0 and k_0 is the number of differential and algebraic equations, respectively.

3.2 Decoupling of index-2 electric networks

We now assume (1) is of index-2 which implies $\mu = 2$ in Equation (2). We first construct basis vectors $(\mathbf{p}_0, \mathbf{q}_0)$ in $\mathbb{R}^n$ with their inversion $(\mathbf{p}_0{}^*, \mathbf{q}_0{}^*)^T$ for the projectors $\mathbf{P}_0$ and $\mathbf{Q}_0$, where $\mathbf{p}_0 \in \mathbb{R}^{n,n_0}$, $\mathbf{q}_0 \in \mathbb{R}^{n,k_0}$. For this case we have two possibilities depending on the spectrum of the matrix pencil $(\mathbf{E}, \mathbf{A})$.

3.2.1 Matrix pencil $(\mathbf{E}, \mathbf{A})$ *with at least one finite eigenvalue*

Here, we assume that the matrix pencil $(\mathbf{E}, \mathbf{A})$ of Equation (1) has at least one finite eigenvalue. Using projector chains $\mathbf{Q}_1, \mathbf{P}_1 \in \mathbb{R}^{n,n}$ leads to a theorem below [5].

Theorem 1 *Let* $\mathbf{P}_{01} = \mathbf{p}_0^{*T}\mathbf{P}_1\mathbf{p}_0$, $\mathbf{Q}_{01} = \mathbf{p}_0^{*T}\mathbf{Q}_1\mathbf{p}_0$. *Then* $\mathbf{P}_{01}, \mathbf{Q}_{01} \in \mathbb{R}^{n_0,n_0}$ *are projectors in* $\mathbb{R}^{n_0}$ *provided the constraint condition* $\mathbf{Q}_1\mathbf{Q}_0 = 0$ *holds.*

Next, we construct another basis vectors $(\mathbf{p}_{01}, \mathbf{q}_{01})$ in $\mathbb{R}^{n_0}$ made of n_{01} independent columns of projector $\mathbf{P}_{01}$ and k_1 independent columns of its complementary projector $\mathbf{Q}_{01}$ such that $n_0 = n_{01} + k_1$. Let the inverse of this basis be denoted by $(\mathbf{p}_{01}^*, \mathbf{q}_{01}^*)^T$. Then Equation (1) can be decoupled as:

$$\xi_p' = \mathbf{A}_p\xi_p + \mathbf{B}_p u, \qquad \xi_p(0) = \mathbf{p}_{01}^{*T}\mathbf{p}_0^{*T} x_0, \tag{5a}$$

$$\xi_{q,1} = \mathbf{A}_{q,1}\xi_p + \mathbf{B}_{q,1}u, \tag{5b}$$

$$\xi_{q,0} = \mathbf{A}_{q,0}\xi_p + \mathbf{B}_{q,0}u + \mathbf{A}_{q,01}\xi_{q,1}', \tag{5c}$$

and the desired solution of (1) is given by:

$$x = \mathbf{p}_0\mathbf{p}_{01}\xi_p + \mathbf{p}_0\mathbf{q}_{01}\xi_{q,1} + \mathbf{q}_0\xi_{q,0}, \tag{6}$$

where $\mathbf{A}_p := \mathbf{p}_{01}^{*T}\mathbf{p}_0^{*T}\mathbf{E}_2^{-1}\mathbf{A}_2\mathbf{p}_0\mathbf{p}_{01} \in \mathbb{R}^{n_{01},n_{01}}$, $\mathbf{B}_p := \mathbf{p}_{01}^{*T}\mathbf{p}_0^{*T}\mathbf{E}_2^{-1}B \in \mathbb{R}^{n_{01},m}$, $\mathbf{A}_{q,1} := \mathbf{q}_{01}^{*T}\mathbf{p}_0^{*T}\mathbf{E}_2^{-1} \times \mathbf{A}_2\mathbf{p}_0\mathbf{p}_{01} \in \mathbb{R}^{k_1,n_{01}}$, $\mathbf{B}_{q,1} := \mathbf{q}_{01}^{*T}\mathbf{p}_0^{*T}\mathbf{E}_2^{-1}B \in \mathbb{R}^{k_1,m}$, $\mathbf{A}_{q,0} := \mathbf{q}_0^{*T}\mathbf{P}_1\mathbf{E}_2^{-1}\mathbf{A}_2\mathbf{p}_0\mathbf{p}_{01} \in \mathbb{R}^{k_0,n_{01}}$, $\mathbf{B}_{q,0} := \mathbf{q}_0^{*T}\mathbf{P}_1\mathbf{E}_2^{-1}B \in \mathbb{R}^{k_0,m}$ and $\mathbf{A}_{q,01} := \mathbf{q}_0^{*T}\mathbf{Q}_1\mathbf{p}_0\mathbf{q}_{01} \in \mathbb{R}^{k_0,k_1}$. Equations (5a), (5b) and (5c) are of dimension n_{01}, k_1 and k_0, respectively, where $n = n_{01} + k_1 + k_0$. System (5a)-(5c) also preserves the dimension and stability of system (1). Since it also be proved that $\sigma(\mathbf{A}_p) = \sigma_f(\mathbf{E}, \mathbf{A})$. For the purpose of stable numerical implementation decoupled system (5a)-(5c) and (6) can be rewritten as:

$$\xi_p' = \mathbf{A}_p\xi_p + \mathbf{B}_p u, \qquad \xi_p(0) = \mathbf{p}_{01}^{*T}\mathbf{p}_0^{*T}x(0), \tag{7a}$$

$$\xi_q = \sum_{i=0}^{1}\mathcal{L}^i\mathbf{A}_q\mathbf{A}_p^i\xi_p + \mathcal{L}\mathbf{A}_q\mathbf{B}_p u + \sum_{i=0}^{1}\mathcal{L}^i\mathbf{B}_q u^{(i)}, \tag{7b}$$

$$x = \mathbf{p}_0\mathbf{p}_{01}\xi_p + \begin{bmatrix}\mathbf{p}_0\mathbf{q}_{01} & \mathbf{q}_0\end{bmatrix}\xi_q, \tag{7c}$$

where $\xi_q = (\xi_{q,1}, \xi_{q,0})^T \in \mathbb{R}^{n_q}$, $\mathbf{A}_q = (\mathbf{A}_{q,1}, \mathbf{A}_{q,0})^T \in \mathbb{R}^{n_q,n_2}$, $\mathbf{B}_q = (\mathbf{B}_{q,1}, \mathbf{B}_{q,0})^T \in \mathbb{R}^{n_q,m}$, $u^{(i)} = \frac{\mathrm{d}^i u}{\mathrm{d}t} \in \mathbb{R}^m$ and

$$\mathcal{L} := \begin{bmatrix}\mathbf{0} & \mathbf{0}\\ \mathbf{A}_{\mathbf{q}_{01}} & \mathbf{0}\end{bmatrix} \in \mathbb{R}^{n_q,n_q} \tag{8}$$

is a strictly lower triangular nilpotent matrix of index 2. Also the decoupled system (7a)-(7c) can be solved in a hierarchical way and the desired solutions of the DAE system can be obtained using (7c).

3.2.2 *Matrix pencil* $(\mathbf{E}, \mathbf{A})$ *with no finite eigenvalues*

We assume that the matrix pencil $(\mathbf{E}, \mathbf{A})$ of Equation (1) has no finite eigenvalues, *i.e.* $\det(\lambda E - A) = c \in \mathbb{R} \setminus \{0\}$, $\forall\lambda \in \mathbb{C}$. Thus Equation (1) can be decoupled as:

$$\xi_{q,1} = \mathbf{B}_{q,1}u, \tag{9a}$$

$$\xi_{q,0} = \mathbf{B}_{q,0}u + \mathbf{A}_{q,01}\xi_{q,1}', \tag{9b}$$

and the desired solution of (1) is given by:

$$x = \mathbf{p}_0\xi_{q,1} + \mathbf{q}_0\xi_{q,0}, \tag{10}$$

where $\mathbf{B}_{q,1} = \mathbf{p}_0^{*T}\mathbf{E}_2^{-1}B \in \mathbb{R}^{n_0,m}$, $\mathbf{B}_{q,0} = \mathbf{q}_0^{*T}\mathbf{P}_1\mathbf{E}_2^{-1}B \in \mathbb{R}^{k_0,m}$ and $\mathbf{A}_{q,01} = \mathbf{q}_0^{*T}\mathbf{Q}_1\mathbf{p}_0 \in \mathbb{R}^{k_0,n_0}$. Here $n = n_0 + k_0$. Still system (9a) and (9b) can simplified to:

$$\xi_q = \sum_{i=0}^{1}\mathcal{L}^i\mathbf{B}_q u^{(i)}, \tag{11a}$$

$$x = \begin{bmatrix}\mathbf{p}_0 & \mathbf{q}_0\end{bmatrix}\xi_q, \tag{11b}$$

where $\xi_q = (\xi_{q,1}, \xi_{q,0})^T \in \mathbb{R}^n$, $\mathbf{B}_q = (\mathbf{B}_{q,1}, \mathbf{B}_{q,0})^T \in \mathbb{R}^{n,m}$, $u^{(i)} = \frac{\mathrm{d}^i u}{\mathrm{d}t} \in \mathbb{R}^m$ and $\mathcal{L} \in \mathbb{R}^{n,n}$ is also a strictly lower triangular nilpotent matrix of index 2 which is defined as (8). We can observe that we do not need initial conditions to solve system (11a) and (11b), in fact the solutions are computed explicitly. We note that the input vector u must be at least $\mu - 1$ times differentiable.

We have discussed that index-1 and -2 electric networks can be decoupled using special basis vectors. These special basis vectors can be computed numerically in an efficient way using the LUQ routine discussed in [8]. Hence this procedure can be applied even on very large electric networks as illustrated in Section 5.2. We call this method the Split-DAE method since it involves splitting the DAE system into differential and algebraic variables.

4 Analysis of decoupled electric networks

In this section, we analyse the solutions of the decoupled systems discussed in the previous section. For simplicity we need to first rewrite the decoupled systems back into the descriptor form:

$$\tilde{\mathbf{E}}\xi' = \tilde{\mathbf{A}}\xi + \tilde{\mathbf{B}}u, \qquad \xi(0) = \xi_0, \tag{12a}$$

$$x = \mathbf{V}\xi, \tag{12b}$$

where $\tilde{\mathbf{E}}, \tilde{\mathbf{A}}, \mathbf{V} \in \mathbb{R}^{n,n}$, $\tilde{\mathbf{B}} \in \mathbb{R}^{n,m}$ and $\xi \in \mathbb{R}^n$ is the projected state space. We note that the solutions of (1) and (12a) and (12b) coincide since the two systems are equivalent to each other. This implies the spectrum of matrix pair $(\mathbf{E}, \mathbf{A})$ and $(\tilde{\mathbf{E}}, \tilde{\mathbf{A}})$ must also coincide. System (12a) and (12b) is much easier to solve than its counter part and moreover it reveals the interconnection or structure of the DAE system. Assume the matrix pencil $(\mathbf{E}, \mathbf{A})$ of (1) has at least one finite eigenvalue then following theorem below holds:

Theorem 2 *If the matrix pair* $(\mathbf{E}, \mathbf{A})$ *has at least one finite eigenvalue, there exist finite* $\lambda \in \mathbb{C}$ *such that*:

$$\det(\lambda\mathbf{I} - \mathbf{A}_p) = 0.$$

This implies that $\sigma_f(\mathbf{E}, \mathbf{A}) = \sigma(\mathbf{A}_p)$.

Thus the stability of the decoupled system (12a) and (12b) depends on the stability of the differential part. Hence if Equation (1) is stable then also (12a) and (12b) must be stable.

Using the decoupled systems in Section 3, we derive the matrices $(\tilde{\mathbf{E}}, \tilde{\mathbf{A}}, \tilde{\mathbf{B}}, \mathbf{V})$ below:

1. Index 1 electric networks

   From decoupled system (3a) and (3b) we obtain:

$$\tilde{\mathbf{E}} := \begin{bmatrix} \mathbf{I}_{n_0} & \mathbf{0} \\ \mathbf{0} & \mathbf{0} \end{bmatrix}, \quad \tilde{\mathbf{A}} := \begin{bmatrix} \mathbf{A}_p & \mathbf{0} \\ \mathbf{A}_q & -\mathbf{I}_{k_0} \end{bmatrix}, \quad \tilde{\mathbf{B}} := \begin{bmatrix} \mathbf{B}_p \\ \mathbf{B}_q \end{bmatrix}, \\ \mathbf{V} := \begin{bmatrix} \mathbf{p}_0 & \mathbf{q}_0 \end{bmatrix} \quad \text{and} \quad \xi := \begin{bmatrix} \xi_p \\ \xi_q \end{bmatrix}. \tag{13}$$

2. Index 2 electric networks
   (i) Matrix pencil $(\mathbf{E}, \mathbf{A})$ with at least one finite eigenvalue

From decoupled system (5a)-(5c) we obtain:

$$\tilde{\mathbf{E}} := \begin{bmatrix} \mathbf{I}_{n01} & \mathbf{0} \\ \mathbf{0} & -\mathcal{L} \end{bmatrix}, \qquad \tilde{\mathbf{A}} := \begin{bmatrix} \mathbf{A}_p & \mathbf{0} \\ \mathbf{A}_q & -\mathbf{I}_{n_q} \end{bmatrix}, \qquad \tilde{\mathbf{B}} := \begin{bmatrix} \mathbf{B}_p \\ \mathbf{B}_q \end{bmatrix},$$

$$\mathbf{V} := \begin{bmatrix} \mathbf{V}_p & \mathbf{V}_q \end{bmatrix}, \quad \text{where } \mathbf{V}_p = \mathbf{p}_0\mathbf{p}_{01}, \mathbf{V}_q = \begin{bmatrix} \mathbf{p}_0\mathbf{q}_{01} & \mathbf{q}_0 \end{bmatrix} \tag{14}$$

$$\text{and } \xi := \begin{bmatrix} \xi_p \\ \xi_q \end{bmatrix}.$$

(ii) Matrix pencil $(\mathbf{E}, \mathbf{A})$ with no finite eigenvalues

From decoupled system (9a) and (9b) we obtain:

$$\tilde{\mathbf{E}} := -\mathcal{L}, \qquad \tilde{\mathbf{A}} := -\mathbf{I}_{n_q}, \qquad \tilde{B} := \mathbf{B}_q, \qquad \mathbf{V} := \mathbf{V}_q = \begin{bmatrix} \mathbf{p}_0 & \mathbf{q}_0 \end{bmatrix}, \qquad \text{and} \quad \xi := \xi_q. \tag{15}$$

We observe that the descriptor form with differential part takes the same form for the case of index-1 and -2 system though $\mathcal{L} = \mathbf{0}$ for index-1 systems. We use this form only for analysis of the solutions of the DAE system (1) but not for solving. For solving one need to solve the decoupled systems derived in the previous section.

4.1 Numerical accuracy of the decoupled system

Suppose that we want to approximate the solution of the DAE system (12a) and (12b). We can do this by discretizing the time interval $t \in [0, T]$ into N subintervals using $h = \frac{T-0}{N}$ time-steps and set $t_n = ih$, $i = 0, 1, \ldots, N-1$. We can then approximate the solutions of DAE system (12a) and (12b) at every point. For simplicity, we consider the implicit Euler method since is the starting point for most high order methods and numerically stable. Applying this method on (12a), we obtain:

$$\tilde{\mathbf{E}}\frac{(\xi_{i+1} - \xi_i)}{h} = \tilde{\mathbf{A}}\xi_{i+1} + \tilde{\mathbf{B}}u_{i+1}. \tag{16}$$

Rearranging Equation (16) we obtain:

$$\xi_{i+1} = (\tilde{\mathbf{E}} - h\tilde{\mathbf{A}})^{-1}[\tilde{\mathbf{E}}\xi_i + h\tilde{B}u_{i+1}]. \tag{17}$$

Thus the approximate solution of Equations (12a) and (12b) can be obtained using the formula: $x_{i+1} = \mathbf{V}\xi_{i+1}$. Then the exact solution of Equations (12a) and (12b) is given by $x(t_i) = \mathbf{V}\xi(t_i)$. The exact solution $\xi(t_i)$ of Equation (12a) satisfies:

$$\tilde{\mathbf{E}}\left[\frac{\xi(t_{i+1}) - \xi(t_i)}{h} + \frac{h}{2}\xi''(\eta)\right] = \tilde{A}\xi(t_{i+1}) + \tilde{\mathbf{B}}u(t_{i+1}).$$

Defining the global error $e_i := \xi(t_i) - \xi_i$ and setting $e_0 = 0$, since the consistent initial condition is known. We obtain an iterative process below:

$$e_{i+1} = (\tilde{\mathbf{E}} - h\tilde{\mathbf{A}})^{-1}\tilde{\mathbf{E}}\left[e_i - \frac{h^2}{2}\xi''(\eta)\right], \quad i \geq 0. \tag{18}$$

We use Equation (18) to analyse the accuracy of the implicit Euler method for the first step, *i.e.* $i = 1$ using the following cases.

1. Index 1 systems

   Using system matrices (13), we obtain:

$$(\tilde{\mathbf{E}} - h\tilde{\mathbf{A}})^{-1}\tilde{\mathbf{E}} = \begin{bmatrix} (\mathbf{I}_{n0} - h\mathbf{A}_p)^{-1} & \mathbf{0} \\ \mathbf{A}_q(\mathbf{I}_{n0} - h\mathbf{A}_p)^{-1} & \mathbf{0} \end{bmatrix}, \qquad \xi''(\eta) = \begin{bmatrix} \xi_p''(\eta) \\ \mathbf{A}_q\xi_p''(\eta) + \mathbf{B}_q u''(\eta) \end{bmatrix}. \tag{19}$$

   Substituting Equation (19) into Equation (18), leads to:

$$e_1 = \frac{h^2}{2}\begin{bmatrix} (\mathbf{I}_{n0} - h\mathbf{A}_p)^{-1}\xi_p''(\eta) \\ (\mathbf{I}_{n0} - h\mathbf{A}_p)^{-1}\xi_p''(\eta) \end{bmatrix} = \mathcal{O}(h^2).$$

   We observe that this corresponds to the local error for implicit Euler method. This implies that the accuracy of the solutions of the decoupled system is determined by the numerical accuracy of the differential part. As illustrated in Example 1.

2. Index 2 systems

   As we have already discussed for the case of index-2 system in Section 3.2, we need to consider two cases:

   (i) Matrix pencil $(\mathbf{E}, \mathbf{A})$ with at least one finite eigenvalue

   Here we use system matrices (14) and obtain:

$$(\tilde{\mathbf{E}} - h\tilde{\mathbf{A}})^{-1}\tilde{\mathbf{E}} = \begin{bmatrix} \mathbf{M}_p & \mathbf{0} \\ \mathbf{M}_L h\mathbf{A}_q\mathbf{M}_p & \mathbf{M}_L\mathcal{L} \end{bmatrix}, \qquad \xi''(\eta) = \begin{bmatrix} \xi_p''(\eta) \\ \xi_q''(\eta) \end{bmatrix}, \tag{20}$$

   where $\mathbf{M}_p = (\mathbf{I}_{n01} - h\mathbf{A}_p)^{-1}$, $\mathbf{M}_L = (h\mathbf{I}_{n_q} - \mathcal{L})^{-1}$ and $\xi_q(\eta)$ is defined by expression (7b). Substituting Equation (20) into (18), leads to:

$$e_1 = -\frac{1}{2}\begin{bmatrix} h^2\mathbf{M}_p\xi_p''(\eta) \\ h^3\mathbf{M}_L\mathbf{A}_q\mathbf{M}_p\xi_p''(\eta) - h^2\mathbf{M}_L\mathcal{L}\xi_q''(\eta) \end{bmatrix} = \mathcal{O}(h), \tag{21}$$

   since $h^3\mathbf{M}_L\mathbf{A}_q\mathbf{M}_p = \mathcal{O}(h^2)$ and $h^2\mathbf{M}_L\mathcal{L} = \mathcal{O}(h)$. We observe that for this case implicit Euler method loses accuracy. We observe that the hidden constraint (7b) destroys the accuracy of the implicit Euler method. We do not face this problem in our method since the algebraic part (7b) is computed explicitly. For our method the accuracy will depend on the method used to solve the differential part (7a). This is illustrated in Example 2.

   (ii) Matrix pencil $(\mathbf{E}, \mathbf{A})$ with no finite eigenvalue

   Using system matrices (15), we obtain:

$$(\tilde{\mathbf{E}} - h\tilde{\mathbf{A}})^{-1}\tilde{\mathbf{E}} = -\mathbf{M}_L\mathcal{L}, \qquad \xi''(\eta) = \xi_q''(\eta), \tag{22}$$

   where $\mathbf{M}_L$ is defined as in the previous case while $\xi_q(\eta)$ is defined by (11a). Substituting Equation (22) into Equation (18), we obtain:

$$e_1 = \frac{1}{2}h^2\mathbf{M}_L\mathcal{L}\xi_q''(\eta) = \mathcal{O}(h).$$

We observe that also this case the implicit Euler method loses local error accuracy. For our method we do not need to use any integration method since the solutions of the decoupled system (11a) are computed ellipticity. This is also illustrated in the Example 3.

Hence implicit Euler method loses accuracy for each increase in the index if used on the DAE system directly. Thus first decoupling the DAE system (1) into differential and algebraic parts is a robust and effective way of solving DAEs. Since on the differential part one can use any numerical integration method and then the solutions of the algebraic part can be obtained explicitly. Hence the Split-DAE method is a very accurate method and very easy to implement as compared to its counterparts discussed in [2].

5 Numerical experiments

In this section, we test our proposed method in Section 3 using problems from electromagnetic community since is the main focus of this paper but it can also be applied to other applications.

5.1 Simple examples

Here, we explicitly discuss how to decouple DAEs into differential and algebraic parts using projectors. We illustrate this using examples below. Example 1 is an index-1 electric network while Example 2 is an index-2 electric network whose decoupled system has a differential part. In Example 3, we illustrate the decoupling of index-2 system without differential part.

Example 1 Consider a linear RLC electric network in Figure 1. We need to find the unknowns $x = [e_1\ e_2\ e_3\ \imath_L\ \imath_V]^T$ in the above electric network. In order to find the unknowns we use the MNA to formulate a mathematical model of the form (1) with system matrices:

$$\mathbf{E} = \begin{bmatrix} 0 & 0 & 0 & 0 & 0 \\ 0 & 0 & 0 & 0 & 0 \\ 0 & 0 & C & 0 & 0 \\ 0 & 0 & 0 & L & 0 \\ 0 & 0 & 0 & 0 & 0 \end{bmatrix}, \quad \mathbf{A} = \begin{bmatrix} -G & G & 0 & 0 & 1 \\ G & -G & 0 & -1 & 0 \\ 0 & 0 & 0 & 1 & 0 \\ 0 & 1 & -1 & 0 & 0 \\ -1 & 0 & 0 & 0 & 0 \end{bmatrix}, \quad \mathbf{B} = \begin{bmatrix} 0 \\ 0 \\ 0 \\ 0 \\ -1 \end{bmatrix}. \tag{23}$$

We observe that det(**E**) = 0, thus the system is a DAEs of dimension $n = 5$. It is solvable since $\det(\lambda E - A) = GCL\lambda^2 + C\lambda + G \neq 0$ for some $\lambda \in \mathbb{C}$. We need to first check the index

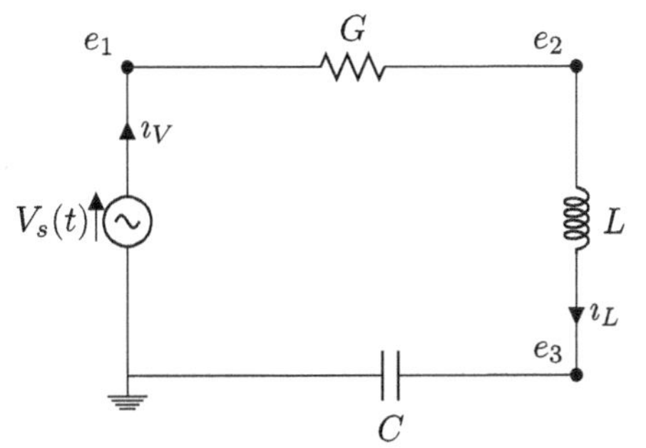

Figure 1 Simple RLC electric network.

of the DAE system. We do this by constructing the matrix and projector chains of matrix pair $(\mathbf{E}, \mathbf{A})$. Setting $\mathbf{E}_0 = \mathbf{E}$, $\mathbf{A}_0 = \mathbf{A}$, we can choose projector $\mathbf{Q}_0$ such that $\operatorname{Im}\mathbf{Q}_0 = \operatorname{Ker}\mathbf{E}_0$, and then compute $\mathbf{E}_1 = \mathbf{E}_0 - \mathbf{A}_0\mathbf{Q}_0$. Thus we obtain the first iterate of matrix and projector chain given by:

$$\mathbf{Q}_0 = \begin{bmatrix} 1 & 0 & 0 & 0 & 0 \\ 0 & 1 & 0 & 0 & 0 \\ 0 & 0 & 0 & 0 & 0 \\ 0 & 0 & 0 & 0 & 0 \\ 0 & 0 & 0 & 0 & 1 \end{bmatrix}, \qquad \mathbf{P}_0 = \begin{bmatrix} 0 & 0 & 0 & 0 & 0 \\ 0 & 0 & 0 & 0 & 0 \\ 0 & 0 & 1 & 0 & 0 \\ 0 & 0 & 0 & 1 & 0 \\ 0 & 0 & 0 & 0 & 0 \end{bmatrix},$$
$$\mathbf{E}_1 = \begin{bmatrix} G & -G & 0 & 0 & -1 \\ -G & G & 0 & 0 & 0 \\ 0 & 0 & C & 0 & 0 \\ 0 & -1 & 0 & L & 0 \\ 1 & 0 & 0 & 0 & 0 \end{bmatrix}. \tag{24}$$

Since $\mathbf{E}_1$ is nonsingular, thus Equation (23) is an index-1 system. Thus the decouple system of (23) will take the form (3a) and (3b). We then construct basis matrices and their respective inverses for projectors $\mathbf{Q}_0$ and $\mathbf{P}_0$ given by:

$$\mathbf{q}_0 = \begin{bmatrix} 1 & 0 & 0 \\ 0 & 1 & 0 \\ 0 & 0 & 0 \\ 0 & 0 & 0 \\ 0 & 0 & 1 \end{bmatrix}, \qquad \mathbf{p}_0 = \begin{bmatrix} 0 & 0 \\ 0 & 0 \\ 1 & 0 \\ 0 & 1 \\ 0 & 0 \end{bmatrix}, \quad \text{and} \quad \mathbf{q}_0^{*T} = \mathbf{q}_0^T, \qquad \mathbf{p}_0^{*T} = \mathbf{p}_0^T. \tag{25}$$

Substituting Equations (23)-(25) into Equations (3a) and (3b) leads to the decoupled system of system (23) given by:

$$\xi_p' = \begin{bmatrix} 0 & \frac{1}{C} \\ -\frac{1}{L} & -\frac{1}{GL} \end{bmatrix} \xi_p + \begin{bmatrix} 0 \\ -\frac{1}{L} \end{bmatrix} u, \qquad \xi_p(0) = \begin{bmatrix} e_3(0) \\ \imath_L(0) \end{bmatrix}, \tag{26a}$$

$$\xi_q = \begin{bmatrix} 0 & 0 \\ 0 & -\frac{1}{G} \\ 0 & 1 \end{bmatrix} \xi_p + \begin{bmatrix} -1 \\ -1 \\ 0 \end{bmatrix} u, \tag{26b}$$

and the desired solution is given by:

$$x = \begin{bmatrix} 0 & 0 & 1 & 0 & 0 \\ 0 & 0 & 0 & 1 & 0 \end{bmatrix}^T \xi_p + \begin{bmatrix} 1 & 0 & 0 & 0 & 0 \\ 0 & 1 & 0 & 0 & 0 \\ 0 & 0 & 0 & 0 & 1 \end{bmatrix}^T \xi_q.$$

The decoupled system (26a) and (26b) can be solved in hierarchical way if we apply $\xi_p(0) = [e_3(0)\ \imath_L(0)]^T$. Observe that $n_0 = 2$, $k_0 = 3 \Rightarrow n = n_0 + k_0 = 5$ and $\sigma_f(\mathbf{E}, \mathbf{A}) = \sigma(\mathbf{A}_p)$ as expected. Thus system (23) is decoupled into 2 and 3 differential and algebraic equations, respectively. We observe that decoupled system (26a) and (26b) can be solved using any

ODE numerical integration methods. Thus system (26a) and (26b) is easier to solve than solving (23).

Example 2 Consider another linear RLC electric network with system matrices:

$$\mathbf{E} = \begin{bmatrix} C & 0 & 0 & 0 \\ 0 & 0 & 0 & 0 \\ 0 & 0 & L & 0 \\ 0 & 0 & 0 & 0 \end{bmatrix}, \quad \mathbf{A} = \begin{bmatrix} -G & G & 0 & 1 \\ G & -G & -1 & 0 \\ 0 & 1 & 0 & 0 \\ -1 & 0 & 0 & 0 \end{bmatrix}, \quad \mathbf{B} = \begin{bmatrix} 0 \\ 0 \\ 0 \\ -1 \end{bmatrix}, \quad x = \begin{bmatrix} e_1 \\ e_2 \\ \imath_L \\ \imath_V \end{bmatrix}. \tag{27}$$

This is also DAE system of dimension $n = 4$ and it is also solvable since $\det(\lambda E - A) = GL\lambda + 1 \neq 0$ for some $\lambda \in \mathbb{C}$. We construct matrix and projector chains by first setting $\mathbf{E}_0 = \mathbf{E}$, $\mathbf{A}_0 = \mathbf{A}$. We choose projector $\mathbf{Q}_0$ such that $\operatorname{Im}\mathbf{Q}_0 = \operatorname{Ker}\mathbf{E}_0$ and its complementary projector $\mathbf{P}_0$:

$$\mathbf{Q}_0 = \begin{bmatrix} 0 & 0 & 0 & 0 \\ 0 & 1 & 0 & 0 \\ 0 & 0 & 0 & 0 \\ 0 & 0 & 0 & 1 \end{bmatrix}, \quad \text{and} \quad \mathbf{P}_0 = \begin{bmatrix} 1 & 0 & 0 & 0 \\ 0 & 0 & 0 & 0 \\ 0 & 0 & 1 & 0 \\ 0 & 0 & 0 & 0 \end{bmatrix}. \tag{28}$$

Then,

$$\mathbf{E}_1 = \mathbf{E}_0 - \mathbf{A}_0\mathbf{Q}_0 = \begin{bmatrix} C & -G & 0 & -1 \\ 0 & G & 0 & 0 \\ 0 & -1 & L & 0 \\ 0 & 0 & 0 & 0 \end{bmatrix}, \quad \mathbf{A}_1 = \mathbf{A}_0\mathbf{P}_0 = \begin{bmatrix} 0 & G & 0 & 1 \\ 0 & -G & 0 & 0 \\ 0 & 1 & 0 & 0 \\ 0 & 0 & 0 & 0 \end{bmatrix}.$$

Since $\mathbf{E}_1$ is singular, the index-1 condition is violet. Thus we need to continue with the process and choose projector $\mathbf{Q}_1$ such that $\operatorname{Im}\mathbf{Q}_1 = \operatorname{Ker}\mathbf{E}_1$ and its complementary projector $\mathbf{P}_1 = \mathbf{I} - \mathbf{Q}_1$:

$$\mathbf{Q}_1 = \frac{1}{C^2+1}\begin{bmatrix} 1 & 0 & 0 & C \\ 0 & 0 & 0 & 0 \\ 0 & 0 & 0 & 0 \\ C & 0 & 0 & C^2 \end{bmatrix}, \quad \text{and} \quad \mathbf{P}_1 = \mathbf{I} - \mathbf{Q}_1. \tag{29}$$

Then,

$$\mathbf{E}_2 = \mathbf{E}_1 - \mathbf{A}_1\mathbf{Q}_1 = \begin{bmatrix} C + \frac{G}{C^2+1} & -G & 0 & -1 + \frac{GC}{C^2+1} \\ -\frac{G}{C^2+1} & G & 0 & -\frac{GC}{C^2+1} \\ 0 & -1 & L & 0 \\ \frac{1}{C^2+1} & 0 & 0 & -\frac{C}{C^2+1} \end{bmatrix}. \tag{30}$$

We can easily check that $\mathbf{E}_2$ is non-singular. Thus Equation (27) is of index-2.

The next step is to decouple the DAE system (27) using the constructed matrix and projector chains. In this step we need to make sure that condition $\mathbf{Q}_1\mathbf{Q}_0 = 0$ holds true. We notice that projectors (28) and (29) don't satisfy this condition. Thus we need to construct other special projectors. Our best choice is the canonical projector suggested by März [4], which is defined as:

$$\tilde{\mathbf{Q}}_1 := -\mathbf{Q}_1\mathbf{E}_2^{-1}\mathbf{A}_1 = \begin{bmatrix} 1 & 0 & 0 & 0 \\ 0 & 0 & 0 & 0 \\ 0 & 0 & 0 & 0 \\ C & 0 & 0 & 0 \end{bmatrix}, \quad \text{and} \quad \tilde{\mathbf{P}}_1 = \mathbf{I} - \tilde{\mathbf{Q}}_1.$$

We can easily check that condition $\tilde{\mathbf{Q}}_1\mathbf{Q}_0 = 0$ is now satisfied. We then use $\tilde{\mathbf{Q}}_1$ instead of $\mathbf{Q}_1$ in the decoupling process. If we repeat step (30), we obtain:

$$\mathbf{E}_2 = \mathbf{E}_1 - \mathbf{A}_1\tilde{\mathbf{Q}}_1 = \begin{bmatrix} C+G & -G & 0 & -1 \\ -G & G & 0 & 0 \\ 0 & -1 & L & 0 \\ 1 & 0 & 0 & 0 \end{bmatrix}, \qquad \mathbf{A}_2 = \mathbf{A}_1\tilde{\mathbf{P}}_1 = \begin{bmatrix} 0 & 0 & 0 & 0 \\ 0 & 0 & -1 & 0 \\ 0 & 0 & 0 & 0 \\ 0 & 0 & 0 & 0 \end{bmatrix}. \tag{31}$$

For convenience we can set $\mathbf{Q}_1 = \tilde{\mathbf{Q}}_1$ and $\mathbf{P}_1 = \tilde{\mathbf{P}}_1$. Since Equation (27) is an index-2 system and its matrix pencil has at least one finite eigenvalues. Thus its decouple system will be in form (5a)-(5c). The next step is to construct basis matrices for projectors and projector products. The basis matrices for projectors $\mathbf{P}_0$ and $\mathbf{Q}_0$ in Equation (28), and their respective inverses are given by:

$$\mathbf{p}_0 = \begin{bmatrix} 1 & 0 \\ 0 & 0 \\ 0 & 1 \\ 0 & 0 \end{bmatrix}, \quad \mathbf{q}_0 = \begin{bmatrix} 0 & 0 \\ 1 & 0 \\ 0 & 0 \\ 0 & 1 \end{bmatrix}, \quad \text{and} \quad \overset{*}{\mathbf{p}}_0 = \begin{bmatrix} 1 & 0 \\ 0 & 0 \\ 0 & 1 \\ 0 & 0 \end{bmatrix}, \quad \overset{*}{\mathbf{q}}_0 = \begin{bmatrix} 0 & 0 \\ 1 & 0 \\ 0 & 0 \\ 0 & 1 \end{bmatrix}. \tag{32}$$

Using Theorem 1, we compute other projectors $\mathbf{P}_{01}$ and $\mathbf{Q}_{01}$ given by:

$$\mathbf{P}_{01} = \overset{*}{\mathbf{p}}_0^T\mathbf{P}_1\mathbf{p}_0 = \begin{bmatrix} 0 & 0 \\ 0 & 1 \end{bmatrix}, \quad \text{and} \quad \mathbf{Q}_{01} = \overset{*}{\mathbf{p}}_0^T\mathbf{Q}_1\mathbf{p}_0 = \begin{bmatrix} 1 & 0 \\ 0 & 0 \end{bmatrix}.$$

The basis matrices for projectors $\mathbf{P}_{01}$ and $\mathbf{Q}_{01}$, and their respective inverses are given by:

$$\mathbf{p}_{01} = \begin{bmatrix} 0 \\ 1 \end{bmatrix}, \quad \mathbf{q}_{01} = \begin{bmatrix} 1 \\ 0 \end{bmatrix}, \quad \text{and} \quad \overset{*}{\mathbf{p}}_{01} = \begin{bmatrix} 0 \\ 1 \end{bmatrix}, \quad \overset{*}{\mathbf{q}}_{01} = \begin{bmatrix} 1 \\ 0 \end{bmatrix}. \tag{33}$$

Substituting Equations (31)-(33) into (7a)-(7c) leads to the decoupled system of (27) given by:

$$\xi_p' = -\frac{1}{GL}\xi_p - \frac{1}{L}u, \qquad \xi_p(0) = \iota_L(0), \tag{34a}$$

$$\xi_q = \begin{bmatrix} 0 \\ -\frac{1}{G} \\ 1 \end{bmatrix}\xi_p + \begin{bmatrix} -1 \\ -1 \\ 0 \end{bmatrix}u + \begin{bmatrix} 0 \\ 0 \\ -C \end{bmatrix}u', \tag{34b}$$

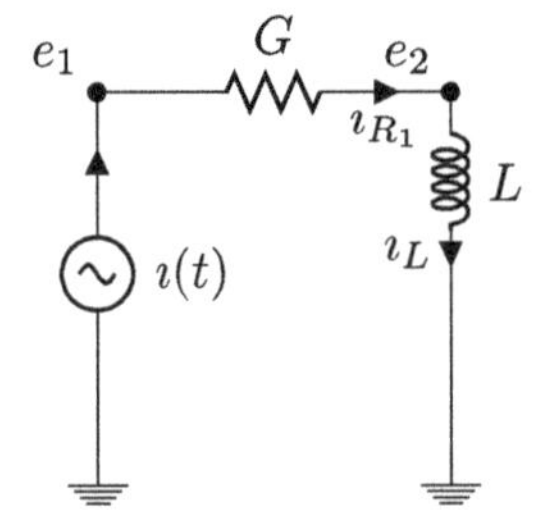

Figure 2 Simple RL network.

and the desired solution is given by:

$$x = \begin{bmatrix} 0 \\ 0 \\ 1 \\ 0 \end{bmatrix} \xi_p + \begin{bmatrix} 1 & 0 & 0 \\ 0 & 1 & 0 \\ 0 & 0 & 0 \\ 0 & 0 & 1 \end{bmatrix} \xi_q. \tag{35}$$

We observe that $n_{10} = 1, k_1 = 1, k_0 = 2 \Rightarrow n = n_{10} + k_1 + k_0 = 4$ and $\sigma_f(\mathbf{E}, \mathbf{A}) = \sigma(\mathbf{A}_p) = \{-\frac{1}{GL}\}$. Equations (34a) and (34b) can be solved if we apply $\xi_p(0) = \imath_L(0)$. Thus the DAE system (27) is decoupled into 1 differential and 3 algebraic equations. The solutions of system (27) and (34a) and (34b), (35) coincides though the later is easier to solve than the former.

Example 3 Consider a simple RL electric network in Figure 2. We need to find the unknowns $x = [e_1, e_2, \imath_L]^T$ in the electric network. Using the Modified Nodal Analysis on this network leads to DAE system of the form (1), where $u = \imath(t)$ with system matrices given by

$$E = \begin{bmatrix} 0 & 0 & 0 \\ 0 & 0 & 0 \\ 0 & 0 & L \end{bmatrix}, \quad A = \begin{bmatrix} -G & G & 0 \\ G & -G & -1 \\ 0 & 1 & 0 \end{bmatrix}, \quad B = \begin{bmatrix} 1 \\ 0 \\ 0 \end{bmatrix}. \tag{36}$$

Since $\det(\lambda E - A) = G$, thus this system is solvable and its matrix pencil $(\mathbf{E}, \mathbf{A})$ has only infinite eigenvalues. Thus its decoupled system must be purely algebraic system. We choose special projectors,

$$\mathbf{Q}_0 = \begin{bmatrix} 1 & 0 & 0 \\ 0 & 1 & 0 \\ 0 & 0 & 0 \end{bmatrix}, \quad \text{and} \quad \mathbf{Q}_1 = \begin{bmatrix} 0 & 0 & L \\ 0 & 0 & L \\ 0 & 0 & 1 \end{bmatrix}, \tag{37}$$

such that $\mathbf{Q}_1\mathbf{Q}_0 = 0$ holds. Then we have,

$$\mathbf{E}_2 = \begin{bmatrix} G & -G & 0 \\ -G & G & 1 \\ 0 & -1 & L \end{bmatrix}. \tag{38}$$

Since the $\mathbf{E}_2$ is non-singular, thus (36) is an index 2 system. The linearly independent columns and their respective inverses of projector $\mathbf{Q}_0$ and its complimentary $\mathbf{P}_0$ are given by;

$$\mathbf{p}_0 = \begin{bmatrix} 0 \\ 0 \\ 1 \end{bmatrix}, \qquad \mathbf{q}_0 = \begin{bmatrix} 1 & 0 \\ 0 & 1 \\ 0 & 0 \end{bmatrix},$$
$$\mathbf{p}_0{}^{*T} = \begin{bmatrix} 0 & 0 & 1 \end{bmatrix}, \qquad \mathbf{q}_0{}^{*T} = \begin{bmatrix} 1 & 0 & 0 \\ 0 & 1 & 0 \end{bmatrix}. \tag{39}$$

Thus, substituting Equations (37)-(39) into (11a) and (11b), we obtain,

$$\xi_q = \begin{bmatrix} 1 \\ G^{-1} \\ 0 \end{bmatrix} \imath(t) + \begin{bmatrix} 0 \\ L \\ L \end{bmatrix} \imath'(t), \tag{40}$$

$$x = \begin{bmatrix} 0 & 1 & 0 \\ 0 & 0 & 1 \\ 1 & 0 & 0 \end{bmatrix} \xi_q. \tag{41}$$

We can observe that system (40) can be computed explicitly without using any numerical integration techniques. Thus instead of solving DAE system (36) is easier to solve (40) and their solutions much coincide.

5.2 Industrial applications

In this section, we test the Split-DAE method on large scale descriptor electromagnetic models from industries. Examples 4 and 5 are index-1 and -2 systems respectively. These examples are multiple-input multiple-output (MIMO) type, *i.e.* we are interested in a few solutions of the DAE system.

Example 4 This is a descriptor model of a large scale power system originating from [9]. It is called bauru5727 power system and can be downloaded from [10]. It is an index-1 DAE system of dimension $n = 40{,}266$ with 2 inputs and 2 outputs. The sparsity of its matrix pencil is shown in Figure 3. We decoupled the DAE system into 5,727 differential equations and 34,639 algebraic equations using the procedure we discussed in Section 3.1, *i.e.* $n_0 = 5{,}727$ and $k_0 = 34{,}639$. The sparsity of the matrix pencil of the decoupled system in the descriptor form is given in Figure 4. We observe that the sparsity in Figure 4 has the same structures as matrix pencil $(\tilde{\mathbf{E}}, \tilde{\mathbf{A}})$ in Equation (13). We solved the decoupled system in Matlab using ODE solver (ode15s) to solve the differential part. We obtained good solutions as shown in Figure 5 and the simulation took 3,486 seconds. We used input function $u(t) = [\cos(t) \;\; \sin(t)]^T$, $t \in [0, \pi]$ and time step $h = 1.5 \times 10^{-4}$. We tried to solve the DAE system directly using he Matlab DAE solver (ode15s) but failed to solve it. It just gave us a warning: 'This DAE appears to be of index greater than 1.'

Example 5 In this example, we consider a RLC network descriptor model of electric power grid originating from [11]. It is called M8OPI system and can also be downloaded

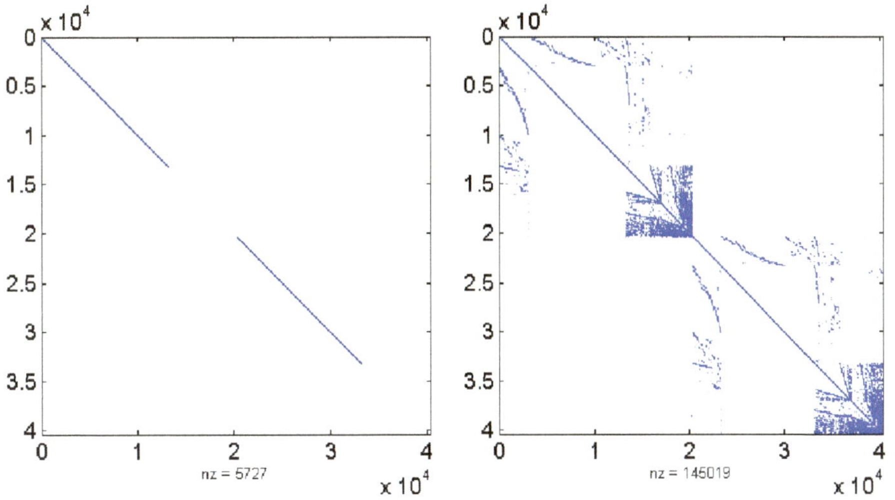

(a) E. (b) A.

Figure 3 Sparsity of matrix pencil (E, A).

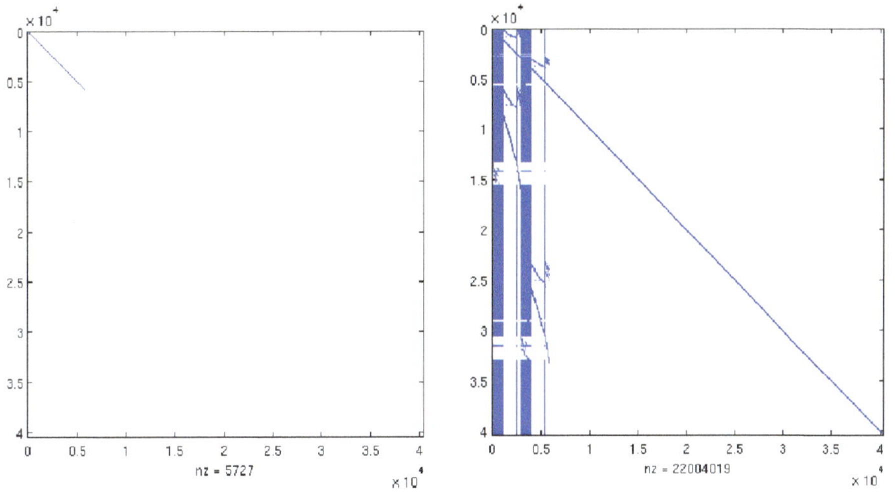

(a) $\tilde{\mathbf{E}}$. (b) $\tilde{\mathbf{A}}$.

Figure 4 Sparsity of matrix pencil ($\tilde{\text{E}}, \tilde{\text{A}}$).

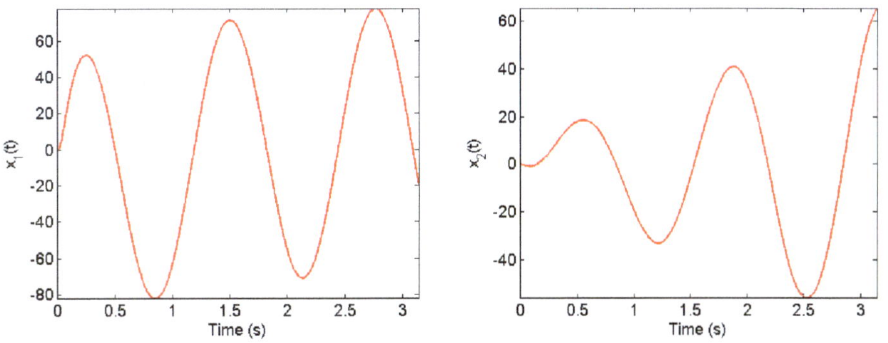

(a) Solution x_1. (b) Solution x_2.

Figure 5 Output solutions of Example 4.

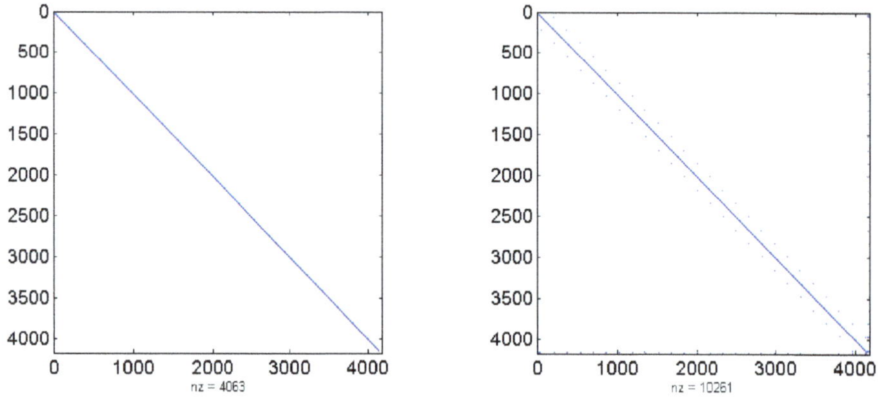

(a) E. (b) A.

Figure 6 Sparsity of matrix pencil (E, A).

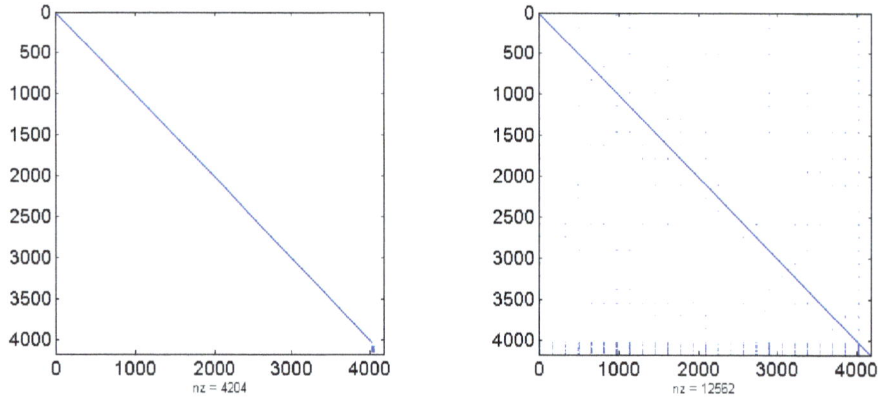

(a) $\tilde{\mathbf{E}}$. (b) $\tilde{\mathbf{A}}$.

Figure 7 Sparsity of matrix pencil ($\tilde{E}, \tilde{A}$).

from [10]. It is an index-2 DAE system of dimension $n = 4{,}182$ with 3 inputs and 3 outputs. The inputs are current signals injected into three buses given by:

$$u = \begin{bmatrix} \sin(\omega_1(t - t_1))U(t - t_1) \\ \sin(\omega_2(t - t_2))U(t - t_2) \\ \sin(\omega_3(t - t_3))U(t - t_3) \end{bmatrix},$$

where $U(t)$ is the step function, $\omega_1 = 2 \times \pi \times 2{,}000$ rad/s, $\omega_2 = 2 \times \pi \times 750$ rad/s and $\omega_3 = 2 \times \pi \times 150$ rad/s. Values $t_1 = 5$ ms, $t_2 = 3$ ms and $t_1 = 1$ ms are time delays associated with the input signals. The desired solutions are voltages $e_1(t)$, $e_2(t)$ and $e_3(t)$ at the three buses. The sparsity of the matrix pencil of this DAE system is shown in Figure 6. We were able to decoupled the DAE system into 4,028 differential equations and 154 algebraic equations using the procedure we discussed in Section 3.2.1, *i.e.* $n_{10} = 4028$, $k_1 = 35$ and $k_0 = 119$. The sparsity of the matrix pencil of the decoupled system in the descriptor form is shown in Figure 7. We solved the differential part of the decoupled system using ode23t solver and we obtained good solutions as shown in Figure 8. The simulation took 39 seconds.

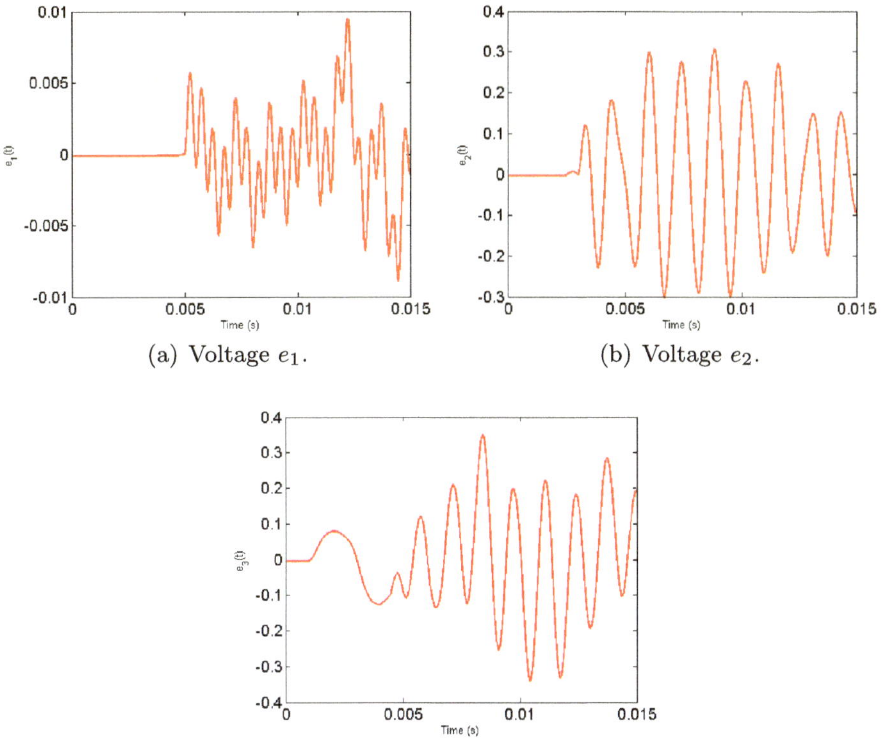

(a) Voltage e_1. (b) Voltage e_2. (c) Voltage e_3.

Figure 8 Output solutions of Example 5.

We could not compare with the Matlab inbuilt solvers since they cannot solve index-2 systems. We validated our results with those presented in [11].

6 Conclusion

In conclusion, splitting DAEs using special projectors leads to decoupled systems which are easier to solve than DAE systems directly. We have tested the Split-DAE method on both large and small problems and proved to be a robust and efficient method. We can now use any numerical integration method to solve DAE systems. Hence this method can be used to simulate electromagnetic descriptor models.

Acknowledgements
This work was supported by The Netherlands Organization for Scientific Research (NWO).

References

1. Tischendorf C: **Topological index calculation of differential-algebraic equations in circuit simulation.** *Surv. Math. Ind.* 1999, **8**:187-199.
2. Brenan K, Campbell S, Petzold L: *Numerical Solution of Initial Value Problems in Differential-Algebraic Equations.* Philadelphia: SIAM; 1996.
3. Differential Algebraic Equations (DAEs) [http://www.lehigh.edu/~wes1/apci/11may00.pdf]
4. März R: **Canonical projectors for linear differential algebraic equations.** *Comput. Math. Appl.* 1996, **31**:121-135.
5. Alì G, Banagaaya N, Schilders W, Tischendorf C: **Index-aware model order reduction for index-2 differential-algebraic equations.** *SIAM J. Sci. Comput.*, in press.
6. Duan GR: *Analysis and Design of Descriptor Linear Systems.* New York: Springer; 2010.
7. Alì G, Banagaaya N, Schilders W, Tischendorf C: **Index-aware model order reduction for differential-algebraic equations.** *Math. Comput. Model. Dyn. Syst.*, in press.
8. Zhang Z, Wong N: **An efficient projector-based passivity test for descriptor systems.** *IEEE Trans. Comput.-Aided Des. Integr. Circuits Syst.* 2010, **29**:1203-1214.

9. Rommes J, Martins N, Freitas F: **Computing rightmost eigenvalues for small-signal stability assessment of large-scale power systems.** *IEEE Trans. Power Syst.* 2010, **25**:929-938.
10. Test Power Systems [http://sites.google.com/site/rommes/software]
11. Freitas F, Martins N, Varrichio S, Rommes J, Véliz F: **Reduced-order transfer matrices from network descriptor models of electric power grids.** *IEEE Trans. Power Syst.* 2011, **26**:1905-1919.

3

An inverse method for color uniformity in white LED spotlights

Corien Prins[1*], Jan ten Thije Boonkkamp[1], Teus Tukker[2] and Wilbert IJzerman[3]

*Correspondence: c.r.prins@tue.nl
[1]Centre for Analysis, Scientific computing and Applications (CASA), Technische Universiteit Eindhoven, P.O. Box 513, 5600 MB Eindhoven, The Netherlands
Full list of author information is available at the end of the article

Abstract

Color over Angle (CoA) variation in the light output of white phosphor-converted LEDs is a common problem in LED lighting technology. In this article we propose an inverse method to design an optical element that eliminates the color variation for a point light source. The method in this article is an improved version of an earlier method by the same authors, and provides more design freedom than the original method. We derive a mathematical model for color mixing in a collimator and present a numerical algorithm to solve it. We verify the results using Monte-Carlo ray tracing.

1 Introduction

LED is a rising technology in the field of lighting. In the past, LEDs were only suitable as indicator lights, but the enormous improvements in energy efficiency, cost and light output now allow the use of LEDs for lighting applications [1]. Additionally, LED lighting benefits from low maintenance cost and long lifetime.

Because LED is a rising technology, companies and researchers are constantly searching for methods to reduce the production cost and increase the efficiency, light output and light quality of LED-based lamps. An important issue for white LED lamps is color variation of the emitted light. This is caused by color variation in the light output of the most common type of white LED, the phosphor-converted LED. This type of LED consists of a blue LED with on top a so-called phosphor layer which converts part of the blue light into yellow and red. The resulting output is white light. The distance that a light ray travels through the phosphor depends on the angle of emission. As a result, the light emitted normal to the LED surface is more bluish, while the light emitted nearly parallel to the surface is more yellowish [2, pp.353-357]. This phenomenon is called Color over Angle (CoA) variation.

A lot of research has been done to reduce this color variation. Introduction of bubbles in the phosphor layer causes scattering of light, reducing the color variation [3]. Another common method is the application of a dichroic coating on the LED [4]. However, these methods reduce the efficiency of the LED and increase the production costs. Wang *et al.* [5] proposed a modification of the optics on the LED to improve the color uniformity. In the case of a spot light, the LED is combined with a collimator. A collimator is an optical component that reduces the angular width of the light emitted by the LED. A common technique is to add a microstructure on top of the collimator. However, this microstructure introduces extra costs in the production process of the collimator, makes the collimator look unattractive and broadens the light beam.

None of the methods mentioned above rigorously solve the problem of color variation, and all methods reduce the efficiency of the optical system. In earlier work [6], we introduced an inverse method to design a specific type of collimator, the so-called TIR (total internal reflection) collimator. The TIR collimator designed with this method mixes light from a point source such that the color variation is completely eliminated. The collimator requires no microstructures nor scattering techniques. However, the inverse method left very little design freedom for optical designers. An optical designer wants to influence the height and width of the collimator, for example, to fit it into the available space in a lamp design. Also, optical designers want a color mixing collimator which resembles a standard collimator as closely as possible. The inverse method introduced in this paper is an improvement of the method introduced in [6]. The collimator has three free surfaces instead of two. As a result, the improved method offers more design freedom, and it is nearly impossible to distinguish the resulting collimator with the naked eye from a collimator without color correction.

The contents of this paper is the following. First we give a thorough introduction to inverse methods for optical systems and the theory of color mixing in Section 2. In Section 3 we explain the improved inverse method. Section 4 describes three examples where the new method is used. Finally, we end with concluding remarks in Section 5.

2 Design of a TIR collimator using inverse methods

A TIR collimator is a rotationally symmetric lens, usually made of a transparent plastic like polycarbonate (PC) or polymethyl methacrylate (PMMA), that is used to collimate the light of an LED into a compact beam. A profile of a TIR collimator can be seen in Figure 1. The design procedure using inverse methods consists of two steps: first we choose a relation between the angles t of rays leaving the LED and the angles θ of rays leaving the collimator, the so-called transfer functions. Subsequently we use these transfer functions to calculate the free surfaces of the TIR collimator such that the light is redirected according to the relation defined by the transfer functions. In Figure 1 these free surfaces are denoted by A, B and C.

2.1 Source and target intensities

The first requirement that determines the choice of the transfer functions is the intensity pattern of the light emitted from the TIR collimator. Let $I(t,u)$ [lm/sr] be the intensity

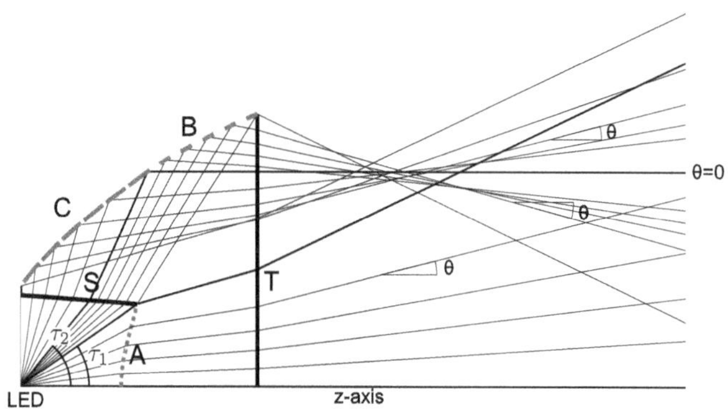

Figure 1 Profile of a TIR collimator. A full TIR collimator can be obtained by rotating the profile around the z-axis. Surface B and C are separated by the ray with angle $\theta = 0$.

distribution of the light source. The unit lm stands for lumen, and is the unit to denote energy flux corrected for the sensitivity of the eye at different wavelengths, and sr stands for steradian, the unit of solid angle. The angle $t \in [0, \pi/2]$ is the angle with respect to the z-axis (inclination), and $u \in [0, 2\pi)$ is the angle that rotates around the z-axis (azimuth). Because of the symmetry of the system, the intensity $I(t, u)$ is independent of u and denoted by $I(t)$. We introduce an effective intensity $\mathcal{I}(t)$, which is the flux per rad through the circular strip $[t, t + \mathrm{d}t]$ on the unit sphere divided by 2π. We calculate $\mathcal{I}(t)$ by integrating $I(t)$ over the angle u:

$$\mathcal{I}(t) = \frac{1}{2\pi} \int_0^{2\pi} I(t) \sin(t) \, \mathrm{d}u = I(t) \sin(t). \tag{1}$$

The effective intensity has unit [lm/rad]. For an LED, the effective intensity is typically positive for $t \in (0, \pi/2)$.

The light emitted from the TIR collimator has a desired pattern in the far field, meaning that the TIR collimator itself can be considered a point source. The desired intensity profile is denoted by $G(\theta, \phi)$ [lm/sr], where $\theta \in [0, \theta_{\max}]$ is the inclination for some maximum inclination angle $0 < \theta_{\max} \leq \pi/2$, and $\phi \in [0, 2\pi)$ is the azimuth. We only consider intensity profiles that are rotationally symmetric and thus independent of ϕ. Integration over the angle ϕ results in an effective intensity $\mathcal{G}(\theta) = \sin(\theta)G(\theta)$ [lm/rad]. A more in-depth discussion of effective intensity distributions can be found in Maes [7]. The target intensity is multiplied by a constant $c > 0$ such that we have conservation of luminous flux for the optical system:

$$\int_0^{\pi/2} \mathcal{I}(t) \, \mathrm{d}t = c \int_0^{\theta_{\max}} \mathcal{G}(\theta) \, \mathrm{d}\theta. \tag{2}$$

The angular space $[0, \pi/2]$ of the light emitted by the LED is partitioned into $N \in \mathbb{N}$ segments $[\tau_{i-1}, \tau_i]$, $i = 1, 2, \ldots, N$. For each segment we define a transfer function $\eta_i : [0, \theta_{\max}] \to [\tau_{i-1}, \tau_i] \subset [0, \pi/2]$. For a certain θ, $\eta_i(\theta)$ gives the emission angle t of the LED in $[\tau_{i-1}, \tau_i]$. We choose each transfer function to be strictly monotonic and thus invertible. The luminous flux emitted from the collimator in the interval $[\theta, \theta + \mathrm{d}\theta]$ must be equal to the sum over i of the luminous fluxes emitted from the source in each interval $[\eta_i(\theta), \eta_i(\theta + \mathrm{d}\theta)]$. This leads to the following relation:

$$\sum_{i=1}^{N} \sigma_i \mathcal{I}\big(\eta_i(\theta)\big) \eta_i'(\theta) = c\mathcal{G}(\theta), \tag{3}$$

where $\sigma_i = -1$ for monotonically decreasing transfer functions and $\sigma_i = 1$ for monotonically increasing transfer functions.

2.2 Color mixing

The second requirement on transfer functions is related to the color of the resulting beam from the collimator. First we give a short introduction to the theory of color perception, then we derive an ordinary differential equation describing the color of the beam.

Color perception is described extensively in [8, 9]. The human perception of a beam of light can be fully described by its luminous flux (in lm) and the two so-called chromaticity coordinates $0 < x, y < 1$. There is a simple rule to calculate the chromaticity coordinates

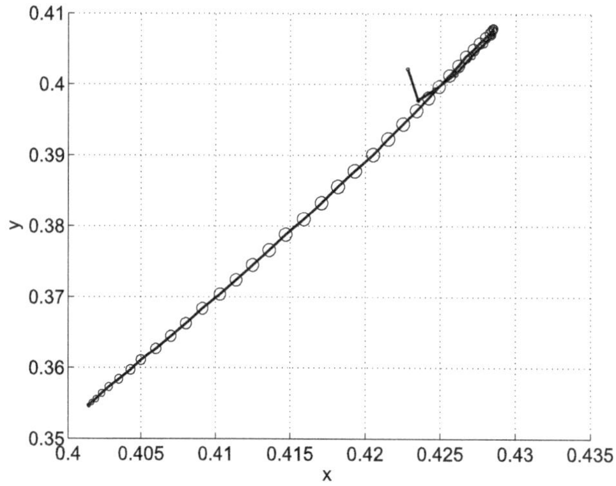

Figure 2 Scatter plot of the measured x and y chromaticity coordinates of the LED used in the numerical experiments in this article. The size of the circles corresponds to the effective intensity. The measured data in the lower left corner correspond small angles of t. The measured data for values of t close to $\pi/2$ around (0.423, 0.4) are unreliable because of the low light intensities, causing the irregularity.

(x, y) of the beam resulting from mixing two beams of light with luminous fluxes L_1 and L_2 and chromaticity coordinates (x_1, y_1) and (x_2, y_2), respectively:

$$x = \frac{x_1 L_1/y_1 + x_2 L_2/y_2}{L_1/y_1 + L_2/y_2}, \tag{4a}$$

$$y = \frac{L_1 + L_2}{L_1/y_1 + L_2/y_2}. \tag{4b}$$

The resulting chromaticity coordinates are weighted averages of the chromaticity coordinates of the original beam with weights L_1/y_1 and L_2/y_2. Note that a point (x, y) is on the straight line segment between (x_1, y_1) and (x_2, y_2).

The chromaticity coordinates of the light emitted from an LED are not constant, but depend on the angle of emission t and are described by functions $x(t)$ and $y(t)$. From measured data we have observed an approximate linear relationship between $x(t)$ and $y(t)$, see Figure 2. From the color mixing rule, we conclude that if we mix light from different angles of the LED into a single beam with color coordinates (x_T, y_T), these coordinates must be on the straight line segment relating $x(t)$ and $y(t)$. Therefore, given y_T, the chromaticity coordinate x_T is fully determined and we only need that the y-coordinate of the mixed light equals a certain constant target value y_T. The light in the interval $[\theta, \theta + \mathrm{d}\theta]$ emitted from the TIR collimator is the sum of beams with intensity $\sigma_i \mathcal{I}(\eta_i(\theta))\,\mathrm{d}\eta_i(\theta)$. The y-coordinate of this light is therefore

$$y_T = \frac{\sum_{i=1}^{N} \sigma_i \mathcal{I}(\eta_i(\theta))\,\mathrm{d}\eta_i(\theta)}{\sum_{i=1}^{N} \sigma_i \mathcal{I}(\eta_i(\theta))/y(\eta_i(\theta))\,\mathrm{d}\eta_i(\theta)}. \tag{5}$$

Using (3) we find the following differential equation:

$$\sum_{i=1}^{N} \sigma_i \frac{\mathcal{I}(\eta_i(\theta))}{y(\eta_i(\theta))} \eta_i'(\theta) = c\mathcal{G}(\theta)/y_T. \tag{6}$$

Figure 3 Geometry of the free surfaces B and C.
The grey arrow shows the 'TIR route' of the light.

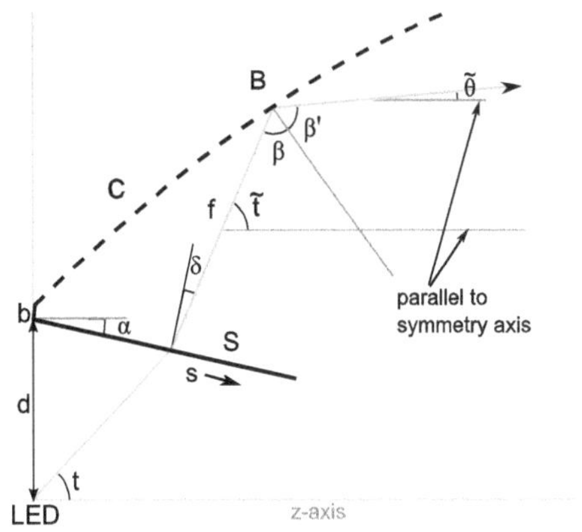

2.3 Free surface calculations

The light paths in the TIR collimator shown in Figure 1 correspond to three transfer functions, one transfer function for each of the free surfaces A, B or C. These free surfaces can be calculated from the transfer functions using the 'generalized functional method' developed by Bortz and Shatz [10, 11]. They derived a differential equation that describes the location of a free surface, given a surface S from which the rays depart with a given angle $\tilde{t}$:

$$\frac{df}{ds} = \frac{d\tilde{t}}{ds}\tan(\beta)f + \left(\tan(\beta)\cos(\delta) + \sin(\delta)\right). \tag{7}$$

Here f is the distance a light ray travels from the surface S to the free surface, s is the arc-length along S, $\tilde{t}$ is the angle of the ray leaving S with respect to the z-axis, and δ is the ray-emission angle measured counterclockwise with respect to the normal of S. The angle β is the angle of incidence on the free surface with respect to the surface normal. For the reflective surfaces B and C, the variables are illustrated in Figure 3. We like to formulate this differential equation in terms of t instead of s. Multiplication by ds/dt gives

$$\frac{df}{dt} = \frac{d\tilde{t}}{dt}\tan(\beta)f + \frac{ds}{dt}\left(\tan(\beta)\cos(\delta) - \sin(\delta)\right). \tag{8}$$

The parameters β and δ depend on t and are derived below.

Light propagates through the collimator by two type of routes. In the 'TIR route', light is refracted by surface S, reflected by surface B or C by total internal reflection and finally refracted by surface T. In the 'lens route', light is refracted by surface A and subsequently refracted by surface T.

First consider the surfaces B and C. These surfaces are on the 'TIR route', which is shown in Figure 3. Surface B is bounded at one side by the rays that leave the source at angle $t = \tau_1$. The boundary between surface B and C is marked by the rays that leave the collimator at angle $\theta = 0$, and we define the angle of this ray when leaving the light source to be $t = \tau_2$. The angles τ_1 and τ_2 are illustrated in Figure 1. First the light is refracted at surface S. Let d be the distance from the left of surface S to the LED and α the clockwise angle of this surface with respect to the symmetry-axis. A ray that leaves the LED at angle t, will hit surface S at $(x_S(t), y_S(t))$ and leave surface S with angle $\tilde{t}(t)$ given by

$$x_S(t) = \frac{d}{\tan(t) + \tan(\alpha)}, \tag{9a}$$

$$y_S(t) = \frac{d\tan(t)}{\tan(t) + \tan(\alpha)}, \tag{9b}$$

$$\tilde{t}(t) = \arccos\left(\frac{\cos(\alpha + t)}{n}\right) - \alpha. \tag{9c}$$

Relation (9c) was derived using Snell's law of refraction. The refractive index of the material of the collimator is denoted by n. From (9a) and (9b) we find s, which is defined to be 0 at $t = \pi/2$. Also we calculate δ:

$$s(t) = \frac{d}{\cos(\alpha)\tan(t) + \sin(\alpha)}, \tag{10a}$$

$$\delta(t) = \tilde{t}(t) + \alpha - \frac{\pi}{2}. \tag{10b}$$

Subsequently, the rays are reflected at surface B or C. For a reflective surface we have [10]

$$\beta(t) = \frac{1}{2}\left(\tilde{t}(t) - \tilde{\theta}(t) - \pi\right). \tag{11}$$

Here $\tilde{\theta}$ is the angle of the rays with respect to the z-axis after reflection. Before the rays leave the TIR collimator, they are once more refracted by surface T. Rays that leave the collimator at angle $\theta = \eta^{-1}(t)$ must enter surface T at angle

$$\tilde{\theta}(t) = \pm\arcsin\left(\sin\left(\eta^{-1}(t)\right)/n\right), \tag{12}$$

where the sign is negative if the rays cross the z-axis, and positive otherwise. Equation (12) is derived using Snell's law. Now we can calculate $f(t)$ by numerically integrating the ODE (8) backwards, starting at $t = \pi/2$. The parameters in (8) are given by (9c), (10a), (10b), (11) and (12). For surface C, a plus sign is chosen in (12) and for surface B a minus sign. The integration for surface C starts with $f(\pi/2) = b \geq 0$, which is usually chosen larger than 0 to prevent a sharp edge of the collimator for manufacturing purposes. At $t = \tau_2$, the final value $f(\tau_2)$ of the calculation of surface C is chosen as starting value for the calculation of surface B. The coordinates of the surfaces B and C can be calculated as

$$x_{B/C}(t) = x_S(t) + f(t)\cos\left(\tilde{t}(t)\right), \tag{13a}$$

$$y_{B/C}(t) = y_S(t) + f(t)\sin\left(\tilde{t}(t)\right). \tag{13b}$$

Now consider surface A, the 'lens route' of the collimator. The light incident on surface A comes from a single point, therefore the arc-length along the source surface is 0, so we take $s(t) = 0$. Furthermore, we have $\tilde{t}(t) = t$. For a refractive surface we need the following expression for β [11]:

$$\tan(\beta) = \frac{\sin(\tilde{\theta} - \tilde{t})}{1/n - \cos(\tilde{\theta} - \tilde{t})}. \tag{14}$$

Using (12), the differential equation (8) is now

$$\frac{\mathrm{d}f}{\mathrm{d}t} = \left(\frac{\sin(\tilde{\theta} - t)}{1/n - \cos(\tilde{\theta} - t)}\right) f, \tag{15}$$

which we solve by numerically integrating backwards subject to the end condition

$$f(\tau_1) = \sqrt{x_S(\tau_1)^2 + y_S(\tau_1)^2}. \tag{16}$$

Surface A can be calculated according to

$$x_A(t) = f(t)\cos(t), \tag{17a}$$

$$y_A(t) = f(t)\sin(t). \tag{17b}$$

3 A TIR collimator with three transfer functions

Our goal is to design a TIR collimator that has a beam with a specified intensity output $\mathcal{G}(\theta)$ and uniform chromaticity coordinates (x_T, y_T). To achieve this, the transfer functions must satisfy (3) and (6). The layout of the TIR collimator as shown in Figure 1 corresponds to three transfer functions, so $N = 3$. For the lens part and surface C, t increases with θ, so $\sigma_1 = \sigma_3 = 1$. For surface B, t decreases for increasing values of θ, and thus $\sigma_2 = -1$. We use the following convention: $\mathcal{I}_i(\theta) = \mathcal{I}(\eta_i(\theta))$ and $y_i(\theta) = y(\eta_i(\theta))$. We now have the following system of differential equations:

$$\begin{pmatrix} \mathcal{I}_1(\theta) & -\mathcal{I}_2(\theta) & \mathcal{I}_3(\theta) \\ \mathcal{I}_1(\theta)/y_1(\theta) & -\mathcal{I}_2(\theta)/y_2(\theta) & \mathcal{I}_3(\theta)/y_3(\theta) \end{pmatrix} \begin{pmatrix} \eta_1'(\theta) \\ \eta_2'(\theta) \\ \eta_3'(\theta) \end{pmatrix} = c\mathcal{G}(\theta) \begin{pmatrix} 1 \\ 1/y_T \end{pmatrix}. \tag{18}$$

The initial and end conditions for the transfer functions follow from the signs σ_i and the boundaries τ_1 and τ_2 between the segments:

$$\eta_1(0) = 0, \qquad \eta_2(0) = \tau_2, \qquad \eta_3(0) = \tau_2, \tag{19a}$$

$$\eta_1(\theta_{\max}) = \tau_1, \qquad \eta_2(\theta_{\max}) = \tau_1, \qquad \eta_3(\theta_{\max}) = \pi/2. \tag{19b}$$

The system (18) is underdetermined, therefore we add an extra equation. We choose an equation which is as simple as possible, has an obvious physical interpretation and yields a regular coefficient matrix for the ODE system. The equation we choose corresponds to the requirement that the intensity resulting from one of the transfer functions contributes a factor $r \in (0,1)$ to the total target intensity. Let j be the index of this transfer function, then we impose

$$\sigma_j \mathcal{I}_j(\theta)\eta_j'(\theta) = rc\mathcal{G}(\theta). \tag{20}$$

For $j = 1$, the coefficient matrix of the system is singular for $\theta = 0$, and for $j = 3$, the coefficient matrix is singular for $\theta = \theta_{\max}$. Such a singular coefficient matrix does not occur for

$j = 2$, so this will be our choice. The ODE system is now

$$\begin{pmatrix} \mathcal{I}_1(\theta) & -\mathcal{I}_2(\theta) & \mathcal{I}_3(\theta) \\ \mathcal{I}_1(\theta)/y_1(\theta) & -\mathcal{I}_2(\theta)/y_2(\theta) & \mathcal{I}_3(\theta)/y_3(\theta) \\ 0 & -\mathcal{I}_2(\theta) & 0 \end{pmatrix} \begin{pmatrix} \eta_1'(\theta) \\ \eta_2'(\theta) \\ \eta_3'(\theta) \end{pmatrix} = c\mathcal{G}(\theta) \begin{pmatrix} 1 \\ 1/y_T \\ r \end{pmatrix}. \tag{21}$$

The system can be inverted, yielding the following explicit system

$$\eta_1'(\theta) = \frac{c\mathcal{G}(\theta)}{\mathcal{I}_1(\theta)} \frac{y_1(\theta)}{y_1(\theta) - y_3(\theta)} \left(1 - r - \frac{y_3(\theta)}{y_T} + r\frac{y_3(\theta)}{y_2(\theta)}\right), \tag{22a}$$

$$\eta_2'(\theta) = -r\frac{c\mathcal{G}(\theta)}{\mathcal{I}_2(\theta)}, \tag{22b}$$

$$\eta_3'(\theta) = \frac{c\mathcal{G}(\theta)}{\mathcal{I}_3(\theta)} \frac{y_3(\theta)}{y_3(\theta) - y_1(\theta)} \left(1 - r - \frac{y_1(\theta)}{y_T} + r\frac{y_1(\theta)}{y_2(\theta)}\right). \tag{22c}$$

The system (22a)-(22c) with boundary conditions (19a) and (19b) has three unknown functions $\eta_1(\theta)$, $\eta_2(\theta)$ and $\eta_3(\theta)$. The functions $\mathcal{I}(t)$ and $y(t)$ are known from measurements on the LED. The function $\mathcal{G}(\theta)$ can be chosen by the optical designer as a finite function on $[0, \theta_{\max}]$. The constants c, y_T and r cannot be chosen freely, we will show that their values follow from conservation of luminous flux, the law of color mixing and the choice of τ_1 and τ_2, respectively. Also the values of τ_1 and τ_2 cannot be chosen freely, we will derive an inequality that guarantees monotonicity of the transfer functions.

Equation (22a) has a removable singularity at $\theta = 0$, because $\mathcal{G}(0) = 0$ and the initial values of the transfer functions imply $\eta_1(0) = 0$ and thus $\mathcal{I}_1(0) = \mathcal{I}(0) = 0$. We calculate $\eta_1'(0)$ using l'Hôpital's rule:

$$\eta_1'(0) = \sqrt{\frac{c\mathcal{G}_+'(0)}{\mathcal{I}_+'(0)} \frac{y_1(0)}{y_1(0) - y_3(0)} \left(1 - \frac{y_3(0)}{y_T}\right)}. \tag{23}$$

We choose the positive sign in front of the square root since $\eta_1'(0)$ should be positive. Here $\mathcal{G}_+'(0)$ and $\mathcal{I}_+'(0)$ are the right derivatives of $\mathcal{G}(\theta)$ at $\theta = 0$ and of $\mathcal{I}(t)$ at $t = 0$, respectively. These right derivatives are positive because $\mathcal{I}(t)$ and $\mathcal{G}(\theta)$ are positive at $t > 0$ and $\theta > 0$. We have $y(t) > 0$ by definition of chromaticity coordinates, and we assume based on measurements that $y(\tau_1) < y(\tau_2)$ and thus $y_1(0) < y_3(0)$. From this we see that we need to choose τ_2 such that $y(\tau_2) > y_T$, so the right hand side of (23) is positive and real.

3.1 The values of c, y_T and r

The system (21) with boundary conditions (19a) and (19b) appears to be overdetermined. However, the system contains three unknown parameters which still need to be chosen. We derive values for three constants c, y_T and r given the boundary conditions and assuming monotonicity of the transfer functions. Later we show that our choice of the constants c, y_T and r imply that three of the boundary conditions are superfluous.

The first unknown value is the constant c. Integration of the first row of (21), using the given boundary conditions and $\mathcal{I}_i(\theta) = \mathcal{I}(\eta_i(\theta))$, yields

$$c\int_0^{\theta_{\max}} \mathcal{G}(\theta)\,\mathrm{d}\theta = \int_0^{\theta_{\max}} \sum_{i=1}^{3} \sigma_i \mathcal{I}_i(\theta)\eta_i(\theta)\,\mathrm{d}\theta = \int_0^{\pi/2} \mathcal{I}(t)\,\mathrm{d}t. \tag{24}$$

The function $\mathcal{I}(t)$ is known from measurements on the LED, the function $\mathcal{G}(\theta)$ is chosen by the optical designer, so from this relation we derive the value of the constant c. This relation corresponds to conservation of luminous flux (equation (2)).

The second unknown is the target chromaticity value y_T. Integration of the second row of (21) using the given boundary conditions and substitution of (24) yields

$$1/y_T \int_0^{\pi/2} \mathcal{I}(t)\,\mathrm{d}t = \int_0^{\pi/2} \mathcal{I}(t)/y(t)\,\mathrm{d}t. \tag{25}$$

This relation shows that y_T is the weighted harmonic average of the y-chromaticity coordinate of the light source. Like $\mathcal{I}(t)$, the function $y(t)$ is known from measurements on the LED, thus we can derive the value of y_T.

The third unknown is r. Integration of the third row of (21) with the given boundary conditions and substitution of (24) yields

$$r \int_0^{\pi/2} \mathcal{I}(t)\,\mathrm{d}t = \int_{\tau_1}^{\tau_2} \mathcal{I}(t)\,\mathrm{d}t. \tag{26}$$

This relations corresponds to conservation of luminous flux for the second transfer function.

3.2 Monotonicity of the transfer functions

The transfer functions calculated from (22a)-(22c) should be monotonic, otherwise they have no physical meaning. From (22b) we can easily see that $\eta_2'(\theta) \le 0$ because $r > 0$, $\mathcal{G}(\theta) \ge 0$ and $\mathcal{I}(t) \ge 0$, thus, $\eta_2(\theta)$ is monotonically decreasing. The monotonicity of η_1 and η_3 is more complicated to show and we need some additional assumptions to derive a sufficient condition for monotonicity.

Theorem 1 *Assume that the chromaticity coordinate function $y(t)$ satisfies the inequalities*

$$\begin{aligned} &0 < y(t_1) < y(\tau_1) < y(t_2) < y(\tau_2) < y(t_3) \\ &\quad \forall t_1, t_2, t_3 \text{ s.t. } 0 < t_1 < \tau_1 < t_2 < \tau_2 < t_3 < \pi/2, \end{aligned} \tag{27}$$

and the transfer functions satisfy the bounds

$$0 \le \eta_1(\theta) \le \tau_1, \qquad \tau_1 \le \eta_2(\theta) \le \tau_2, \qquad \tau_2 \le \eta_3(\theta) \le \pi/2. \tag{28}$$

If

$$\frac{\int_{\tau_1}^{\tau_2} \mathcal{I}(t)\,\mathrm{d}t}{\int_0^{\pi/2} \mathcal{I}(t)\,\mathrm{d}t} = r < \min\left(\frac{y(\tau_1)}{y_T}, \frac{y(\tau_2)/y_T - 1}{y(\tau_2)/y(\tau_1) - 1}, \frac{1 - y(\tau_1)/y_T}{1 - y(\tau_1)/y(\tau_2)}\right), \tag{29}$$

then $\eta_1(\theta)$ and $\eta_3(\theta)$ are monotonically increasing.

Proof We need to prove that the derivatives of η_1 and η_3 are positive. From (22a), using assumptions (27) and (28), we find that η_1 is monotonically increasing if

$$1 - r - \frac{y_3(\theta)}{y_T} + r\frac{y_3(\theta)}{y_2(\theta)} \le 0,$$

and likewise η_3 is monotonically increasing if

$$1 - r - \frac{y_1(\theta)}{y_T} + r\frac{y_1(\theta)}{y_2(\theta)} \geq 0.$$

Subtracting the second inequality from the first we obtain

$$t - \frac{1}{y_T}\big(y_3(\theta) - y_1(\theta)\big) + \frac{r}{y_2(\theta)}\big(y_3(\theta) - y_1(\theta)\big) \leq 0.$$

Using (27) and (28) we find from the assumption $r < y(\tau_1)/y_T$ that

$$\frac{r}{y_2(\theta)} \leq \frac{1}{y_T}.$$

Define

$$M(\theta) = \frac{1-r}{1/y_T - r/y_2(\theta)}.$$

Using $\frac{r}{y_2(\theta)} \leq \frac{1}{y_T}$, the inequality for the monotonicity of the first transfer function can be rewritten as $M(\theta) \leq y_3(\theta)$ and the inequality for the third transfer function as $y_1(\theta) \leq M(\theta)$. We combine these two results to obtain

$$y_1(\theta) \leq M(\theta) \leq y_3(\theta).$$

The function $M(\theta)$ it is monotonically increasing because $\eta_2(\theta)$ is monotonically decreasing. Therefore, if $y(\tau_1) \leq M(0)$, we find using (27) and (28) that $y_1(\theta) \leq M(\theta)$ for all θ. Similarly, if $M(\theta_{\max}) \leq y(\tau_2)$, then $M(\theta) \leq y_3(\theta)$ for all θ. The inequalities $y(\tau_1) \leq M(0)$ and $M(\theta_{\max}) \leq y(\tau_2)$ are equivalent to the second and third inequality in (29). □

Figure 4 shows a scatter plot of values of τ_1 and τ_2 for which (29) is satisfied for an LED which was also used in the numerical experiments. The acceptable values of τ_1 and τ_2 are bounded by the lines $\tau_1 = t_{\text{av}}$ and $\tau_2 = t_{\text{av}}$, where t_{av} is the such that $y(t_{\text{av}}) = y_T$. In this case, the value of t_{av} is unique. From (29) we see that $\tau_1 > t_{\text{av}}$ results in $y(\tau_1) > y_T$, and thus $r < 0$. Therefore we cannot guarantee the monotonicity of the transfer functions. Using (22c) we can verify that indeed the third transfer function is not monotonic at $\theta = \theta_{\max}$. Similarly, $\tau_2 < t_{\text{av}}$ results in $r < 0$, and we can verify using (23) that the first transfer function is not monotonic at $\theta = 0$.

3.3 The initial value problem

The ODE-system (21) with the boundary conditions (19a) and (19b) can be solved as an initial value problem. We remark that solving the system as an end value problem has no advantages or disadvantages. We discard the end conditions and solve the initial value problem using a Runge-Kutta method. The end conditions are satisfied as a result of our choices of c, y_T and r.

Theorem 2 *Assume monotonicity of the transfer functions. The solution of the initial value problem defined by the ODE system* (21) *and the initial conditions* $\eta_1(0) = 0$, $\eta_2(0) = \eta_3(0) = \tau_2$ *satisfies the end conditions* $\eta_1(\theta_{\max}) = \eta_2(\theta_{\max}) = \tau_1$, $\eta_3(\theta_{\max}) = \pi/2$.

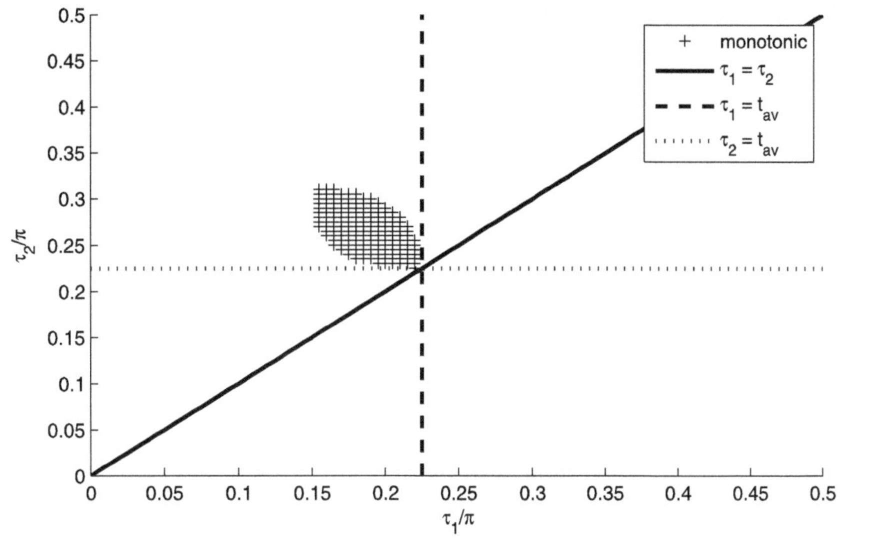

Figure 4 Scatter plot of the values of τ_1 and τ_2 for LED16 that guarantee monotonic transfer functions. t_{av} is such that $y(t_{av}) = y_T$.

Proof First we show that $\eta_2(\theta_{max}) = \tau_1$ is satisfied. Integration of the last row of (21) from $\theta = 0$ to $\theta = \theta_{max}$ using (26) gives

$$-\int_0^{\theta_{max}} \mathcal{I}_2(\theta)\eta_2'(\theta)\,d\theta = -\int_{\tau_2}^{\eta_2(\theta_{max})} \mathcal{I}(t)\,dt = rc\int_0^{\theta_{max}} \mathcal{G}(\theta)\,d\theta = \int_{\tau_1}^{\tau_2} \mathcal{I}(t)\,dt.$$

So we find

$$\int_{\tau_2}^{\eta_2(\theta_{max})} \mathcal{I}(t)\,dt + \int_{\tau_1}^{\tau_2} \mathcal{I}(t)\,dt = 0.$$

Because $\mathcal{I}(t) > 0$ for all t except for two points at the boundary, we can conclude $\eta_2(\theta) = \tau_1$. Note that this implies $\tau_1 \le \eta_2(\theta) \le \tau_2$.

Using the monotonicity of the transfer functions, we can integrate the left hand side of the first row of (21):

$$\int_0^{\theta_{max}} \left(\mathcal{I}_1(\theta)\eta_1'(\theta) - \mathcal{I}_2(\theta)\eta_2'(\theta) + \mathcal{I}_3(\theta)\eta_3'(\theta)\right) d\theta$$
$$= \int_0^{\eta_1(\theta_{max})} \mathcal{I}(t)\,dt - \int_{\tau_2}^{\tau_1} \mathcal{I}(t)\,dt + \int_{\tau_2}^{\eta_3(\theta_{max})} \mathcal{I}(t)\,dt.$$

For the right hand side we have due to (24)

$$c\int_0^{\theta_{max}} \mathcal{G}(\theta)\,d\theta = \int_0^{\pi/2} \mathcal{I}(t)\,dt.$$

By subtracting the last two relations we find

$$\int_{\tau_1}^{\eta_1(\theta_{max})} \mathcal{I}(t)\,dt + \int_{\pi/2}^{\eta_3(\theta_{max})} \mathcal{I}(t)\,dt = 0. \qquad (*)$$

Similarly, we can derive from the second row of (21) using equation (25)

$$\int_{\tau_1}^{\eta_1(\theta_{\max})} \frac{\mathcal{I}(t)}{y(t)}\,\mathrm{d}t + \int_{\pi/2}^{\eta_3(\theta_{\max})} \frac{\mathcal{I}(t)}{y(t)}\,\mathrm{d}t = 0.$$

The functions $\mathcal{I}(t)$ and $1/y(t)$ are continuous, and $\mathcal{I}(t)$ does not change sign in the interval $(0, \pi/2)$. Using the expanded first mean-value theorem for integrals [12, p.487], we find that for some $t_1 \in (\eta_1(\theta_{\max}), \tau_1)$ and some $t_3 \in (\eta_2(\theta_{\max}), \pi/2)$ we have

$$\frac{1}{y(t_1)} \int_{\tau_1}^{\eta_1(\theta_{\max})} \mathcal{I}(t)\,\mathrm{d}t + \frac{1}{y(t_3)} \int_{\pi/2}^{\eta_3(\theta_{\max})} \mathcal{I}(t)\,\mathrm{d}t = 0. \qquad (**)$$

The equations (*) and (**) form a linear system for the integrals of $\mathcal{I}(t)$ over $[\tau_1, \eta_1(\theta_{\max})]$ and $[\pi/2, \eta_3(\theta_{\max})]$. If $t_1 < \tau_2$, we find from the assumption $y(t_1) < y(t_3)$ that

$$\det \begin{pmatrix} 1 & 1 \\ 1/y(t_1) & 1/y(t_3) \end{pmatrix} \neq 0,$$

and thus

$$\int_{\tau_1}^{\eta_1(\theta_{\max})} \mathcal{I}(t)\,\mathrm{d}t = 0,$$

$$\int_{\pi/2}^{\eta_3(\theta_{\max})} \mathcal{I}(t)\,\mathrm{d}t = 0.$$

Because $\mathcal{I}(t) > 0$ for all t except at the boundary points, we conclude

$$\eta_1(\theta_{\max}) = \tau_1,$$

$$\eta_3(\theta_{\max}) = \pi/2. \qquad \square$$

4 Numerical procedure and results

We solved the mathematical model described in the previous section to design three different TIR collimators. The collimators were designed for two different LEDs, which we refer to as LED16 and LED02. Both of them are Luxeon Rebel IES white LEDs without a dichroic coating, and have a larger than usual CoA variation. The intensity and chromaticity-coordinates of the LEDs were measured, and the measured data were interpolated. The interpolation polynomials have been used to approximate $\mathcal{I}(t)$ and $y(t)$ in (22a)-(22c). The first two collimators were designed for LED16 and have a Gaussian-shaped target intensity profile. The two collimators differ in their values for τ_1 and τ_2. The third collimator was designed for LED02 and has a block-shaped target intensity profile. The collimators were evaluated using the LightTools software package [13].

4.1 Modelling of the LEDs

The LEDs were measured using a goniophotometer [14]. A goniophotometer is a device that measures intensity, chromaticity coordinates and many other characteristics of light at different solid angles. Our LEDs were measured at 46 different angles t and 4 different angles u. For each LED, the chromaticity values were averaged and the intensities were

Table 1 **Coefficients from the linear least squares fits**

LED	i	0	2	3	4	5	6	7
16	C_i	0	−181.8279	76.0797	221.9411	−624.0461	499.8253	−127.0787
	D_i^x	0.4013	0.0578	−0.0367	0.0271	−0.0564	0.0444	−0.0116
	D_i^y	0.3546	0.0988	−0.0557	0.0255	−0.0371	0.0196	−0.0034
02	C_i	0	−148.8533	−60.3797	520.5934	−913.6760	636.8623	−153.5863
	D_i^x	0.4591	0.0452	−0.0590	0.0916	−0.1399	0.1030	−0.0271
	D_i^y	0.3792	0.0892	−0.1790	0.3596	−0.4585	0.2823	−0.0658

summed over the angle u. These data have been interpolated with a least squares fit using the following polynomials:

$$I(t) = \sum_{i=2}^{7} C_i\left(t^i - (\pi/2)^i\right),$$

$$x(t) = D_0^x + \sum_{i=2}^{7} D_i^x t^i,$$

$$y(t) = D_0^y + \sum_{i=2}^{7} D_i^y t^i.$$

The polynomial for the intensity was chosen because it equals 0 at $t = \pi/2$ and has zero derivative at $t = 0$, both properties are characteristic for the intensity distribution of an LED. The effective intensity equals $\mathcal{I}(t) = \sin(t)I(t)$. The polynomials for the chromaticity coordinates were chosen because their derivative equals 0 at $t = 0$. The coefficients for the two LEDs can be found in Table 1.

In the LightTools software package, two three-dimensional models were built to simulate the LEDs. The range $(0, \frac{\pi}{2})$ of the angle t was discretized into 46 different subintervals, labeled $k = 1, 2, \ldots, 46$. For $k = 1$ we have the interval $(0, \frac{\pi}{180})$, for $k = 2, \ldots, 45$ we have $(\frac{(2k-3)\pi}{180}, \frac{(2k-1)\pi}{180})$ and finally for $k = 46$ we have $(\frac{89\pi}{180}, \frac{\pi}{2})$. The intensity and chromaticity coordinates of the LED models in these subintervals correspond to the measured data of the real LEDs at the angles $t = \frac{(2k-2)\pi}{180}$. The size of the LED model was reduced to 0.01 mm by 0.01 mm to simulate a point light source. A comparison of the measured data, the least squares fit and the raytracing results of the LightTools model of LED16 without collimator can be seen in Figures 5 and 6. A scatter plot of the measured x and y chromaticity coordinates for this LED was shown earlier in Figure 2. The plot shows the near-linear relationship between x and y, indeed.

4.2 Computation of the transfer functions

Three example collimators have been calculated. The first collimator was designed for a Gaussian target intensity [15] with full width at half maximum (FWHM) [16] at $\pi/9$. This yields the following effective target intensity:

$$\mathcal{G}(\theta) = \sin(\theta)\exp\left(-4\ln(2)\left(\frac{\theta}{\theta_{\mathrm{FWHM}}}\right)^2\right), \tag{33}$$

with $0 \le \theta \le 1.25\theta_{\mathrm{FWHM}} = \theta_{\max}$, $\theta_{\mathrm{FWHM}} = \pi/9$. The collimator was designed for LED16. The choice of τ_1 and τ_2 is restricted by (29). This relation is highly nonlinear. A scatter plot of values of τ_1 and τ_2 that satisfy (29) for LED16 is shown in Figure 4.

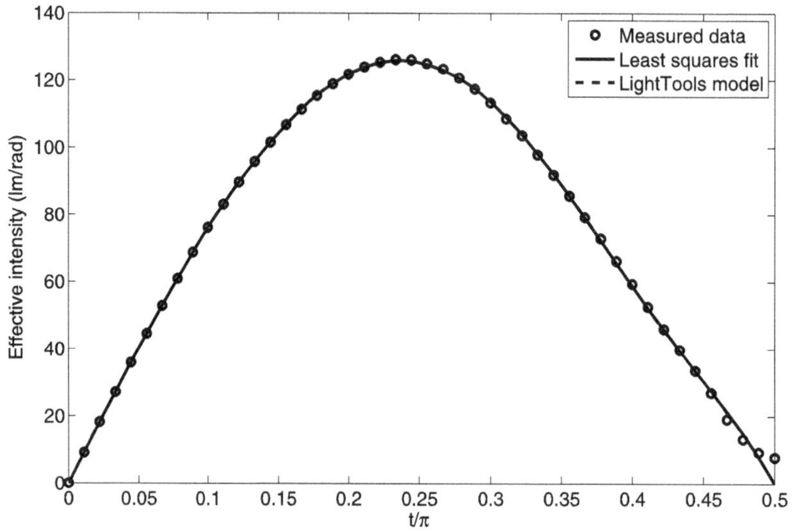

Figure 5 Comparison of the measured effective intensity $\mathcal{I}$, the least squares fit and the LightTools model of LED16. The graph of the LightTools model is not visible because it is hidden behind the least squares fit.

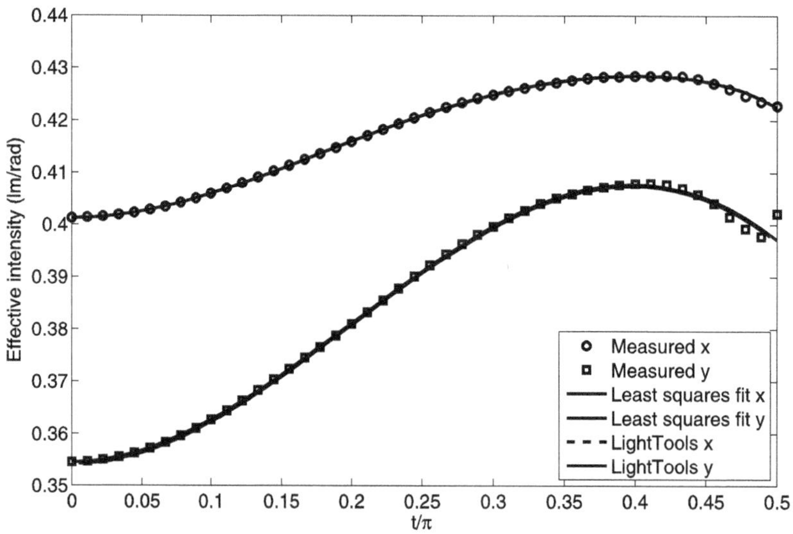

Figure 6 Comparison of measured chromaticity coordinates *x* and *y*, the least squares fit and the LightTools Model of LED16. The graph of the LightTools model is not visible because it is hidden behind the least squares fit.

We chose $\tau_1 = 0.2\pi$ and $\tau_2 = 0.25\pi$. The second collimator was designed for the same LED and target intensity, but this time we chose $\tau_1 = 0.16\pi$ and $\tau_2 = 0.3\pi$, which gives a larger second segment. The third collimator was designed for LED02. The target intensity was chosen to be a block function, yielding the effective intensity $\mathcal{G}(\theta) = \sin(\theta)$, with $0 \leq \theta \leq \theta_{\max} = \pi/9$. We chose $\tau_1 = 0.2\pi$ and $\tau_2 = 0.25\pi$, which satisfies (29). An overview of the values chosen and calculated for the three collimators is shown in Table 2.

The ODE system (22a)-(22c) with initial conditions (19a) was solved using the ODE-solver `ode45` in Matlab. The calculation times were a few seconds on a laptop computer

Table 2 Parameter values and characteristics for the three different collimators

Collimator	LED	τ_1	τ_2	r	y_T	c	$\theta_{\max}$
Gaussian, small 2nd segment	LED16	0.20π	0.25π	0.1630	0.3862	5,585.9	11.25π
Gaussian, large 2nd segment	LED16	0.16π	0.30π	0.4417	0.3862	5,585.9	11.25π
Block profile	LED02	0.20π	0.25π	0.1612	0.3986	4,843.3	9π

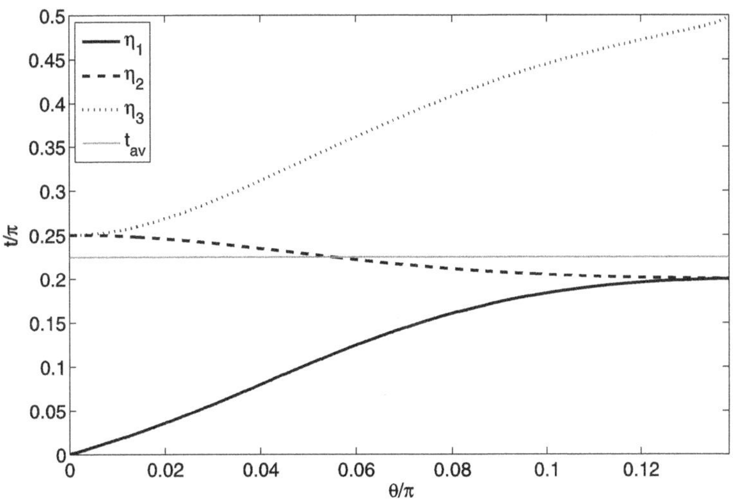

Figure 7 Transfer functions for the collimator with Gaussian profile and a small second segment.

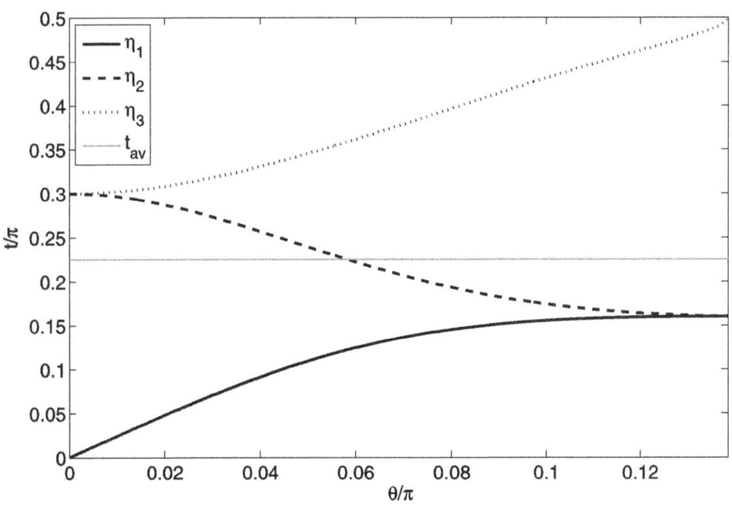

Figure 8 Transfer functions for the collimator with Gaussian profile and a large second segment.

with a 2.4 GHz processor and 4 GB RAM. The calculated transfer functions are shown in Figure 7, 8 and 9. The transfer functions are indeed monotonic, as expected. Also, $\eta_1(\theta_{\max}) = \eta_2(\theta_{\max}) = \tau_1$ and $\eta_3(\theta_{\max}) = \pi/2$, as anticipated.

4.3 Performance of the TIR collimators

Subsequently, a TIR collimator was designed for each set of transfer functions, and evaluated using LightTools. We chose for all the collimators $d = 4$ mm, $b = 0.4$ mm and

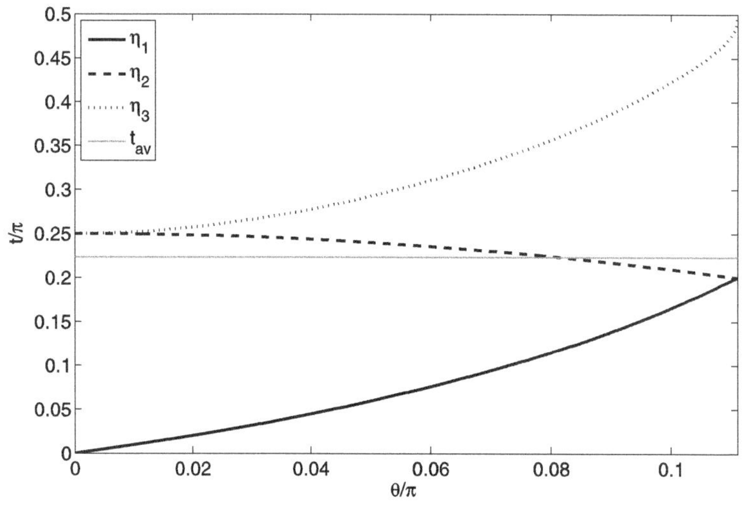

Figure 9 Transfer functions for the collimator with block profile.

Figure 10 LightTools model of the first collimator.

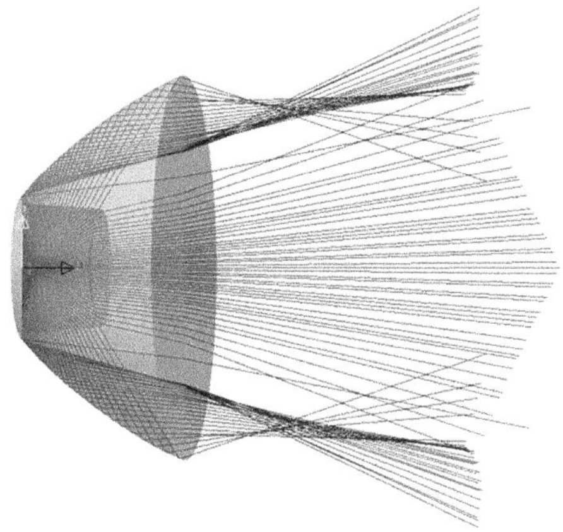

$\alpha = 4\pi/180$. For every collimator, each free surface was discretized using 500 points and converted into a LightTools model. A screenshot of the LightTools model of the first collimator can be seen in Figure 10. Results of the simulations can be seen in Figures 11, 12 and 13. In these figures, we see the expected profiles of the effective intensity and chromaticity. Figures 11 and 12 look very similar, because the first and second collimator were designed with the same output specifications. In these figures, an irregularity is visible in the chromaticity coordinates near $\theta = \theta_{max}$. This can be explained as follows. Every bar in the graph corresponds to a range of one degree ($\pi/180$ rad). We chose $\theta_{max} = 25\pi/180$, and thus the flux at this angle should be zero. Due to small errors in the free surfaces, a small number of rays exits the collimator at angles larger than θ_{max}. This happens at surface C, and therefore the chromaticity coordinates at $\theta > \theta_{max}$ are larger than the target values. Because the luminous flux of this light is very small, the irregularity is not visible. A similar irregularity is visible for the collimator with the block profile, only with a smaller chromaticity difference.

Apart from this small irregularity, the variation in chromaticity is very small. The maximum difference between the average chromaticity of the LEDs and chromaticity coordi-

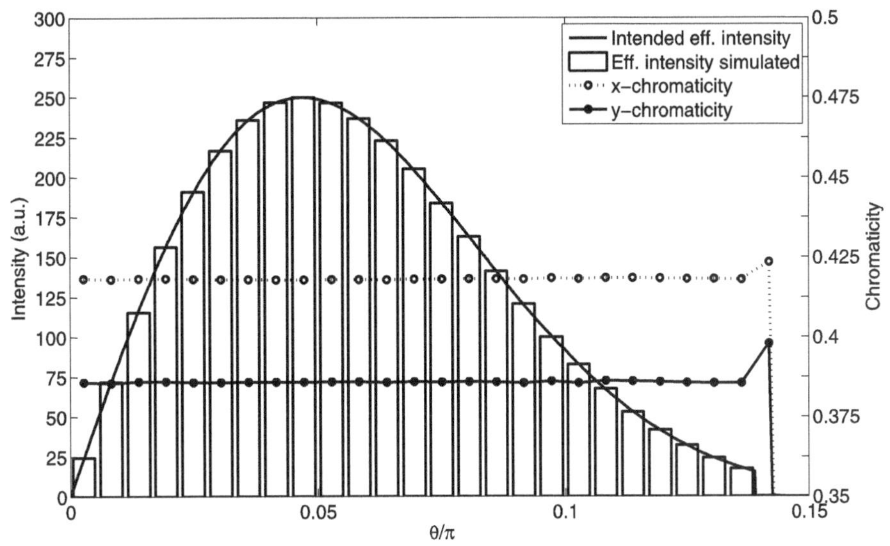

Figure 11 LightTools simulation results for the collimator with Gaussian profile and a small second segment.

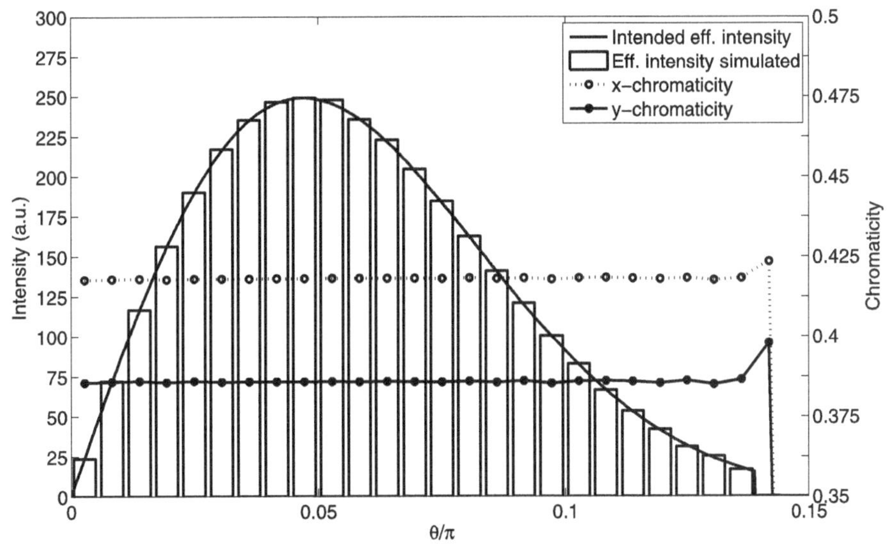

Figure 12 LightTools simulation results for the collimator with Gaussian profile and a large second segment.

nates in the simulations are shown in Table 3. A color difference of 0.003 is considered very good by optical designers and is invisible for the human eye [8, 17]. The measured color differences in the simulations are comfortably below this value, thus the color variation in the beam is eliminated.

5 Conclusions

We introduced an inverse method to design a TIR collimator that eliminates CoA variation for a point light source. This method improves the method introduced earlier in [6] by producing collimators that closely resemble standard collimators and at the same time

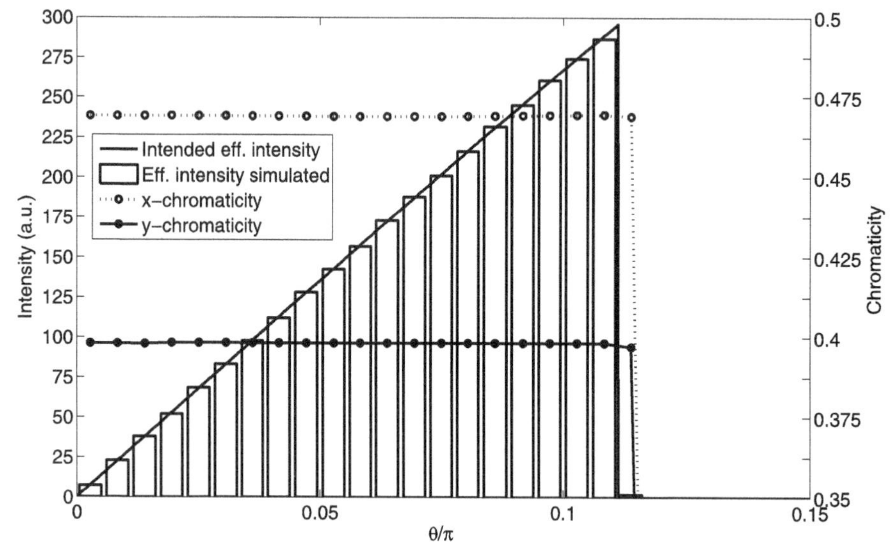

Figure 13 LightTools simulation results for the collimator with Block profile for LED02.

Table 3 Average chromaticity coordinates of the LEDs and the maximum difference with the chromaticity coordinates in the simulations

Collimator	LED	Target x	$\max(\lvert\Delta x\rvert)$	Target y	$\max(\lvert\Delta y\rvert)$
Gaussian, small 2nd segment	LED16	0.4181	$4 \cdot 10^{-4}$	0.3862	$7 \cdot 10^{-4}$
Gaussian, large 2nd segment	LED16	0.4181	$4 \cdot 10^{-4}$	0.3862	$1 \cdot 10^{-3}$
Block profile	LED02	0.4691	$4 \cdot 10^{-4}$	0.3985	$5 \cdot 10^{-4}$

have more parameters for optical design. In Section 3 we discussed which choices for these design parameters give meaningful results. In Section 4 we tested the method and verified the resulting collimators with Monte-Carlo raytracing using the software package LightTools. The simulations show color variations that are not visible with the human eye.

Unfortunately, LEDs are too large to be treated as a point light source. In future research, we would like to extend this method to take the finite size of the light source into account using iterative methods such as described in for example [18, 19]. This point source method will be an important building block in such an iterative method.

Competing interests

The authors declare that they have no competing interests.

Authors' contributions

The main idea in this paper was proposed by CP, who also prepared the manuscript initially. The proofs were elaborated by CP and JtTB. Technical assistance was provided by TT. All authors provided feedback on the manuscript, and read and approved the final manuscript.

Author details

[1]Centre for Analysis, Scientific computing and Applications (CASA), Technische Universiteit Eindhoven, P.O. Box 513, 5600 MB Eindhoven, The Netherlands. [2]Philips Research, High Tech Campus 44, 5656 AE Eindhoven, The Netherlands. [3]Philips Lighting, High Tech Campus 44, 5656 AE Eindhoven, The Netherlands.

References

1. Haitz R, Jeffrey YT: **Solid-state lighting: 'The case' 10 years after and future prospects.** *Phys Status Solidi A* 2011, **208**:17-29.
2. Schubert F: *Light-Emitting Diodes.* 2nd edition. Cambridge: Cambridge University Press; 2006.

3. Wu H, Narendran N, Gu Y, Bierman A: **Improving the performance of mixed-color white LED systems by using scattered photon extraction technique.** In *Light-Emitting Diodes*; 2007. [*Proc. SPIE*, vol. 6669.]
4. Mueller GO: **Luminescent ceramic for a light emitting device.** United States patent US7361938 B2.
5. Wang K, Wu D, Chen F, Liu Z, Luo X, Liu S: **Angular color uniformity enhancement of white light-emitting diodes integrated with freeform lenses.** *Opt Lett* 2010, **35**:1860-1862.
6. Prins C, ten Thije Boonkkamp C, Tukker T, IJzerman W: **An inverse method for the design of TIR collimators to achieve a uniform color light beam.** *J Eng Math* 2013, **81**:177-190.
7. Maes M: **Mathematical methods for reflector design.** *PhD thesis*. University of Amsterdam; 1997.
8. Malacara D: *Color Vision and Colorimetry. Theory and Applications*. Bellingham: SPIE; 2002.
9. Wyszecki G, Stiles W: *Color Science. Concepts and Methods, Quantitative Data and Formulae*. 2nd edition. New York: Wiley; 2000.
10. Bortz J, Shatz N: **Generalized functional method of nonimaging optical design.** In *Nonimaging Optics and Efficient Illumination Systems III*. Edited by Winston R, Benitez P; 2006. [*Proc. SPIE*, vol. 6338.]
11. Bortz J, Shatz N: **Mathematical relationships between the generalized functional, edge-ray and sms design methods.** In *Nonimaging Optics and Efficient Illumination Systems III*; 2010. [*Nonimaging Optics: Efficient Design for Illumination and Solar Concentration*, vol. 7785.]
12. Stöcker H: *Taschenbuch mathematischer Formeln und moderner Verfahren*. Thun: Verlag Harri Deutsch; 1993.
13. **ORA LightTools product website** [http://www.opticalres.com/lt/ltprodds_f.html] [Accessed May 31, 2013].
14. **Instrument Systems - Light measurement** [http://www.instrumentsystems.com/products/led-measurement-accessories/ledgon-goniophotometer/] [Accessed July 10, 2012].
15. Weisstein E: **Gaussian function** [http://mathworld.wolfram.com/GaussianFunction.html]
16. Weisstein E: **Full width at half maximum** [http://mathworld.wolfram.com/FullWidthatHalfMaximum.html]
17. MacAdam D: **Visual sensitivities to color differences in daylight.** *J Opt Soc Am* 1942, **32**:247-274.
18. Zhang W, Liu Q, Gao H, Yu F: **Free-form reflector optimization for general lighting.** *Opt Eng* 2010, **49**(6):063003.
19. Cassarly WJ: **Iterative reflector design using a cumulative flux compensation approach.** *Opt Eng* 2010, **7652**:76522.

4

Parallel solutions of static Hamilton-Jacobi equations for simulations of geological folds

Tor Gillberg[1,2*], Are Magnus Bruaset[1], Øyvind Hjelle[2] and Mohammed Sourouri[1]

*Correspondence: torgi@simula.no
[1]Simula Research Laboratory, Martin Linges vei, Oslo, Norway
[2]Kalkulo AS, Martin Linges vei, Oslo, Norway

Abstract

Two new algorithms for numerical solution of static Hamilton-Jacobi equations are presented. These algorithms are designed to work efficiently on different parallel computing architectures, and numerical results for multicore CPU and GPU implementations are reported and discussed. The numerical experiments show that the proposed solution strategies scale well with the computational power of the hardware. The performance of the new methods are investigate for tow types of static Hamilton-Jacobi formulations are investigated, the isotropic eikonal equation and an anisotropic formulation used to simulate different types of geological folding. Simulations of geological folding is a key component in an industrial software used in oil and gas exploration. In particular, our experiments indicate that the new algorithms would be capable of accelerating an existing industrial simulator substantially. Direct comparison with the current industry code shows that computing times can be reduced from several minutes to a few seconds. The new methods are now being migrated to the industrial software.

Keywords: parallel computing; general purpose GPU computing (GPGPU); multicore CPU; eikonal equation; static Hamilton-Jacobi equations; front propagation; geological folding; generalised distance computing; CUDA; OpenMP

1 Static Hamilton-Jacobi equations

This paper concerns methods for simulating monotonically propagating fronts modelled by static Hamilton-Jacobi equations. A general formulation of such equations is

$$\begin{aligned} H(\mathbf{x}, \nabla T(\mathbf{x})) &= 1, \\ T(\mathbf{x}) &= g(\mathbf{x}) \quad \forall \mathbf{x} \in \Gamma, \end{aligned} \tag{1}$$

where $g(\mathbf{x})$ represents initially given values at points in the set Γ, and the Hamiltonian H is here assumed Lipschitz continuous and convex in ∇T. The unknown entity, T, can be thought of as a generalised distance field, as well as the time of arrival of a monotonically expanding front. Usually, $g(\mathbf{x}) = 0$ which means that Γ represents the object from which the generalised distance should be computed. Static Hamilton-Jacobi equations are known to often have multiple-valued solutions, in that a point in the domain may have several values simultaneously. The viscosity solution of the static Hamilton-Jacobi equations is also the minimal distance value, that is the first time of arrival [1]. In some applications, the multiple-valued solutions can be important [2]. However, in many applications it is enough to solve for the viscosity solution of the equation. In this paper we solve solely for the viscosity solution.

1.1 A mathematical framework for propagating fronts

The solution T of (1) can be thought of as the time of arrival of a wavefront that propagates through a domain. More generally, the t-isosurface of T shows the position of the front at time t. The t-isosurface can also be regarded as the generalised t-distance from Γ, that is, solution values are increasing away from Γ. In a front propagation setting, equation (1) is often formulated as $V\|\nabla T\| = 1$, where the scalar V is the speed at which the front is moving in the direction normal to the front. At every point, V is greater or equal to 0, with equality implying that the front never reaches the point. If the speed is independent of the direction the front is moving in, the front propagation is said to be isotropic. The isotropic eikonal equation is important in many applications and is formulated as

$$F(\mathbf{x})\left\|\nabla T(\mathbf{x})\right\| = 1. \tag{2}$$

For the special case of $F \equiv 1$ and $g(x) \equiv 0$, the viscosity solution of the eikonal equation is the minimal Euclidean distance to Γ.

The problem is anisotropic if the rate of increase of T values changes with the direction it is measured in. There is a clear causality interpretation of a propagating front since the current motion does not depend on future events. In other words, smaller values of T are independent of larger T values, which imply that the discrete solution representing T should be computed in increasing order. In the isotropic eikonal case this causality observation can be directly translated to discrete nodal values. However, for the more general anisotropic case a similar discrete interpretation is not valid [3], and anisotropic problems are therefore more complicated to solve. The algorithms developed in this paper have been applied to both isotropic eikonal formulations and to anisotropic problems of the form (3).

Most formulations of static Hamilton-Jacobi equations are nonlinear, and the computations needed for updating the solution in a grid node are expensive in terms of both floating-point operations and logical branching. The amount of computations needed can be decreased by updating nodes in different orderings.

In the following section we give a brief overview of solution algorithms. For a more complete algorithmic overview the reader is to [4] and references therein.

1.2 Applications

Many physical phenomena can be described by a propagating front. A front can for example describe an interface between different objects or fluids (multi-phase flow) [5], a shock wave [6], or the arrival of a wave [7]. In this work, the front is assumed monotonically expanding, like a wildfire spreading only to unburned grounds. Monotonically expanding fronts can be entirely described by the time of arrival of the front to all points in the domain. Applications requiring fast simulations of monotonic front propagations include medical tomography [8], simulation of cardiac activation times [9], segmentation of images [10], and seismic wave propagation [11]. Shorter computing times may result in faster and more accurate models since one can afford to better explore parameter ranges, and thereby the model uncertainty. As a result, faster methods can provide further insight into the underlying physical problems. In this paper, we focus on the development of new efficient algorithms. For a proper introduction to front propagation and a range of applications, see [1]. The application initiating the research presented in this paper is from a software project developed by Statoil ASA and Kalkulo, and is presented in the following paragraph.

The compound earth simulator

Structural restoration and reconstruction is a methodology for building and improving structural models of subsurface geology. The Compound Earth Simulator (CES) is a software in which large three dimensional models can be transformed between different structural states [12]. Generalised distance fields are needed in the transformations of the geology and to deposit the layers. The generalised distance fields are given from the following static Hamilton-Jacobi equation,

$$\begin{aligned} F\left\|\nabla T(\mathbf{x})\right\| + \psi\left(\mathbf{a}\cdot\nabla T(\mathbf{x})\right) &= 1, \quad \mathbf{x}\in\Omega, \\ T(\mathbf{x}) &= t_0 \quad \forall\mathbf{x}\in\Gamma. \end{aligned} \tag{3}$$

Here, Ω is the computational domain, Γ_0 is the surface of a geological layer (horizon), and $\mathbf{a}$ is a unit vector. Geological volumes are transformed repeatedly in the workflow [13]. For the software to be user friendly, the distance fields must be computed rapidly while still being accurate.

In this paper, software is said to be interactive if the delay due to computations is of the order of a few seconds at the most. Some of the available algorithms are fast enough for interactive use in two spatial dimensions. This is not the case in three dimensions, since the number of nodes and the nodal update cost increase significantly. A case study of geological folding based on structures extracted from offshore seismic data, is included in the numerical experiments reported in this paper. The methods presented in this study enables interactive high-resolution restorations in three dimensions.

The ray equations

Characteristic curves are paths along which 'particles' on the front are pushed forward, as well as lines along which distances are measured. Let $\mathbf{x}(s) = (x,y,z)(s)$ denote a characteristic curve parameterised by the scalar s. Characteristic curves are often referred to simply as characteristics. The solution T is strictly increasing along characteristics that are perpendicular to the moving front. The characteristic equations can be formulated on basis of (1) as

$$\frac{d\mathbf{x}}{ds} = \nabla_{\mathbf{p}} H(\mathbf{p},\mathbf{x}), \tag{4}$$

$$\frac{d\mathbf{p}}{ds} = \nabla_{\mathbf{x}} H(\mathbf{p},\mathbf{x}), \tag{5}$$

where $\mathbf{p} = \nabla T$, $\nabla_{\mathbf{x}} = (\partial/\partial x, \partial/\partial y, \partial/\partial z)^T$, and $\nabla_{\mathbf{p}} = (\partial/\partial T_x, \partial/\partial T_y, \partial/\partial T_z)^T$ [14, 15]. For the Hamiltonians (2) and (3), the respective characteristic curves $\mathbf{x}_e(s)$ and $\mathbf{x}_f(s)$ are given by

$$\mathbf{x}_e(s) = \mathbf{x}_e(0) + sF\mathbf{n}, \tag{6}$$

$$\mathbf{x}_f(s) = \mathbf{x}_f(0) + s(F\mathbf{n} + \psi\mathbf{a}), \tag{7}$$

where $\mathbf{n} = \nabla T / \|\nabla T\|$ is the unit normal vector to the propagating front, and $x(0)$ marks the starting location of the curve. The velocity parameters F and $\psi\mathbf{a}$ are assumed to be constant in a small neighbourhood, which results in locally linear characteristic curves along which $\mathbf{p}$ is constant. On a discrete grid, the parameters are commonly assumed to be constant within the stencil area. An investigation of the solution value along a characteristic curve shows that

$$\frac{dT}{ds} = \mathbf{p} \cdot \frac{d\mathbf{x}}{ds} = \mathbf{p} \cdot \nabla_{\mathbf{p}} H = 1 \quad \text{for (2) and (3)} \tag{8}$$

$$\implies \quad T\big(\mathbf{x}(s)\big) = T\big(\mathbf{x}(0)\big) + s. \tag{9}$$

That is, T is increasing linearly in s along characteristics. Because of this one-dimensional dependency, solution values can be extrapolated with high accuracy along characteristic curves. An understanding of characteristics is therefore important when creating numerical stencils, as well as designing new algorithms for front propagation problems. The solutions T_e and T_f for equations (2) and (3) change explicitly according to

$$T_e(s\mathbf{n}) = T_e(\mathbf{0}) + \frac{s}{F}, \tag{10}$$

$$T_f\left(s\frac{F\mathbf{n} + \mathbf{a}}{\|F\mathbf{n} + \mathbf{a}\|}\right) = T_f(\mathbf{0}) + \frac{s}{\|F\mathbf{n} + \mathbf{a}\|} \tag{11}$$

along characteristics. In particular, T_e increases at a constant rate since the front moves with a speed that is independent of the normal and its direction. This is not the case for T_f, which increases with a rate that depends on the direction in which the front moves. From these formulations the increased complexity for the anisotropic formulation is obvious. The anisotropic dependencies in the fold equation complicate the computations, and the computational cost is significantly higher than for the isotropic problem.

1.3 Numerical solution methods

The solution can be built by computing the local solution along a set of characteristic curves, and thereafter interpolating the travel time between the curves. This approach is traditionally referred to as the method of characteristics, or ray tracing. In seismology, the term 'rays' is often used in lieu of characteristics. The methods developed in this paper assumes a fixed grid of nodal values. Grid-based methods use stencils that are strongly related to ray-tracing methods. Instead of tracing values along specific characteristics, the grid-based tracking methods update nodal values by interpolating solution values between neighbouring nodes organised in stencils. Commonly, solution values are interpolated linearly along edge-connected nodes in the grid [16], implying a planar modelling of the front. In Appendix 1, we present an efficient stencil formulation for the eikonal equation. A smaller value corresponds to an earlier arrival time, and thus a better estimate of the sought viscosity solution. The solution at every point is therefore strictly decreasing. A too small value will not be enlarged, but instead it generates an error that will spread throughout the domain. It is therefore important that new estimates are based on upwind information, that is, the characteristic originates from values that are upwind of the node being updated and therefore already passed by the front.

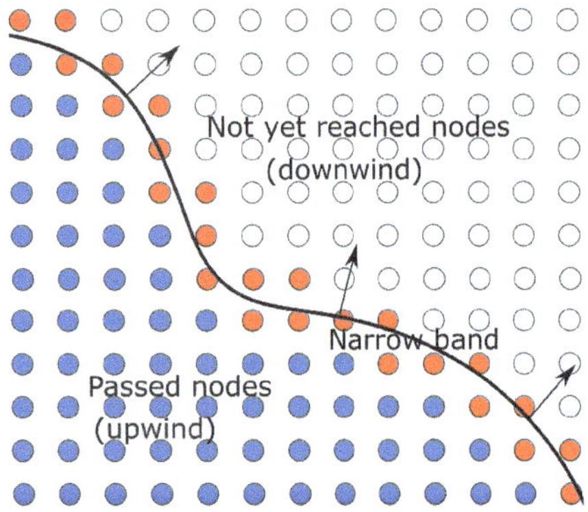

(a) Tracking order.

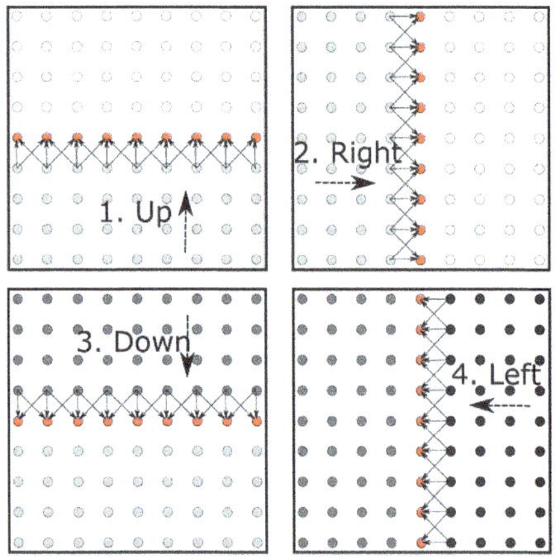

(b) Parallel marching sweep.

Figure 1 Illustration of update order of algorithms, where circles are nodes on the grid. (**a**) Tracking order. Red nodes belong to the narrow band, describing the current front position. One node at a time is moved from the narrow band to the collection of already passed nodes (in blue). Nodes that have yet to be assigned a first value are empty circles. (**b**) The four sub-sweeps of the parallel marching method. Red nodes are being computed using values of nodes connected by arrows, which points in the direction of the sub-sweep. Nodes are coloured darker after each time their value has been updated.

Tracking methods

Methods that compute nodes in the same order as they are reached by the front, are referred to as front *tracking* methods. Tracking algorithms classify a node as passed by the front only once, and are often referred to as one-pass algorithms. Therefore, these algorithms compute solution estimates only a few times per unknown node. The current position of the front is described by a set of nodes known as the *narrow band*. Figure 1(a) is an illustration of the classification of nodes in tracking methods. To know which value is the to be considered passed by front, the narrow band nodes are stored in a min-heap data structure that is kept sorted [17, 18]. With N being the total number of nodes, the worst case computational cost of the algorithm is $O(N \log(N))$, where the $\log(N)$ factor is due to the min-heap sorting. It is important to note that a direct application of the causality principle to discrete nodal values does not hold for anisotropic problems [3, 19], and the original tracking method sometimes fails to converge [20]. An algorithmic extension for tracking solvers, is to use adaptive (also called dynamic) stencil shapes [20, 21]. When updating a node, the stencil shape then varies depending on the dynamics of the front, and nodes covering a wider area must contribute to the update step. The footprint of the stencil is often determined on basis of the anisotropy coefficient [20], which is a measure of the degree of anisotropy in the chosen problem.

The need for a sorted data structure makes tracking methods conceptually sequential. However, there are some parallel implementations of tracking-like methods. The group marching method [22] classifies several nodes as passed by the front in every iteration. Since several nodes are passed simultaneously, the method allows a certain degree of parallel processing. However, a parallel implementation is not straightforward since all neighbours should be updated according to two different orderings to ensure convergence. The entire domain can also be decomposed into subdomains in which separate processors update nodes with tracking methods [23, 24]. Subdomains must then be padded with extra layers of nodes, referred to as *ghost nodes* or *ghost points*. In order for rays to travel between subdomains, the resulting algorithms are not strictly of tracking type, since already computed subdomains can be recomputed several times.

Iterative methods

Sweeping methods are iterative methods based on the observation that the minimal distance follows along a unique direction. By sweeping through the grid in one direction at a time, the minimal distances in that direction is computed for all nodes [25]. Computed nodes are used shortly after they have been assigned new values. The characteristic curves are extended in the same direction as the sweep moves through the domain. Moreover, sweeping methods are almost directly applicable to anisotropic problems [26], and can be adapted to parallel architectures with good performance on multicore systems with a domain-division approach [8, 27]. Still, the parallelism of traditional sweeping methods is coarse grained. This is caused by neighbouring nodes in a sweep depending on each other's values. Nodes on 'slices' of the domain do not share dependencies, and can therefore be computed in parallel [28]. Slices are small in corners of the domain, and parallel processing is then not beneficial. Another disadvantage is that memory accesses of nodes on a slice are incoherent, which impedes the performance of GPU implementations [29].

The parallel marching methods also sweep through the domain, but in slightly modified directions and with alternative stencil shapes [29, 30]. The alternative stencil shapes remove dependencies between neighbouring nodes, so that they can be updated simultaneously. Figure 1(b) give an illustration of the sweeping process of the two dimensional parallel marching method. A full sweep of the grid consists of four sub-sweeps, during which the front is propagated up, right, down and in the left direction. In two spatial dimensions, lines of nodes can be updated simultaneously, and in three dimensions an entire layer of nodes can be updated in parallel [30, 31]. The method scales well when implemented on GPUs. Sweeping methods are faster than tracking methods when the domain structure and the velocity formulation are simple. For complicated problems, many sweeps are needed for convergence and the methods become slow [32, 33]. Similarly, when anisotropy is dominant, more iterations are often needed for convergence [30]. Traditional sweeping methods need to sweep the domain in 2^k directions in k dimensions [34] to span the solution space. Interestingly, only $2k$ sweeps are needed for extensions of the parallel marching method to higher dimensions. However, in practice more full sweeps of the domain might be needed by the parallel marching method than by the traditional sweeping methods [29].

Label correcting methods

There are algorithms that combine iterative and tracking methods by tracing a front with a less strict ordering. A common name for such methods is *label correcting*, emanating from shortest path methods in graph networks [35, 36]. Tracking methods are referred to as *label setting*, since nodes are labelled as 'passed' only once. Since label correcting methods use a less strict ordering than tracking methods, they can sometimes be implemented on parallel architectures. For instance, the massive marching method [37] updates an entire list of nodes in parallel on a multicore CPU. If a node receives a new value during the update, all direct neighbours with larger solution values are collected in a new list. The procedure is repeated until there are no nodes to be added to a list. Massive marching is very similar to the fast iterative method [38] (FIM), which has also been used to solve problems with anisotropy [10]. GPU FIM have been used in several applications [8, 10], and can be extended to compute geodesic distances on triangulated surfaces [39]. Large velocity contrasts may cause the list to grow long and nodes to be recomputed many times, thereby increasing the computing time significantly [38]. The amount of needed computations is reduced if nodes are updated in an order more similar to the actual position of the expanding front. The two-queue method [40] and semi-ordered FIM (SOFI) [3] are more uniform in their performance since they enforce a stricter ordering of the updates. In these methods, only selected neighbours are directly added to the new list, leaving for later computations the ones too far ahead of the front. As a result, the lists in SOFI and the two-queue method follow the isocurves of the solution more closely than FIM.

2 Developing fast parallel solvers

For a front propagation method to perform efficiently, the algorithm must make use of the causality information embedded in the problem. Brute force iterative and sweeping approaches are too costly in terms of computations, and methods that are strictly tracking the front are too sequential. For instance, the parallel marching algorithms work well on GPUs, but their performance on multi-core devices is not impressive when compared to the algorithms presented in this paper. Since the entire grid is updated repeatedly, such sweeping-like methods cannot perform well enough due to the high amount of computations. Tracking methods are stable, and each node is recomputed only a few times [33]. Since tracking methods are conceptually sequential, the computing time increases noticeably when the grid size increases. Extensions for problems with anisotropy are complicated and sometimes not applicable [19]. In contrast, the more general class of label correcting methods can solve anisotropic problems [3] and can sometimes be implemented on parallel computing architectures [37, 41]. Such methods are therefore particularly interesting to investigate further.

2.1 Algorithmic inspiration

Several recent contributions to the field have inspired the design of the algorithms presented in this paper. In this section we discuss the relation between our algorithms and other existing methods.

Two-scale methods

For problems where the characteristics are straight lines, sweeping methods converge quickly and are often faster than tracking methods. However, problems with highly nonlinear characteristic curves often require many sweeps to reach convergence, and ordered methods then perform better. For most problem formulations, the characteristics bend smoothly and can be approximated well as piecewise linear. With these properties in mind, Chacon et al. [35] design several two-scale algorithms for isotropic problems. In the two-scale methods, the domain is split into subdomains and an ordering of the subdomains is created, for instance with a FMM solver acting at the subdomain level. Every subdomain is thereafter computed with a sweeping method, one subdomain at a time in the determined order. Alternatively, subdomains can enter a heap repeatedly in the *heap-cell method* (HCM). The local sweeping method converges fast since characteristic curves do not bend much within a neighbourhood, such as a subdomain. Recently, the same authors introduced a parallel two-scale method with the *parallel Heap-Cell Method* (pHCM) for eikonal equations [34]. The pHCM targets specifically multicore architectures, and show impressive speedups when compared with other algorithms. Every thread has its own heap with non-overlapping subdomains awaiting computations. In this paper we consider more general anisotropic front propagations.

Domain decomposition
Two subdomains can be updated independently of each other as long as they do not share a boundary. On shared boundaries there is a risk of read-write memory interferences [42]. To avoid any collision between read and write operations, all subdomains are padded with a layer of ghost nodes [43]. Ghost nodes are local copies of nodes belonging to subdomains that are immediate neighbours. After a subdomain has been computed, ghost nodes in surrounding subdomains must be updated with the most recently computed values. This is done in a *synchronisation* step that follows the computation of a subdomain. The boundary condition for a subdomain is then based on the values that are synchronised from the neighbouring subdomains. Subdomains surrounded by ghost nodes are further discussed in Appendix 2. Any algorithm can be used to compute new values within a subdomain.

Locks
The concept of locking is a methodology for reducing the number of unnecessary computations, which was introduced at the nodal level in the locked-sweeping method [40]. The solution value of a node needs to be updated only if values on neighbouring nodes have changed since the last update. Therefore, a node can be locked for computation right after an update, and be kept locked until relevant changes are observed in its vicinity. A lock can be considered a label that can be reset several times. Consequently, algorithms using locks are of the label correcting type. We have extended the idea of locks to the subdomain level. In our algorithms, we use a compute lock (`CL`) to prevent unnecessary computations of subdomains. After computing all values in a subdomain, it is locked for computation. This compute lock is unlocked whenever the subdomain receives a new value in the synchronisation step. Adopting the terminology used in FIM, we refer to a subdomain with an open compute lock as *active*.

2.2 Two new algorithms

In this paper, we propose two new algorithms that process sets of subdomains in parallel. We refer to the sets of subdomains as *schedules*. The order in which subdomains are computed depends on which subdomains are put into schedules. A schedule is represented as a list of indices to subdomains. The new algorithms differ mainly in the way they build their respective schedules. All subdomains in the schedule are computed and synchronised repeatedly until most subdomains have been locked for computations, and a new schedule is built. In the development of these algorithms also a third algorithm was investigated, called the *two-scale parallel marching method* [4]. Since this method is outperformed by the list-based methods, a presentation is omitted from this paper.

In the reported implementations, we compute new nodal values in a subdomain with the *three-dimensional parallel marching method* (3D PMM) [30], of which details are given in Appendix B.2. Since 3D PMM is a parallel method, the resulting algorithm has two levels of parallelism. Hardware support for multiple layers of parallelism exists for instance in GPUs. More advanced subdomain solvers (FMM, FIM or SOFI) may increase performance, especially when implemented on devices with lesser processing capabilities. However, such investigations are beyond the scope of this paper. The remainder of this section presents the two new algorithms in detail.

Algorithmic framework

The computed values $T_{i,j,k}$ approximating T in the nodal positions (i, j, k) are stored in an array corresponding to a rectangular grid in three spatial dimensions. The domain is decomposed in (s_x, s_y, s_z) subdomains in the x, y, and z, directions, respectively. Every subdomain consists of (b_x, b_y, b_z) nodes, making the number of nodes in the x direction of the rectangular domain $b_x s_x$. All subdomains are surrounded by a layer of ghost nodes on either sides. The ghost nodes are stored physically in memory, resulting in a total storage of $(b_x + 2)s_x$ nodal values along the x-axis. Similarly there are $(b_y + 2)s_y$ and $(b_z + 2)s_z$ nodes along the two other axes when ghost nodes are included. The grid may have a nonuniform spacing defined by the parameters dx, dy, and dz, but the same spacing values apply to all subdomains. When algorithmic details are discussed on a nodal basis, nodes are indexed with (i, j, k) while subdomains are indexed with (ii, jj, kk). For increased readability, subdomains are sometimes indexed by the triple `sd` $= (ii, jj, kk)$. The compute lock `CL` is implemented as an array with one binary state value for each subdomain, `Open` or `Locked`.

Similar to most front propagation methods, the proposed algorithms assume monotone convergence of the solution. All nodes are first assumed never to be reached by the front, $T_{i,j,k} = \infty$. A set of nodes with initial values is assumed known initially. These values constitute the discretised version of the boundary condition in (3). The subdomains that receive at least one initial value are unlocked for computations during the initialisation phase. As new T values are computed, they are only accepted if they are smaller than the previous value, thus assuring monotone convergence from above. The solution is bounded from below by the smallest initialised value since the solution T is increasing away from Γ_0. Note that solution values will not be increased during the iterations once accepted, and too small values should therefore never be accepted. Too small approximations can be avoided in upwind stencils if the used solution values lie upwind of the node being updated. Further details on the construction of stencils for front propagation problems are discussed in Appendix 1, where an efficient eikonal stencil is presented.

Generic solver

The proposed algorithms have a top level structure given by the template in Algorithm 2. As discussed below, each algorithm have individual specifications of the conditions governing the loops marked Ⓐ and Ⓑ. After initiating the problem, a schedule of computing tasks is created. In this context, a task corresponds to the computations needed for a single subdomain. The scheduled tasks are computed and the ghost nodes updated repeatedly through synchronisations with neighbouring subdomains, before a new schedule is created.

The compute and synchronisation steps are similar for both algorithms while the actual construction of the schedule differs. After a subdomain `sd` has been computed, the compute lock is closed, `CL(sd)` $\leftarrow$ `Locked`. In ComputeSchedule(), all active scheduled subdomains are computed in parallel as shown in Algorithm 3.

Algorithm 2 SOLVER()

comment: Generic driver for the proposed algorithms.

```
INITIATEPROBLEM()
BUILDSCHEDULE(Schedule)
while Schedule is not empty (A)
  do { while Repeat condition fulfilled (B)
         do { COMPUTESCHEDULE(Schedule)
              SYNCFROMSCHEDULE(Schedule)
       BUILDSCHEDULE(Schedule)
```

Algorithm 3 COMPUTESCHEDULE(Schedule)

comment: Parallel execution of computations for all scheduled subdomains sd.

```
for each sd in Schedule in parallel
  do { if CL(sd) is Open
         then { COMPUTESUBDOMAIN(sd)
                CL(sd) ← Locked
```

Algorithm 4 SYNCFROMSCHEDULE(Schedule)

comment: Communication between subdomains updating the values of locally stored ghost nodes.

```
for dir ∈ 1, −1
  do { for each sd in Schedule in parallel
         do SYNCiVALUES(sd, dir)
       for each sd in Schedule in parallel
         do SYNCjVALUES(sd, dir)
       for each sd in Schedule in parallel
         do SYNCkVALUES(sd, dir)
```

The COMPUTESUBDOMAIN() function updates the solution for all nodes and ghost nodes in the subdomain sd. Because of the presence of the ghost nodes, the computations rely only on nodal values stored locally for the subdomain, thus eliminating the need for subdomain communication during the computations. After computing new values for the active subdomains in the schedule, nodal values must be communicated between adjacent subdomains to assure that all ghost nodes have correct values. This is done in SYNCFROMSCHEDULE(), shown in Algorithm 4.

To avoid any memory interference, the nodal values of neighbouring subdomains are updated in parallel in one direction at a time. First, the ghost nodes belonging to subdomains with larger i values are updated in parallel, and thereafter ghost nodes in subdomains with larger j, k, and smaller i, j, k indices are updated. Note that no computations are performed during the synchronisation, rather the nodal values are set to match their copies in the other subdomain. A nodal value is only changed if the value of its counter-node is smaller.

Algorithm 5 BUILDSCHEDULELAS(L,CL)

comment: Construction of the computational schedule,
tailored for the LAS method.

```
noSched ← 0
for sd ← index to all subdomains
  do { if CL(sd) is Open
       then {L[noSched++] ← (sd)
return (noSched)
```

If a node in subdomain sd receives a new value, the subdomain is activated by opening the compute lock, CL(sd) ← Open. Further details of the SYNC□VALUES() functions are given in Appendix B.1. The algorithms finish when all subdomains are locked for computations and the schedule is empty.

2.2.1 The list of active subdomains method

The list of active subdomains (LAS) algorithm schedules all active subdomains for computation by storing their indices in a list. As the algorithm iterates through the grid, the list follows a front of solution estimates as it travels through the domain, similarly to the lists in the massive marching and FIM methods [37, 38]. Subdomains that have already been computed may at a later stage be added to a new version of the list. Pseudocode for building the schedule (list) for the LAS method is shown in Algorithm 5. The returned variable noSched is the number of subdomains in the list L.

After computing and synchronising the list of subdomains, the number of active subdomains left in the list are computed, here referred to as noActive. The schedule is computed again if the density of active subdomains is higher than a given fraction. Let noSched denote the length of the list. After some experimentation with Ⓑ in Algorithm 2, we have chosen to repeat the computations while $\frac{1}{64} < \frac{\texttt{noActive}}{\texttt{noSched}}$. This threshold gives good performance for most examples. Moreover, let noC denote the number of subdomains that can be computed simultaneously by the current computational platform. If there are fewer active subdomains than what can be computed simultaneously, i.e. noActive < noC, a new schedule is created. This optional condition to improve load balancing is included in the LAS pseudocode given in Algorithm 6, which replaces the generic SOLVER() in Algorithm 2.

The LAS method has some similarities with FIM, but there are differences to notice: For FIM, only the GPU implementation computes subdomains, and no compute locks are used. In FIM, the scheduled tasks are computed only once before a new schedule is built, whereas in LAS the schedule is reused until at most (around 98%) scheduled subdomains are locked for computation. Any method can be used to compute a subdomain in LAS, whereas in FIM all nodes of a subdomain are computed iteratively until convergence. This brute force approach of FIM would likely reduce the performance on multicore systems.

2.2.2 The semi-ordered list of active subdomains method

In the LAS method, previously active subdomains may be reactivated at later stages of the algorithm. In some front propagation problems, this reactivation can force the list to grow long. A similar observation holds for FIM and SOFI, although SOFI performs

Algorithm 6 SolverLAS()

comment: Specific driver for LAS, variant of algorithm 2.1.

```
InitiateProblem()
noSched ← BuildScheduleLAS(L, CL)
noActive ← noSched
while L is not empty
do  { while 1/64 < noActive/noSched
        do { ComputeSchedule(L)
             SyncFromSchedule(L)
             noActive ← Number of active subdomains in L
             if noActive < noC
               then noActive ← 0
      noSched ← BuildScheduleLAS(L, CL)
      noActive ← noSched
```

more uniformly thanks to the enforced heuristic ordering of updates [3], which makes the behaviour of the list more similar to the narrow band of tracking methods. The semi-ordered list of active subdomains (SOLAS) method similarly schedules subdomains in lists that in a sense stay close in shape to isosurfaces of the solution. If a subdomain is too far ahead of the others, it is not added to the schedule.

In order to measure the approximate location of a subdomain, each subdomain has an associated scalar value stored in a global array structure, here referred to as `SD`. The `SD` value for a subdomain is the smallest new T value that activates the subdomain, and is set during the synchronisation procedure. This minimal T value is similar to the 'rollback state' of the domain decomposition FMM [23]. Details on handling the `SD` values in the synchronisation is given in Appendix B.1.

The `SD` value is an approximate position of the current front as it reaches a subdomain. Let `Av` denote the current average `SD` value of all active subdomains, while `oldAv` denotes the same average of the previous iteration. All subdomains with a `SD` value smaller than a cutoff value `cutT` are scheduled for computation. In our implementation the `cutT` value is set to be $\texttt{Av} + 0.4\max(0, \texttt{Av} - \texttt{oldAv})$. Several methods of creating this cut-off value was tested, but the chosen 0.4 relaxation from the average value gives good performance for many problems. With the enforced ordering of subdomains, the list will mimic the behavior of the solution's isosurfaces.

If the list is made too short, some processing units may be idling since there are not enough computations to perform. Since the list is computed repeatedly, neighbouring subdomains may be reactivated repeatedly, and if the neighbouring subdomain is not in the list it will not be computed until a new list is created. It is therefore not beneficial to restrict the list when the number of active subdomains is low. A cut-off is therefore only created if there are more active subdomains than an empirically chosen `MinAct` value. With the variable `noC` representing the total number of subdomains that the targeted platform can compute simultaneously, the choice $\texttt{MinAct} = \max(2\texttt{noC}, \frac{3}{2}\max(s_x s_y, \texttt{noSched}))$ seems to give a good overall performance, and has been used in all reported results. The `noSched` bound assures that the list is not limited if there are much fewer active subdo-

---

Algorithm 7 BUILDSCHEDULESOLAS(L,CL)

---

comment: Construction of the computational schedule,
tailored for the SOLAS method.

```
noActive ← 0
sumSD ← 0
cutT ← ∞
for sd ← index to all subdomains
  do { if CL(sd) is Open
       then { noActive++
              sumSD ← sumSD + SD(sd)
Av ← ∞
if noActive > MinAct
  then { Av ← sumSD / noActive
         cutT ← Av + 0.4 max(0, Av − oldAv)
         oldAv ← Av
noSched ← 0
for sd ← index to all subdomains
  do { if CL(sd) is Open and SD(sd) < cutT
         then { L[noSched++] ← sd
       MinAct = max(2noC, 2/3 max(s_x s_y, noSched))
return (noSched)
```

---

mains than was scheduled in the previous list. The $s_x s_y$ bound is included so that the list is shortened only if the current front is rather large (or thick).

Pseudocode for creating a task schedule in the SOLAS method is given in Algorithm 7. The returned variable noSched is the number of scheduled tasks that have been added to the list L.

The pseudocode for the SOLAS driver is identical to the driver SOLVERLAS() in Algorithm 6, except that calls to BUILDSCHEDULELAS() are replaced by calls to the SOLAS-specific procedure for constructing the subdomain schedule, that is, BUILDSCHEDULESOLAS() defined in Algorithm 7.

2.3 Algorithmic observations and implementation details

In this section we discuss implementation specific details, and some of the optimisations that were tested. The gain of a particular optimisation depends on its interplay with other imposed optimisations. In general, if an optimisation is said to give a 'significant' performance boost, the reduction in computational time was lowered with more than 10%. From an implementation perspective, LAS is easier to implement. However, SOLAS requires only small modifications of the LAS code. The workload and number of iterations of LAS and SOLAS differ significantly with the underlying physical problem and boundary conditions. A performance modelling analysis is therefore difficult to perform.

Initial computations

An expanding front passes a node only once and reaches nodes further away at later times. Because of this causal dependency, the parts of the solution holding larger values depend on nodes with smaller values. Numerical errors introduced close to the origin will propagate to the whole domain. It is therefore important that the computed solution is particularly accurate close to the initial condition. To ensure that all initiated subdomains fully converge, they are computed with one sweep of 3D PMM in the beginning of the algorithms.

Blocking for cache
In our initial implementation, the numerical grid was stored in row-major order. Nodes within one subdomain are stored in several rows, and every row of data continues into other subdomains. Therefore, some of the data read into cache will belong to other subdomains, and will not be used in computations. When data that are reused often are stored physically close in memory, more relevant data are loaded at once to the faster cache memory. Nodes within a subdomain were therefore ordered contiguously in memory. This optimisation improves the data locality, and is sometimes referred to as *blocking for cache* [44]. The GPU performance was improved significantly with this optimisation, since the access time between global memory and shared memory differ more on GPUs than on CPUs. On GPUs, rows of two and three dimensional data structures are stored in memory allocations of 'pitch' bytes. The pitch is chosen by CUDA to ensue best performance when accessing the row addresses [45]. In the GPU implementation, all nodes in one subdomain lie within one pitch so that the memory accesses are aligned for both the synchronisation and the computation kernels. The CPU performance was also improved, but only by a couple of per cent.

Sliding window
If the entire subdomain is read into the shared memory on the GPU, the performance is limited by the high use of shared memory. This was implemented as a first attempt, but led to unacceptably low speedup. Instead, we use a *sliding window* approach [46]. When two adjacent nodes are updated in 3D PMM, they share several values in the bottom layer of the pyramid stencils, see Appendix B.2. To reduce the number of reads from global memory, the entire bottom layer of nodes is first loaded to a shared memory structure, from where cache loads are faster. The GPU performance was improved significantly.

A similar sliding window optimisation is used in the CPU implementation, since the entire bottom layer of nodes is loaded to a cached data structure. This increases the performance slightly, and has been used in the numerical examples in Section 3.

List building in GPU implementations
Both the SOLAS and the LAS methods create lists. This is difficult to do efficiently on a GPU. Instead, a boolean structure indicating whether or not a subdomain should be present in the list is created on the GPU. This structure is transferred to the CPU where the list creation can be performed fast. Unfortunately, this requires the GPU to wait while the list is being built on the CPU. Stalling communications between the CPU and the GPU is known to be a bottleneck of GPU algorithms [47].

Kernel timeout on GPUs
In our application, we sometimes experienced difficulties launching kernels for the anisotropic experiments in double precision on laptop GPUs. These devices were simultaneously used by the operating system to display the graphical desktop environment. It

appeared that when launching a kernel on many subdomains, we received the error message CUDA_ERROR_LAUNCH_TIMEOUT. According to the CUDA Driver API [48], this error is returned when a kernel takes too long to execute. To avoid the device making the operating system halt, Nvidia has implemented a specific time limit on the execution time of kernels. If this time limit is exceeded by the kernel, the kernel will simply not launch or in some cases be terminated once being launched.

Our solution is to limit the number of launched kernels. If there are too many active subdomains in the list, the list is sliced into chunks on which kernels are launched. The size of each chunk is chosen so that the device can execute the kernel with maximum occupancy. In cases where the number of active subdomains will not lead to a kernel timeout, the chunking is simply omitted.

3 Numerical verification

In this section, we review a series of numerical experiments conducted in order to study the performance of the new solvers proposed in this paper. These experiments include isotropic and anisotropic problems for which analytic solutions can be derived, and isotropic problems with changing velocity and obstacles that are impermeable or very slowly permeable. Our final case is that of simulating a geological fold, based on seismic field data.

Definitions

For any scalar field ξ represented by values in every grid node (i, j, k), we will apply the discrete L_2 norm

$$\|\xi\|_{L_2} = \sqrt{\frac{\sum_{i,j,k} \xi_{i,j,k}^2}{N}}, \tag{12}$$

where the summation is performed over all N nodes in the domain. We refer to the grid as uniform only when $dx = dy = dz$.

Computing platforms

The numerical experiments reported herein have been conducted on a 16 core CPU referred to as *SandyBridge*, and a Nvidia Tesla K20 GPU, hereafter called *K20*. The capabilities of these platforms are summarised in Tables 1 and 2, respectively.

All computing times presented in this section are from implementations of the algorithms, one multicore CPU variant using OpenMP [49], and one GPU variant using CUDA [50]. For all test cases, we report on the fastest solution extracted from at least three executions of the chosen algorithm on the chosen platform. In general, we have observed that the computing times have only insignificant variations from one run to another.

Sizing subdomains

In general, 3D PMM is an efficient algorithm with significant parallel capabilities. For each subdomain, one needs to store both the internal nodes and the ghost nodes to the subdomain. The domain decomposition is slightly overlapping, considering that the ghost nodes are inner nodes of other subdomains. Therefore, the smaller the subdomains get, the larger will the fraction of ghost nodes relative to inner nodes become. From a global

Table 1 Specifications for the Intel Xeon E5-2650 2 × 8-core CPU (one of two identical NUMA domains).

SandyBridge: Intel Xeon E5-2650	
CPU frequency	2.0 GHz
Number of cores	2×8
Single precision	256 GFLOP/s
Double precision	512 GFLOP/s
Shared L3	2×20 MB
Memory	32 GB DDR3
Memory bandwidth	2×51.2 GB/s

Table 2 Specifications for the NVIDIA Tesla K20 GPU.

K20: NVIDIA Tesla K20	
CUDA cores	2496
Single precision	3520 GFLOP/s
Double precision	1170 GFLOP/s
Memory	5120 MB GDDR5
Memory bandwidth	208.0 GB/s

perspective, a reduction of the subdomain size will increase the memory footprint and the communication overhead due to storing and handling a growing number of ghost nodes. Still, a relatively small subdomain size may offer an advantage in that the list can better mimic the front propagation, thus leading to a more efficient exploitation of the underlying causality. Moreover, nodal values of smaller subdomains are more likely to fit in fast memory locations, thereby increasing the computational speed per subdomain.

We experimented with different cubic sizes of subdomains for a range of problems. For the CPU implementation, the optimal subdomain size depends on the exact problem formulation and the total number of nodes. A change in subdomain size have only small impact on the performance of the CPU implementation. We have therefore focused on the GPU implementation when choosing the subdomain size for our computations. Subdomains with 14^3 nodes lead to the best GPU performance in terms of wall clock time used for the solution. Since the computations include updates of ghost nodes, a total of $(14+2)^2$ threads are used to compute a subdomain, see Appendix B.2 for details. The GPUs used for the analysis all have a warp size of 32, that is, a minimum of 32 identical operations per instruction is performed in a Single Instruction Multiple Data (SIMD) fashion [47]. If the number of threads is not a multiple of the warp size, some GPU threads will be idling while the others performs computations. Having analysed our implementation with the CUDA Device Occupancy calculator [51], we have concluded that the subdomain size of 14^3 allows for both the best occupancy and the shortest computing time. This subdomain size is kept constant throughout all experiments reported in this paper.

3.1 Synthetic examples

The LAS and SOLAS algorithms proposed in Section 2.2 have been tested on a range of problems that expose the solvers to several challenges. Here, we briefly summarise our findings by presenting five of those examples. For all these examples, we have computed solutions for grids with 84, 168, 336 and 504 nodes along each axis, grouped accordingly in 6^3, 12^3, 24^3 and 36^3 subdomains. That is, these grids represent from 592, 704 to 128, 024, 064 unknowns. The computations have been conducted on the SandyBridge (CPU)

Figure 8 Isosurfaces of the anisotropic solution with 13 point sources located irregularly in the rectangular domain as boundary condition (Ex_A).

Ex_A: 13 point sources.

and the K20 (GPU) platforms. In all experiments, the solvers stop when all subdomains are locked for computations, and the schedules are empty.

The reported performance numbers bellow are based on computations in double precision. Due to specific hardware features of GPUs, a transition from CPU to GPU is far more attractive for computations in single precision. For the type of calculations done in this paper, single precision arithmetics can be used without noticeable loss of accuracy. This topic is further discussed in Section 3.2.

3.1.1 Example with analytic solution

The first example uses the anisotropic formulation (3), with constant velocity parameters $F = 1.4$, and $\psi\mathbf{a} = (0.9, -0.75, -0.07)$. As boundary condition we used 13 point sources spread out irregularly in the rectangular domain, defined by $(0,0,0) \leq \mathbf{x} \leq (10,13,9)$. The ratio between the maximum and minimum velocities, referred to as the anisotropy coefficient, is 11.2 for this problem. A visualization of the anisotropic solution is given in Figure 8.

As shown for the Ex_A case in Figure 9(a), the GPU implementations always outperform its CPU counterparts. Notice that SOLAS performs slightly better than LAS for large grids across both platforms. The difference between these two solvers is more prominent on the CPU than in the GPU-based computations.

The efficiency[a] for eight and 16 cores is 83-88% for SOLAS, whereas it drops down to 65-75% for LAS. Many subdomains are active simultaneously in Ex_A, so the additional ordering of SOLAS is worthwhile. These efficiency ratios are good, taken into account the sequential construction of schedules and that threads are synchronised between the COMPUTESCHEDULE() and SYNCFROMSCHEDULE() procedures. Going from CPU to GPU architecture, we observe 6.9 and 8.5 times faster computations for the Ex_A case when employing the SOLAS and LAS solver on the largest grids, respectively. We also run the same problem with an isotropic setting, $\psi\mathbf{a} = 0$. For the isotropic case, LAS is instead slightly faster than SOLAS on large grids. Compared to the isotropic counterpart, the anisotropic computing time increases with a factor of 7-12 on the CPU and 5-8 on the GPU. The anisotropic stencil is significantly more complicated than the isotropic, and the solution dependencies in the anisotropic case increase the total number of times that subdomains are activated.

In Figure 9(b), we have repeated the computing times from Figure 9(a), using logarithmic scales on both axes. The thin dotted lines illustrate linear relations to N. That is, the figure indicates that all the tested methods have a computational cost that scales with the

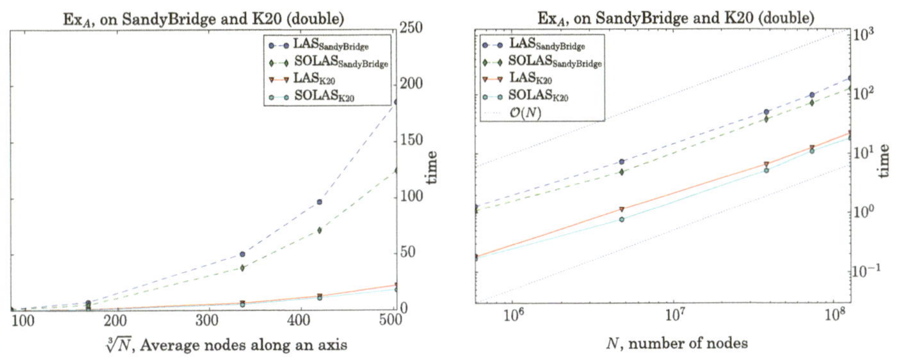

(a) Ex_A: computing times.

(b) Ex_A: logarithmically scaled computing times.

Figure 9 Performance on the SandyBridge and K20 platforms for the isotropic test case with 13 irregularly located point sources (Ex_A). (**a**) Observed computing times (seconds) for 16 cores. (**b**) Observed computing times presented in logarithmic scales.

grid size close to $O(N)$, or even *superlinearly*, that is, the computing time seem to increasing less than linearly with growing problem size. Similar behaviour is observed for all our numerical investigations. There are several factors that can help explain this observation. Small data sets cause more computing units to be idle, thereby reducing the observed computational speed. When the data set grows in size, the level of parallelism increases and a better load balance between the processing units is achieved. The time needed to transfer data to and from the GPU is included in the time measurements. The overhead associated with (synchronous) data transfers between the CPU and GPU decreases as the data set gets larger. The cost of data transfer is therefore relatively larger for smaller data sets. Although we have not formally analysed the computational cost of the new algorithms, they will likely have linear complexity due to being semi-ordered and of sweeping type. The superlinear behaviour is thus caused by other artefacts as explained above.

Accuracy and convergence This example has analytical solutions [30], making it possible to study how the numerical solution converges towards the true solution. Let h_k be the edge length of the kth grid in a sequence of increasingly finer grids used for numerical solution of the problems. Denoting the corresponding error $e^{(k)}$, the estimated rate of convergence is

$$p = \frac{\log(\|e^{(k)}\|_{L_2}/\|e^{(k+1)}\|_{L_2})}{\log(h_k/h_{k+1})}. \tag{13}$$

For the grid sizes in this study, our convergence estimates are 0.70, 0.74, 0.76 and 0.77, for both the anisotropic and isotopic results. Convergence estimates for single and double precision solutions are for all practical purposes identical. To reach first order convergence, the amount of nodes initially assigned analytic values should be kept constant [16]. The estimated convergence rates are therefore satisfactory.

3.1.2 *Examples with curved characteristics*

Iterative algorithms are sensitive to curved characteristics [32]. It is therefore common to investigate how algorithms perform for problems with obstacles that force the characteristics to turn sharply [18, 29, 38].

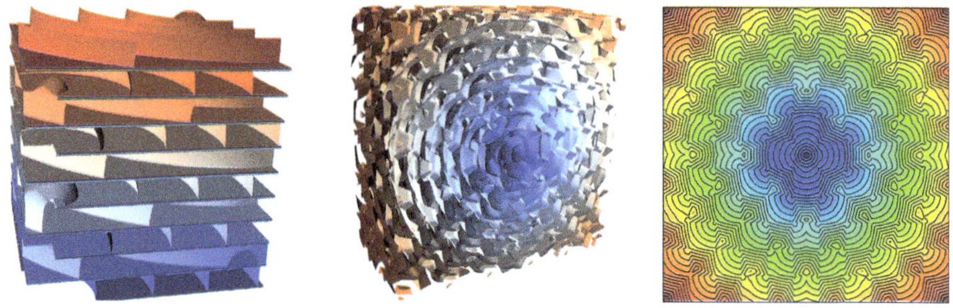

(a) (Ex_C)Ex_B: (Im)permeable walls. (b) Ex_D: Velocity-cubes. (c) Ex_D: Velocity-cubes cut.

Figure 10 Illustrations of solutions to examples with curved characteristics. (**a**) The distance field in presence of semi-permeable obstacles organised as nine walls (examples Ex_B and Ex_C). (**b**) Three-dimensional visualisation of the distance field obtained by solving the chequerboard velocity problem (example Ex_D). (**c**) Two-dimensional planar cut through the velocity-cubes distance field (example Ex_D).

In the test cases Ex_B and Ex_C, we consider a cubic volume divided by nine walls with small openings in either the upper left or lower right corners. The walls are one node layer thick, and are either semi-permeable ($F = 0.03dz$ for Ex_B) or impermeable ($F = 0$ for Ex_C). Outside these walls, the isotropic velocity is set to $F = 1$. The fastest path from the starting point in a corner of the domain to the opposite corner is by zigzagging through the small openings in the walls. A sample solution is visualised in Figure 10(a) [38, 52].

In the final example, Ex_D, the computational domain is divided into a set of equally sized cubes. The velocity is constant within each cube, having the value of either 1 or 2. For a cube with velocity value 1, all adjacent cubes sharing a side with it will have a velocity equal to 2, and vice versa. We have placed in total 11^3 such velocity cubes within a computational domain measuring 10 length units along each axis. The boundary condition is one point source at the center of the domain. Figure 10(b) shows a volumetric view of the computed solution, while Figure 10(c) shows a planar cut of the solution through the center of the domain [34, 40].

Performance for the obstacle problems The examples Ex_B and Ex_C, having barriers embedded in the computational domains, are challenging since the front is forced to travel along irregularly shaped paths. Figures 11 and 12 summarise the computing times and the speedup factors for these two problems.

The barriers are only one node thick for all grids, and the fraction of nodes that are obstacles reduces as the grid gets finer. The front is small since it only propagates between two layers of walls. For larger grids, more subdomains fit between two barriers, and the parallel processing possibilities increases as the front activates more subdomains. Still, many schedules are built in the computation of these examples. The changing dynamics of the problems causes the compute times to change rather peculiarly with grid sizes, especially LAS since the extra ordering in SOLAS reduce the effects. At a large enough grid size, the LAS solvers changes to such an extent that the solve time decreases for a larger grid size. The irregular behaviour of LAS is present for both the CPU and the GPU implementations. Although the variations for each platform is at the same level when seen relative to the platform's computational power, this issue is most pressing in the CPU environment due to generally longer computing times.

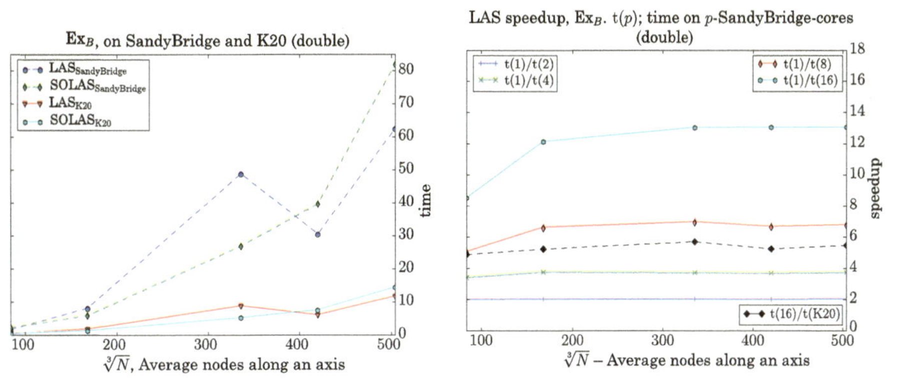

(a) Ex$_B$: computing times.

(b) Ex$_B$: LAS scaling.

Figure 11 Performance on the SandyBridge and K20 platforms for the example with semi-permeable barriers (Ex$_B$). (**a**) Observed computing times (seconds) for 16 cores. (**b**) Observed speedup of LAS when moving from one ($t(1)$) to many ($t(p)$, $p = 2,4,8,16$) cores, and for 16 cores relative to the GPU.

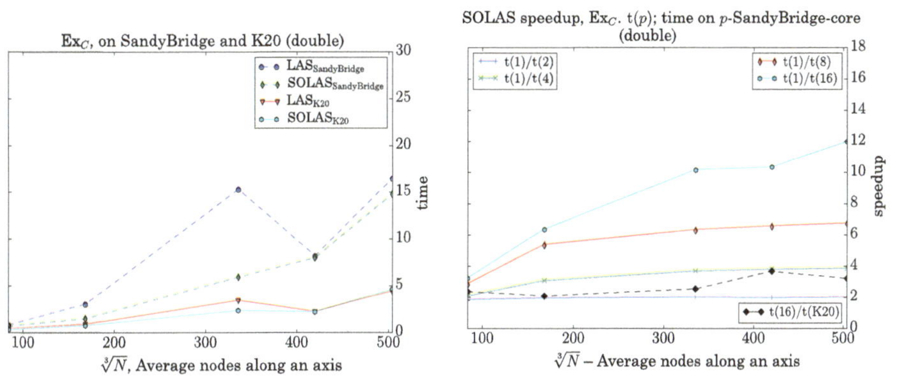

(a) Ex$_C$: computing times.

(b) Ex$_C$: LAS scaling.

Figure 12 Performance on the SandyBridge and K20 platforms for the example with impermeable barriers (Ex$_C$). (**a**) Observed computing times (seconds) for 16 cores. (**b**) Observed speedup of SOLAS when moving from one ($t(1)$) to many ($t(p)$, $p = 2,4,8,16$) cores, and for 16 cores relative to the GPU.

For Ex$_B$, where the barriers are semi-permeable, LAS turns out to be the most efficient solver for large N, regardless of platform. For the impermeable barriers in Ex$_C$, SOLAS would be the preferred method, although the behaviour comes very close to that of LAS when the grid gets large. For the 504^3 grid, the GPU versions of LAS and SOLAS compute the solution in 4.5 seconds, while the CPU implementations require about 15 seconds. It is noted that the semi-permeable case requires about two to three times as long computing time as the impermeable problem, regardless of solvers and platforms.

Figure 11(b) shows that LAS applied to the Ex$_B$ problem gives a very good speedup, even on eight and 16 cores for which the efficiency is 82% and 84%, respectively. Moving on to the GPU, the LAS method runs 5.5 times faster than the 16 CPU cores. The speedup of SOLAS on eight cores is the same as for LAS, but there is a drop to 77% efficiency when going up to 16 cores. Similar to the observation we made for Ex$_A$, this drop of speedup does not propagate over to the GPU, where SOLAS performs at same level as LAS. As shown in Figure 12(b), the computing times obtained for applying LAS and SOLAS to

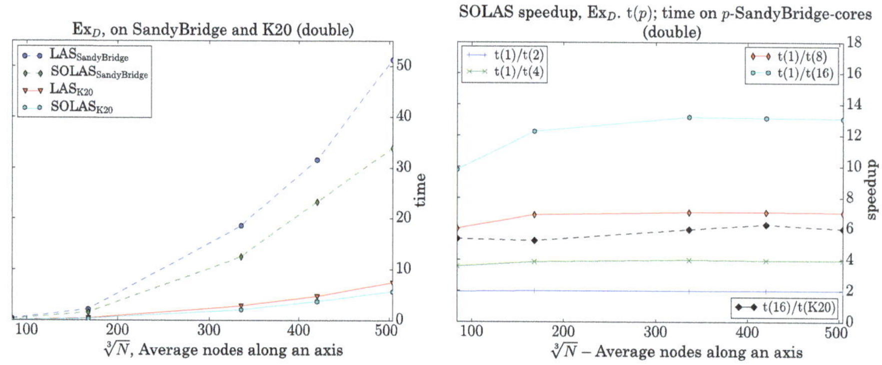

(a) Ex$_D$: computing times.

(b) Ex$_D$: SOLAS scaling.

Figure 13 Performance on the SandyBridge and K20 platforms for the example semi-permeable barriers (Ex$_D$). (**a**) Observed computing times (seconds) for 16 cores. (**b**) Observed speedup when moving from one ($t(1)$) to many ($t(p)$, $p = 2, 4, 8, 16$) cores, and for 16 cores relative to the GPU.

Ex$_C$ scale with the number of cores quit similar to the factors observed for Ex$_B$. However, the performance gain of moving from 16 cores to the GPU is less for the impermeable problem in Ex$_C$, where the speedup factor is in the range 3.2-3.8.

Performance for the velocity-cubes problem For the velocity-cubes problem Ex$_D$, Figure 13 illustrates smooth behaviour of all algorithms with respect to the computing time as a function of the grid size. As for the two examples with embedded barriers, SOLAS proves to be an efficient solver. It performs consistently better than LAS, in particular on the CPU platform. On the GPU, the solution can be computed for the largest grid in 5.7 seconds, compared to the 33.9 seconds required by the 16 CPU cores. Both LAS and SOLAS demonstrate excellent speedup on the CPU, delivering 80%-88% of the theoretical potential for eight and 16 cores. For both methods, the GPU calculations runs 6-7 times faster than the 16 cores.

Overall performance for the synthetic examples Combining the observations from the synthetic test cases discussed above, we conclude that the proposed methods are well suited for efficient computations on multicore CPUs and GPUs. Even for complicated problems these algorithms offer computational efficiency that can support interactive software applications, also when handling finely spaced grids. This behaviour is due to the methods being designed to efficiently exploiting the causality of a propagating front.

For anisotropic problems, the GPU implementations of LAS and SOLAS spend on the average around 96% of the time on computing, 3-4% on synchronisation, and less than 1% on list building. For isotropic problems, the computations take around 90% of the total time on the GPUs. Computations dominate even more on CPUs where 98-99% of the total time is spent on subdomain computations and the remaining time is used for synchronisation. List creation is then less than 0.1% of the total time, and is therefore negligible. As to be expected, the synchronisation of values is slightly more costly for SOLAS than for the LAS algorithm.

Regardless of the test cases, the GPU implementations offer consistently faster computations than the CPU implementation running on up to 16 cores. The speedup factors when moving from 16 cores to the GPU ranges from 3 to 8 with an average of 6. For single precision computations, this speedup factor varies from 6 to 16 with an average of 10. LAS is preferred for the isotropic or mildly anisotropic problems, whereas SOLAS is the best choice for problems with complicated velocity profiles.

3.2 Single versus double precision arithmetics

The difference in peak performance between single and double precision of the Sandy-Bridge CPU assumes a perfectly vectorised code. Unfortunately, the large amount of logical branching prohibits a vectorisation of the computationally dominating update step. By hardware design, GPUs compute substantially faster in single precision than in double precision. In particular, Nvidia's Kepler architecture, such as our K20 platform detailed in Table 2, is able to perform three times as many operations per second in single precision than in double precision. In addition to the raw processing capabilities, values represented in single precision need only half the memory space that values stored in double precision need. Therefore, one can fit larger data sets into the limited memory of a GPU if single precision is used. Data-intensive applications will thereby get an additional improvement of performance due to better utilisation of fast memory and less shuffling of data between the GPU and the host computer. Across the range of synthetic problems investigated in Section 3.1, we have observed an average speedup of 2.1 when comparing single and double precision computations on the K20 platform. In comparison, the corresponding speedup of 1.2 for the 16-cores SandyBridge CPU is rather modest. On both platforms, these speedup factors are independent of the solvers. This observation agrees with the fact that the total time spent by the algorithms is heavily dominated by arithmetic operations, rather than memory transfers and synchronisation.

Given the advantage of single precision arithmetics, it is relevant to investigate whether computations in single precision can provide a satisfactory level of numerical accuracy when solving problems of the form (3). Empirically, we have observed that the single precision solution, T_f, is practically identical to the double precision solution, T_d. For the largest grid, $N = 504^3$, we have observed that $\|T_d - T_f\|_{L_2} = O(10^p)$, where $p = -7$ for Ex_A, $p = -6$ for Ex_A and Ex_D, and $p = -4$ for the complicated cases Ex_B and Ex_C. By the triangle inequality, we have

$$\|T - T_f\|_{L_2} = \|T - T_f + T_d - T_d\|_{L_2} \tag{14}$$

$$\leq (1 + \varepsilon)\|T - T_d\|_{L_2}, \tag{15}$$

where $\varepsilon = \|T_d - T_f\|_{L_2}/\|T - T_d\|_{L_2}$. That is, if $\varepsilon < 1$, the single precision error will be of the same order as for the computations in double precision. We have estimated ε empirically for Ex_A, which have analytic solutions. Depending on the problem and the grid size, this entity stays in the range from 10^{-5} to 10^{-3} and is thus significantly smaller than one. Moreover, we have computed the ratio $\beta = \|T_d - T_f\|_{L_2}/\|T_d\|_{L_2}$, which is obtainable also for problems that can not be solved analytically. We have then observed that β stays between 10^{-5} and 10^{-8} for Ex_A-Ex_D across all grid sizes, and that β gets smaller as N increases, thus indicating convergence.

Based on these empirical studies, we conclude that single precision computations provide high enough accuracy when using the algorithms proposed in this paper with first order stencils to simulate front propagation based on (3). More subdomains are reactivated for computation when using double precision arithmetics than in the case of single

(a) (b)

Figure 14 Illustrations of a fold simulation. (**a**) A surface extracted in the CES application, with two cross sections through a three-dimensional offshore seismic volume. (**b**) The folded layers simulated on basis of the extracted surface and an isotropic formulation.

precision. A subdomain is only reactivated when it receives a new value. Since the smallest noticeable change in single precision is effectively 10^{-7}, the subdomains with smaller changes will not be reactivated in single precision, but only in double precision computations. The increase of reactivations contributes to the increase of time needed by the solvers in double precision compared to single precision. Note that a subdomain could be reactivated only if the new value t_{new} differs from the old value t_{old} more than a chosen threshold-limit in order to limit the amount of reactivations. In other words, to reduce the number of subdomain computations, a subdomain can be activated only if $t_{\text{old}} - t_{\text{new}} > c$ for some empirically chosen value c. In all our numerical experiments we have used $c = 0$.

3.3 Fold simulation

Using the CES software, a user can easily segment structured surfaces from seismic data sets by autotracking. Figure 14(a) shows an autotracked surface in the CES software, extracted from a three-dimensional seismic volume as indicated by the transparent seismic cross sections. The physical domain measures 6700, 4975, and 11055 meter along the x, y and z axes, respectively. Values in the nodes immediately above and below the extracted horizon were computed in CES for grids with 256^3 and 512^3 nodes, respectively. These values were then used as boundary conditions in our numerical experiments. Parallel folded layers were simulated by solving the isotropic eikonal equation for the different grids. An example of the computed solution is visualised in Figure 14(b).

We have compared our new algorithms with the CES solver. The fold simulation solver in CES is based on a tracking algorithm, and is therefore sequential. When measuring the performance of the CES solver, we used a 4 core, 3.5 GHz Intel Ivy Bridge Core i7-3770K CPU with 8MB L3 cache. This is the fastest platform available that can run the CES software. Applying the new algorithms on K20 and SandyBridge, we have measured the computing times for both single and double precision arithmetics. Because of the sizing of subdomains, the grids used by the new algorithms are slightly larger than the grids used by the CES solver. The new grids consist of 266^3 and 518^3 nodes, respectively. These grids represent from 343,000 to more than 138 million unknowns. Table 3 shows the computing time for double and single precision arithmetics, together with the speedup achieved by replacing the Ivy-based CES solver. The speedup factors are prefixed with the symbol '×.'

Table 3 Comparison of computing times (in seconds) for the CES solver on Ivy and LAS and SOLAS solver on SandyBridge and K20, including the speedup factors ('×') relative to CES.

Solver	Device	Double precision				Single precision			
		266^3		518^3		266^3		518^3	
CES	Ivy	27.68	×1.0	354.9	×1.0	24.91	×1.0	319.42	×1.0
LAS	K20	0.87	×32.0	6.2	×57.3	0.44	×57.2	2.99	×107.0
SOLAS	K20	0.84	×33.1	6.0	×59.6	0.43	×58.6	2.98	×107.2
LAS	SandyBridge	5.22	×5.3	36.0	×9.9	4.31	×5.8	28.46	×11.2
SOLAS	SandyBridge	4.66	×5.9	32.8	×10.8	3.81	×6.5	26.67	×12.0

The CES time for single precision computations has been empirically estimated as 90% of the CES time for double precision.

Reviewing the results, we observe that the difference between solution times for LAS and SOLAS are negligible when run on the same platform and with the same arithmetic resolution. We also observe that the speedup is higher for the GPU platforms than for the CPU platforms. As noted in Section 3.2, there is no difference in solution quality between single and double precision computations. However, the computing times are considerably lower in single precision computations, especially on the GPU. The computing time for the GPU is roughly halved when moving from double to single precision computations. For the largest grid in single precision, the K20 implementation of SOLAS is more than 100 times faster than the CES solver on Ivy.[b] We do not expect the speedup to be as impressive in all fold simulations. However, the user experience is significantly improved when fold simulations are performed with the new algorithms. Even on the largest grids, the computational time is a few seconds instead of minutes when LAS or SOLAS uses K20 instead of the current solution method. Based on this and several other numerical experiments with seismic field data, the proposed algorithms are now being implemented in the industrial code.

The task of simulating a folded volume is rather different from the synthetic examples discussed initially. For fold simulations, the initial geological surface defining the boundary condition will contain a significant number of subdomains. Because of this, the list used for subdomain scheduling will be quite long already from the start. When this list is long, it is easier to make full use of the available computing resources, thus resulting in a substantial effect of parallelisation. For simpler problems, where the boundary condition is represented by one or a few point sources, only a few subdomains will be scheduled as active. Thus, some of the available computing resources will be idling until later stages of the solution process when the lists grow longer. Since GPUs have an intrinsic capability of processing a large number of subdomains simultaneously, the simulation of geological folds based on LAS or SOLAS is particularly suitable for GPUs.

4 Conclusions

Two new parallel solvers are presented, together with several numerical experiments conducted on both multicore CPUs and GPUs. All solvers use a domain decomposition approach, where each subdomain is surrounded by a layer of ghost nodes. A subdomain needs to be processed only if its local boundary condition has changed since the previous computation. Due to this observation, a computed subdomain is locked for further computations until its boundary condition changes, and the subdomain is reactivated.

The presented algorithms differ mainly in the way subdomains are scheduled for computations. LAS places all active subdomains in a list, and updates the subdomains in this list in parallel. The GPU implementation of LAS builds the lists on the CPU, but performs all computations on the GPU. SOLAS is very similar to LAS, but creates the lists slightly differently to enforce a more causal ordering of the subdomain computations.

From empirical studies, the sequential solution times seem to scale linearly with the total number of nodes. The parallel solvers appears to scale somewhat superlinearly, accredited to the increase in parallel processing efficiency obtained for larger grids. Some algorithmic details are crucial for achieving good performance. For instance, if a schedule is only used once by LAS and SOLAS, many more and longer lists are built. For anisotropic cases, repeated activations between neighbouring subdomains are more common than for isotropic cases. It is therefore important not to enforce a too strict ordering of subdomains. For some examples the LAS ordering is not very different from the SOLAS ordering, and the extra overhead of the SOLAS ordering is then not worthwhile.

We have investigated the potential for accelerating an industrial software for simulation of geological folding by introducing new algorithms for numerical solution of static Hamilton-Jacobi equations. The computing times are reduced from several minutes to seconds, enabling the software to be used interactively even for large three-dimensional data sets.

Possible extensions of the presented work

In our implementation, the computing and synchronisation procedures are entirely completed before the list building starts. Future implementations can overlap the building of new schedules with the processing of previous schedules. That is, one processor can build the new list while other processors reuse the old list. This will result in reduced overhead of list building and processor synchronisation. On the GPU, the copying of data to and from the device can be done asynchronously with the computations, and the utilisation of the CPU will increase.

The ghost nodes assure that subdomains can be updated simultaneously by different processors. No data is communicated between subdomains during the update procedure. Therefore, a parallel implementation iterates through the grid in the same way as a sequential implementation. The computed solutions are independent of how many processors are used, and the computed solution is always of good quality.

Ghost nodes carry the drawback of needing a synchronisation procedure. Before a synchronisation of nodal values, all computations must be finalised and the processing threads must be synchronised. The pHCM method avoids synchronisations of threads by allowing memory to be shared between different threads [34], resulting in so called *false sharing*. Because of the stable upwind stencils, the pHCM method converge to the correct solution. However, the algorithmic behaviour changes noticeably depending on by which processor, and in which order, memory is accessed. Similarly to pHCM and FIM, the ghost nodes can be avoided in LAS and SOLAS. Some minor changes of the reactivation system is then needed, and the synchronization can be entirely avoided. Values are instead 'synchronised' between adjacent subdomains directly when computed. We expect the performance when running on a few threads to improve significantly, but the number of iterations will change with the number of threads.

Other methods than 3D PMM should be investigated for the nodal updates within each subdomain. It is difficult to foresee which type of methods would be better suited. Iterative methods such as SOFI, FIM and Locked Sweeping [40] might perform well since the

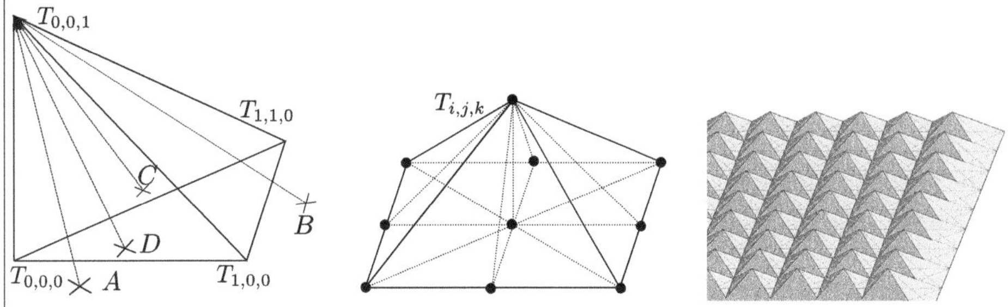

(a) Stencil tetrahedron. (b) One stencil shape. (c) Independent updates.

Figure 15 Illustration of numerical stencil and nodal update procedure of 3D PMM. (**a**) A stencil shaped as a tetrahedron, with four characteristics marked with *A*, *B*, *C* and *D*. Only the solution along characteristic *D* is acceptable. (**b**) The pyramid-shaped stencil used in 3D PMM, consisting of eight tetrahedrons. Nodal values from the bottom layer is used to update the value of node $T_{i,j,k}$. (**c**) A layer of pyramid stencils. All top nodes in the layer can be updated in parallel, since none of the computed values affects the values of other nodes being updated.

subdomains are rather small. Second order stencils can be implemented with almost no algorithmic alterations, thereby increasing accuracy further.

Appendix 1: Efficient formulation of the eikonal 'pyramid' stencil

The most common way to propagate a front numerically is through use of semi-Lagrangian [53] or conditional upwind [16, 30] discretisations. The semi-Lagrangian approach searches through all possible characteristics for the one yielding the minimal distance. The conditional upwind stencils first solve for a solution, and thereafter accepts the new estimate only if the associated characteristic originates from within the spatial grid element, defined as the convex hull of the nodes supporting the stencil. Both approaches are identical on eikonal formulations but may differ in anisotropic cases [30]. If the characteristic curve originate from outside of the element, the new estimate is not based on upwind values, and should not be accepted. Due to the isotropic nature of the eikonal problem one can often formulate upwind conditions for the acceptance of a new estimate [16]. These conditions can significantly reduce the amount of computations needed to update a nodal value.

In this appendix we present an efficient stencil for the eikonal equation, derived as a conditional upwind stencil. The stencil element is shaped as a tetrahedon defined by the corner nodes $\mathcal{X}_{i,j,k}$ with values $T_{i,j,k} = T(\mathcal{X}_{i,j,k})$, as visualised in Figure 15(a). The value $T_{0,0,1}$ for the top node is to be updated using the values $T_{0,0,0}$, $T_{1,0,0}$ and $T_{1,1,0}$. Four characteristic curves (dashed arrows) are shown with their entrance points on the *bottom* surface defined by nodes $\mathcal{X}_{i,j,0}$. Only the solution associated with the characteristic curve D would be acceptable since the other curves originate from outside the stencil element. When updating a node in 3D PMM, the tetrahedron shape is used eight times in a configuration that resembles a *pyramid*, shown in Figure 15(b).

In line with general consensus, the velocity F is assumed to be constant within the element. When updating $T_{0,0,1}$, as in Figure 15(a), only nodes with strictly smaller T values than $T_{0,0,1}$ are used. A larger value implies that the front moves away from $\mathcal{X}_{0,0,1}$, and towards the position of the larger value. Initially we assume that no new value is found, that is

$$t_{\text{new}} \leftarrow T_{0,0,1}. \tag{16}$$

From the characteristic equations (6) we know that the characteristic coincides with the gradient $\mathbf{n}$, and since $\mathbf{n} = \nabla T / \|\nabla T\| = F \nabla T$, we get the following estimates

$$\mathbf{n}_x \leftarrow F\frac{T_{1,0,0} - T_{0,0,0}}{dx}, \qquad \mathbf{n}_y \leftarrow F\frac{T_{1,1,0} - T_{1,0,0}}{dy}. \tag{17}$$

These estimates and the assumption of constant velocity correspond to modelling an arriving front as planar [29]. If $\mathbf{n}_y > 0$, the ray reaching $\mathcal{X}_{0,0,1}$ cuts the bottom surface outside of the tetrahedron, in areas with negative i offset (ray A). The shortest distance from within the tetrahedron then originates from the side with $i = 0$, that is, the new estimate is given from a two-dimensional diagonal stencil solution $\mathtt{s}_{2D}(\mathcal{X}_{0,0,0}, \mathcal{X}_{1,0,0})$ [16]. A similar observation holds for $\mathbf{n}_x$; in order for the characteristic to cut below the line connecting $\mathcal{X}_{0,0,0}$ and $\mathcal{X}_{1,1,0}$, the condition $\mathbf{n}_x\, dy < \mathbf{n}_y\, dx$ must hold. Otherwise (ray C), the minimal new estimate is $\mathtt{s}_{2D}(\mathcal{X}_{0,0,0}, \mathcal{X}_{1,1,0})$. These conditions results in the following upwind conditions

$$\mathtt{if}\ (\mathbf{n}_y > 0)\left\{t_{\text{new}} \leftarrow \min\left(t_{\text{new}}, \mathtt{s}_{2D}(\mathcal{X}_{0,0,0}, \mathcal{X}_{1,0,0})\right)\right\}, \tag{18}$$

$$\mathtt{else\ if}\ (\mathbf{n}_x\, dy > \mathbf{n}_y\, dx)\left\{t_{\text{new}} \leftarrow \min\left(t_{\text{new}}, \mathtt{s}_{2D}(\mathcal{X}_{0,0,0}, \mathcal{X}_{1,1,0})\right)\right\}. \tag{19}$$

Next, $\mathbf{n}_z$ is estimated using the unit argument of the normal, $\mathbf{n}_z^2 = 1 - \mathbf{n}_x^2 - \mathbf{n}_y^2$. To make sure that the ray originates from a location with $i \leq 1$ (to remove ray B), the following upwind condition is needed

$$\mathtt{else\ if}\ \left(\mathbf{n}_z^2\, dx^2 < \mathbf{n}_x^2\, dz^2\right)\left\{t_{\text{new}} \leftarrow \min\left(t_{\text{new}}, \mathtt{s}_{2D}(\mathcal{X}_{1,0,0}, \mathcal{X}_{1,1,0})\right)\right\}. \tag{20}$$

This condition also assures that $\mathbf{n}_z^2 > 0$. If all upwind conditions are true, the characteristic curve that cuts node $\mathcal{X}_{0,0,1}$ enters the tetrahedron through its base (ray D), and the new arrival time estimate is

$$\mathtt{else}\ \left\{t_{\text{new}} \leftarrow T_{0,0,0} + \mathbf{n}_z \frac{dz}{F}\right\}. \tag{21}$$

Details for the anisotropic stencil

The anisotropic stencil for equation (3) uses a direct conditional upwind discretisation in our implementation. Note that nodes upwind of the new solution point can be used in the anisotropic stencil, as long as the point where the characteristic curve enters the stencil lies upwind of the updated node [30].

Appendix 2: Subdomains

Subdomains are surrounded by a layer of virtual nodes referred to as *ghost nodes*. These ghost nodes are copies of nodes in adjacent subdomains. Since each ghost node exists at multiple locations, their values must be kept equal at all locations. We refer to the procedure of comparing such nodal values as synchronisation. Figure 16 is a 2D illustration

Figure 16 Two neighbouring subdomains, without and with ghost nodes, top and lower illustrations. Ghost nodes are coloured in green. Columns of nodes that exist in both subdomains have been marked with waves or stripes. Arrows show the direction in which nodal values are copied between the subdomains. Notice that two layers of nodes are synchronised between subdomains.

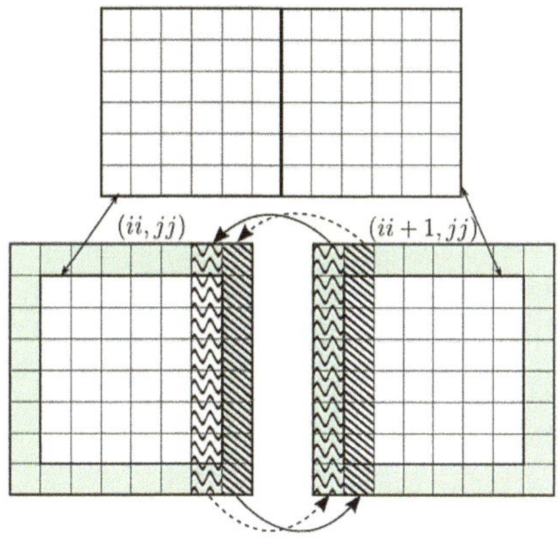

of subdomains in their logical grid (top) and with the ghost nodes marked by light green (bottom). Columns of nodes that exists in both subdomains have been marked with waves and stripes in the lower figure.

Nodes belonging to a subdomain, including the ghost nodes, will during computations only be accessed, and written to, by the thread(s) updating that subdomain. Two subdomains can be computed simultaneously without any risk of memory interference since all data is private to the updating thread(s). Ghost nodes have been used and discussed in relation to front propagation problems in other works [23, 24]. Our use is different, and details regarding the synchronisation of ghost nodes are given in Section B.1. The computation of a subdomain using 3D PMM is discussed in Section B.2.

B.1 Synchronisation of nodal values

Consider the two-dimensional case when comparing values from nodes in subdomain (ii, jj) to values in subdomain $(ii+1, jj)$, as shown with the two lower arrows in Figure 16. When a computed nodal value, t_s in (ii, jj) is compared to its copy t_t in $(ii+1, jj)$, the copy is only changed if its value is larger than the computed nodes, that is $t_t = \min(t_t, t_c)$. If at least one node of the inner part of $(ii+1, jj)$ (lower solid arrow in Figure 16) receives a new value, the subdomain is activated by releasing the compute lock, $\texttt{CL}(ii+1, jj) \leftarrow \texttt{Open}$. A subdomain that receives new values only on its out-most boundary (dashed arrow) does not need to be activated.

In SOLAS we need the minimum nodal value that activates a subdomain. The minimum activation value is computed efficiently in our GPU implementation using parallel reduction. Note that if a tetrahedron (diagonal) stencil is used, the corners will be shared and updated by eight (four) blocks in three (two) spatial dimensions. However, by synchronising values in one direction at a time, the corner nodes receive correct solution values.

B.2 Computation of a subdomain

In all our numerical experiments, 3D PMM has been used for the computation of new nodal values in the subdomains. Here we give a brief description of the method [30, 54].

A specific *pyramid* shape of the stencil is used, as shown in Figure 15(b), where the value $T_{i,j,k}$ for the top node of the pyramid is updated. 3D PMM computes *layers* of nodes, as illustrated in Figure 15(c). Nodes in a layer have one index (i, j or k) in common. All nodes in one layer can be updated entirely in parallel since there are no dependencies between the computed nodes, since the value of a top node does not depend on other top

Algorithm 17 *i*SUBSWEEP(sd,dir)

comment: Subsweep in a subdomain using in 3D PMM.
STENCIL($B_{\mathcal{X}}$) returns an estimate using values on nodes in the set $B_{\mathcal{X}}$.

```
T ← T values in subdomain sd
if dir is 1
  then order ← 0 to b_x
if dir is −1
  then order ← b_x + 1 downto 1
for i ← order
  do { for each j ← 0 to b_y + 1 and k ← 0 to b_z + 1, in parallel
       do { B_X ← { All X_{i,j±a,k±b}, a ∈ {0,1}, b ∈ {0,1} within sd }
            t_new ← STENCIL(B_X)
            if t_new < T_{i+dir,j,k}
              then T_{i+dir,j,k} = t_new
```

Algorithm 18 COMPUTESUBDOMAIN(sd)

comment: Computation of nodal values in subdomain sd with 3D PMM.

```
T ← T values in subdomain sd
if CL(sd) is Open)
  then { CL(sd) ← Locked
         for dir ← 1 and −1
           do { iSUBSWEEP(sd, dir)
                jSUBSWEEP(sd, dir)
                kSUBSWEEP(sd, dir)
```

nodes, see Figure 15(c). After processing a layer, the computed values are used as bottom-pyramid values to compute the next layer of nodes. An iteration through the domain for an increasing or decreasing common index is referred to as a *subsweep*. A set of all six subsweeps, increasing and decreasing *i*, *j*, *k* indices, are referred to as a *sweep*.

Conceptually, the ghost nodes in a subdomain define the boundary condition of the subdomain. In our implementation we compute new values also for the ghost nodes. When the ghost nodes are updated, no values from other subdomains are used in the computations. Therefore, 'half'-pyramids are used at the sides, and 'quarter'-pyramids in the corners. Pseudocode for a subsweep in the *i* direction is given in Algorithm 17, where the layers defined by $i = 0$ and $i = b_x + 1$ contains only ghost nodes.

A subdomain that is activated receives (at least) one surface with two layers of nodes in the synchronisation. Assuming that these 'thick' boundary values are correct, the characteristic curves are extrapolated to the rest of the subdomain in the sweeps as new solution values are computed. These 'thick' boundary conditions explain why the solution converges fast. On a local scale, the characteristic curves are straight lines, or only slightly curved, and the solution within a subdomain will often converge in only one sweep. The final solution remains the same in all our examples whether one, two, or more sweeps are used when computing subdomains. Since the total computational time is minimal when only one sweep is used, we have used only one sweep in the numerical experiments.

After processing a subdomain, the compute lock is closed. As explained above, it can be opened again during a synchronisation. Algorithm 18 presents pseudocode for the computation of new nodal values in subdomain `sd`, using 3D PMM. This algorithm, named COMPUTESUBDOMAIN(), is called from the procedure in Algorithm 3.

Competing interests
The authors declare that they have no competing interests.

Authors' contributions
The main idea of this paper was proposed by TG, based on earlier research by ØH. TG and MS contributed with the implementation and numerical experiments. TG and AM performed the analysis, and prepared the manuscript initially. All authors contributed to the manuscript, and have read and approved the final manuscript.

Acknowledgements
The presented work was funded by Statoil ASA through the Akademia program, and the Research Council of Norway under grant 202101/I40. The work has been conducted at Kalkulo AS, a subsidiary of Simula Research Laboratory.

Endnotes

[a] Here, the term 'efficiency' refers to the ratio between observed speedup and the theoretically possible speedup. For instance, the LAS speedup of 6.0 for $\mathbf{Ex}_A$ running eight cores gives an efficiency of $6.0/8 \approx 75\%$.

[b] The CES software is only available on the Ivy platform running in double precision. Experiments indicate that single precision computations on Ivy runs at about 90% of the time used in double precision. Therefore, the CES times for single precision experiments are constructed by multiplying the observations for double precision with a factor 0.9.

References

1. Sethian JA: *Level Set Methods and Fast Marching Methods: Evolving Interfaces in Computational Geometry, Fluid Mechanics, Computer Vision, and Materials Science.* 2nd edition. Cambridge: Cambridge University Press; 1999.
2. Rawlinson N, Hauser J, Sambridge M: **Seismic ray tracing and wavefront tracking in laterally heterogeneous media.** *Adv. Geophys.* 2009, **49**:203-267.
3. Gillberg T: **A semi-ordered fast iterative method (SOFI) for monotone front propagation in simulations of geological folding.** In *MODSIM2011, 19th International Congress on Modelling and Simulation*; 2011:641-647.
4. Gillberg T: **Fast and accurate front propagation for simulation of geological folds.** *PhD thesis.* University of Oslo; 2013.
5. Natvig JR, Lie K-A: **Fast computation of multiphase flow in porous media by implicit discontinuous Galerkin schemes with optimal ordering of elements.** *J. Comput. Phys.* 2008, **227**(24):10108-10124.
6. Kornhauser ET: **Ray theory for moving fluids.** *J. Acoust. Soc. Am.* 1953, **25**(5):945-949.
7. Podvin P, Lecomte I: **Finite difference computation of traveltimes in very contrasted velocity models: a massively parallel approach and its associated tools.** *Geophys. J. Int.* 1991, **105**:271-284.
8. Li S, Mueller K, Jackowski M, Dione D, Staib L: **Physical-space refraction-corrected transmission ultrasound computed tomography made computationally practical.** In *Medical Image Computing and Computer-Assisted Intervention - MICCAI 2008.* Berlin: Springer; 2008:280-288. [*Lecture Notes in Computer Science*, vol. 5242.]
9. Wallman M, Smith NP, Rodriguez B: **A comparative study of graph-based, eikonal, and monodomain simulations for the estimation of cardiac activation times.** *IEEE Trans. Biomed. Eng.* 2012, **59**(6):1739-1748.
10. Jeong W-K, Fletcher PT, Tao R, Whitaker R: **Interactive visualization of volumetric white matter connectivity in DT-MRI using a parallel-hardware Hamilton-Jacobi solver.** *IEEE Trans. Vis. Comput. Graph.* 2007, **13**(6):1480-1487.
11. Rawlinson N, Sambridge M: **Multiple reflection and transmission phases in complex layered media using a multistage fast marching method.** *Geophysics* 2004, **69**(5):1338-1350.
12. Petersen SA, Hjelle Ø, Hustoft S, Haubiers M: **Process based data-restoration and model-reconstruction workflow for seismic interpretation and model building.** In *EAGE 74th Conference & Exhibition, Extended Abstracts*; 2012.
13. Petersen SA, Hjelle Ø: **Earth recursion, an important component in shared earth model builders.** In *EAGE 70th Conference & Exhibition, Extended Abstracts*; 2008.
14. Cerveny V: *Seismic Ray Theory.* Cambridge: Cambridge University Press; 2001.
15. Hjelle Ø, Petersen SA: **A Hamilton-Jacobi framework for modeling folds in structural geology.** *Math. Geosci.* 2011, **43**(7):741-761.
16. Gillberg T, Hjelle Ø, Bruaset AM: **Accuracy and efficiency of stencils for the eikonal equation in earth modelling.** *Comput. Geosci.* 2012, **16**(4):933-952.
17. Chopp D: **Recent advances in the level set method.** In *Handbook of Biomedical Image Analysis.* Edited by Micheli-Tzanakou E, Suri JS, Wilson DL, Laxminarayan S. New York: Springer; 2005:201-256.
18. Tsitsiklis JN: **Efficient algorithms for globally optimal trajectories.** *IEEE Trans. Autom. Control* 1995, **40**(9):1528-1538.
19. Cacace S, Cristiani E, Falcone M: **Requiem for local single-pass methods solving stationary Hamilton-Jacobi equations?** 2013 [arXiv:1301.6775]
20. Sethian JA, Vladimirsky A: **Ordered upwind methods for static Hamilton-Jacobi equation: theory and algorithms.** *SIAM J. Numer. Anal.* 2003, **41**(1):325-363.

21. Hjelle Ø, Petersen SA, Bruaset AM: **A numerical framework for modeling folds in structural geology.** *Math. Geosci.* 2013, **45**(3):255-276.
22. Zhang J, Huang Y, Song L-P, Liu Q-H: **Fast and accurate 3-D ray tracing using bilinear traveltime interpolation and the wave front group marching.** *Geophys. J. Int.* 2011, **184**(3):1327-1340.
23. Herrmann M: **A domain decomposition parallelization of the fast marching method.** In *Annual Research Briefs 2003.* Stanford: Center for Turbulence Research, Stanford University; 2003:213-225.
24. Tugurlan MC: **Fast marching methods - parallel implementation and analysis.** *PhD thesis.* Louisiana State University and Agricultural and Mechanical College; 2008.
25. Zhao H-K: **A fast sweeping method for eikonal equations.** *Math. Comput.* 2004, **74**(250):603-627.
26. Qian J, Zhang Y-T, Zhao H-K: **A fast sweeping method for static convex Hamilton-Jacobi equations.** *J. Sci. Comput.* 2007, **31**(1-2):237-271.
27. Zhao H-K: **Parallel implementations of the fast sweeping method.** *J. Comput. Math.* 2007, **25**(4):421-429.
28. Detrixhe M, Gibou F, Min C: **A parallel fast sweeping method for the eikonal equation.** *J. Comput. Phys.* 2013, **237**:46-55.
29. Weber O, Devir YS, Bronstein AM, Bronstein MM, Kimmel R: **Parallel algorithms for approximation of distance maps on parametric surfaces.** *ACM Trans. Graph. (TOG)* 2008, **27**(4):104:1-104:16.
30. Gillberg T, Sourouri M, Cai X: **A new parallel 3D front propagation algorithm for fast simulation of geological folds.** *Proc. Comput. Sci.* 2012, **9**:947-955. Proceedings of the International Conference on Computational Science, ICCS 2012.
31. Gillberg T, Hjelle Ø, Bruaset AM: **A parallel 3D front propagation algorithm for simulation of geological folding on GPUs.** In *EAGE 74th Conference & Exhibition, Extended Abstracts*; 2012.
32. Gremaud PA, Kuster CM: **Computational study of fast methods for the eikonal equation.** *SIAM J. Sci. Comput.* 2006, **27**(6):1803-1816.
33. Hysing S-R, Turek S: **The eikonal equation: numerical efficiency vs. algorithmic complexity on quadrilateral grids.** In *Proceedings of ALGORITMY*; 2005.
34. Chacon A, Vladimirsky A: **A parallel heap-cell method for eikonal equations**; 2013 [arXiv:1306.4743]
35. Chacon A, Vladimirsky A: **Fast two-scale methods for eikonal equations.** *SIAM J. Sci. Comput.* 2012, **34**(2):A547-A578.
36. Glover F, Klingman D, Phillips N: **A new polynomially bounded shortest path algorithm.** *Oper. Res.* 1985, **33**(1):65-73.
37. Dejnozkova E, Dokladal P: **A parallel algorithm for solving the eikonal equation.** In *IEEE International Conference on Acoustics, Speech, and Signal Processing, 2003. Volume* 3; 2003:325-328. IEEE.
38. Jeong W-K, Whitaker RT: **A fast iterative method for eikonal equations.** *SIAM J. Sci. Comput.* 2008, **30**(5):2512-2534.
39. Fu Z, Jeong W-K, Pan Y, Kirby RM, Whitaker RT: **A fast iterative method for solving the eikonal equation on triangulated surfaces.** *SIAM J. Sci. Comput.* 2011, **33**(5):2468-2488.
40. Bak S, McLaughlin J, Renzi D: **Some improvements for the fast sweeping method.** *SIAM J. Sci. Comput.* 2010, **32**(5):2853-2874.
41. Jeong W-K, Whitaker RT: **A fast iterative method for a class of Hamilton-Jacobi equations on parallel systems.** *Technical report.* University of Utah; 2007.
42. Tanenbaum AS: *2. Modern Operating Systems.* 3rd edition. Upper Saddle River: Prentice Hall; 2007.
43. Kjolstad FB, Snir M: **Ghost cell pattern.** In *Proceedings of the 2010 Workshop on Parallel Programming Patterns.* ParaPLoP '10. New York: ACM; 2010:4-149.
44. Rivera G, Tseng C-W: **Tiling optimizations for 3D scientific computations.** In *Proceedings of the 2000 ACM/IEEE Conference on Supercomputing.* Supercomputing '00. Washington: IEEE Computer Society; 2000:1-23.
45. Nvidia: **CUDA C programming guide**; 2012 [http://docs.nvidia.com/cuda/cuda-c-programming-guide/index.html]
46. Michéa D, Komatitsch D: **Accelerating a three-dimensional finite-difference wave propagation code using GPU graphics cards.** *Geophys. J. Int.* 2010, **182**(1):389-402.
47. Nvidia: **CUDA C Best Practices Guide**; 2012 [http://docs.nvidia.com/cuda/cuda-c-best-practices-guide/index.html]
48. Nvidia: **CUDA driver API**; 2013 [http://docs.nvidia.com/cuda/cuda-driver-api/index.html]
49. OpenMP.org: **OpenMP application program interface**; 2013 [http://openmp.org/]
50. Nvidia: **What is CUDA**; 2013 [https://developer.nvidia.com/what-cuda]
51. Nvidia: **CUDA occupancy calculator**; 2012.
52. Kadlec B, Dorn G: **Leveraging graphics processing units (GPUs) for real-time seismic interpretation.** *Lead. Edge* 2010, **29**(1):60-66.
53. Cristiani E, Falcone M: **Fast semi-Lagrangian schemes for the eikonal equation and applications.** *SIAM J. Numer. Anal.* 2007, **45**(5):1979-2011.
54. Sourouri M: **A parallel front propagation method: simulating geological folds on parallel architectures.** *Master's thesis.* University of Oslo; 2012.

5

An accuracy comparison of polynomial chaos type methods for the propagation of uncertainties

Florian Augustin[1*], Albert Gilg[2], Meinhard Paffrath[2*], Peter Rentrop[1], Manuel Villegas[1*] and Utz Wever[2*]

*Correspondence: augustin@ma.tum.de; meinhard.paffrath@siemens.com; villegas@ma.tum.de; utz.wever@siemens.com
[1]Department of Mathematics, Technische Universität München, Boltzmannstr. 3, Garching, 85748, Germany
[2]Corporate Technology, Siemens AG, Otto-Hahn-Ring 6, Munich, 81730, Germany

Abstract

In (Augustin *et al.* in European J. Appl. Math. 19:149-190, 2008) we considered the Polynomial Chaos Expansion for the treatment of uncertainties in industrial applications. For many applications the method has been proven to be a computationally superior alternative to Monte Carlo evaluations. In the current overview we compare the accuracy of Polynomial Chaos type methods for the propagation of uncertainties in nonlinear problems and verify them on two examples relevant for industry. For weakly nonlinear time-dependent models, the generalized Kalman filter equations define an efficient method, yielding good approximations if the quantities of interest are restricted to the first two moments of the solution. Secondly, stochastic collocation is discussed. The method is applied to delay differential equations and random ordinary differential equations. Finally, a generalized PC method is discussed which is based on a subdivision of the random space. This approach is even suitable for highly nonlinear models.

1 Introduction

Uncertainty quantification in industrial applications is an interesting and important field of research [1–4]. Due to uncertainties in system parameters, measurements and in the modeling of physical processes itself, deterministic approaches for the simulation of those processes are not appropriate. Uncertainty quantification became a vast field of research over the past decades and many methods have been proposed to deal with random data [5–10]. Amongst these methods, approaches based on Polynomial Chaos expansions have proven to be very efficient in dealing with uncertainties, see [3, 10, 11].

Those methods revealed to be efficient alternatives to Monte-Carlo (MC) simulations. They were successfully applied to solve stationary problems *e.g.* in the field of stochastic finite elements [10] and to solve differential equations with uncertain parameters and initial values [12]. The suitability of PC for industrial applications was considered in [1]. Additionally, an extensive discussion of the Polynomial Chaos approaches, applied to linear differential equations, was presented. In the current paper we consider numerical methods for the propagation of uncertainties through nonlinear differential equations. The main part is concerned with a comparison of Kalman filter techniques for state estimation and Polynomial Chaos Galerkin techniques. The methods themselves are known from literature, the contribution of this work is the comparison of these methods in selected examples relevant for industry. Our main focus is on the accuracies of the methods. Based on these results, we try to answer the question what is the numerical effort in order to achieve a low, medium or high accuracy while, in terms of function evaluations, gaining at least one order compared to a crude Monte Carlo method.

- The most widespread method for treating differential equations with uncertainties are techniques of Kalman filtering. While the original Kalman filter was developed to identify noisy observations with linear differential equations [13], here we use it only to integrate the underlying system with uncertainties. A direct generalization to nonlinear equations is given by the extended Kalman filter [14]; it is based on the sensitivities of the equation. The unscented Kalman filter, a nonlinear filter which does not use derivative information, was developed in [15]. It directly approximates the covariance matrix and is, in this context, used just for state estimation.
- Stochastic collocation methods (SCMs) are introduced by [16, 17]. The first practical application of the methods are discussed in [10]. Here, we apply the methods to nonlinear differential equations with delay, which present a more complex behavior than the delay-free counterparts. Stochastic collocation methods have the advantage of using the differential equation as a black box.
- In [1] we presented a theory of the stochastic Galerkin method (SGM), see [3, 10], for ordinary differential equations with uncertain parameters. Nevertheless, this method suffers from the lack of convergence, especially when time evolves [18]. Thus, we have to consider generalized approaches to scope with this problem. Several methods have been proposed in literature, see for example the multi resolution method [9, 19] and the Multi-Element generalized Polynomial Chaos (MEgPC) method [20]. In [21] a rigorous convergence theory for MEgPC and a fully adaptive scheme in time and random space has been developed. In the current paper we compare this MEgPC approach to the other PC type methods mentioned above.

In Section 2 we introduce two examples from industrial applications, which are the base for the comparison of the discussed methods. The first example concerns crack propagation, while the second considers the quorum sensing in biofilms. Next, we present a summary of the Kalman filter equations for state estimation in Section 3. Especially, nonlinear filters in the context of uncertainty propagation of state are discussed. In Section 4 we consider the SCM, which we apply to the random delay differential equations of quorum sensing. We discuss the multi-element approach of the stochastic Galerkin method in Section 5. Finally we compare the three described methods on the base of the two applications, Sections 6.2 and 6.3. In Section 7 we close this article by drawing our conclusions.

2 Test sets of (delay) differential equations

In this section, we introduce the applications considered for the comparison of the methods of the Kalman filter, the SCM and the SGM.

- The first application deals with the growth of cracks, which is important in the context of life cycle analysis. Industrial applications are for example crack growth in turbine blades or in train wheelsets. Here the task is to define appropriate inspection schemes in order to prevent severe damages. The crack growth is modelled by a highly nonlinear ordinary differential equation.
- The second example is a model of the quorum sensing of biofilms, a delay differential equation. The control of biofilms requires a deep understanding of the structured community of microorganisms living on inert surfaces. Industrial applications range

from sewer systems, fresh water systems to corrosion prevention in cooling flow networks of (electric) power plants.

We restrict ourselves to a low number of uncertain variables. Nevertheless, due to the non-linearities and the delay, differences between the methods with respect to their achieved accuracies will become visible. A more complex model of crack growth, for the evaluation of inspection schemes for train wheelsets, has been computed with the unscented approximation in [22]. The biofilm model, in particular the physical background, has been studied in more detail in [23].

2.1 Crack growth example

For the modeling of a growing crack we use a simplified model of fracture mechanics, see Harris [24], where a center cracked plate in tension is studied. The crack growth relation is based on a modified form of the Forman relation [25]. The crack size a is governed by the ordinary differential equation in the abstract form

$$\frac{\partial a}{\partial N} = f(a, K_{Ic}, \Delta_\sigma, C_F), \quad a(0) = a_0, \ 0 \leq N \leq N_L \tag{1}$$

where N denotes the number of load cycles and N_L the life cycle. The parameters are the stress intensity factor K_{Ic}, the stress range Δ_σ, the fatigue growth coefficient C_F and the initial crack size a_0. In more detail the crack growth differential equation reads [24]

$$\frac{\partial a}{\partial N} = C_F\left(\frac{(1-f)\Delta K}{1-R}\right)^{n_f} \frac{(1 - \frac{4}{\pi}\frac{\Delta K_0}{\Delta K}\tan^{-1}(1-R))^p}{(1 - \frac{\Delta K}{(1-R)K_{Ic}})^q}, \tag{2}$$

where

$$\Delta K = \Delta_\sigma \sqrt{\pi a}\, Y\left(\frac{2a}{h}\right), \qquad Y(\alpha) = \frac{1 - 0.5\alpha + 0.370\alpha^2 - 0.044\alpha^3}{\sqrt{1-\alpha}}$$

and

$$f = \begin{cases} A_0 + A_1 R + A_2 R^2 + A_3 R^3 & \text{for } R \geq 0, \\ A_0 + A_1 R & \text{for } -1 < R < 0, \end{cases}$$

$$A_0 = \left(0.825 - 0.34\alpha' + 0.05\alpha'^2\right)\left[\cos(\pi s/2)\right]^{1/\alpha'},$$

$$A_1 = \left(0.415 - 0.071\alpha'\right)s,$$

$$A_2 = 2A_0 + A_1 - 1,$$

$$A_3 = 1 - (A_0 + A_1 + A_3).$$

R is the stress ratio between maximum and minimum stress, h denotes the width of the plate and n_f, ΔK_0, p, q, α', s are curve-fitting constants.

2.2 The biofilm model with quorum sensing

The model for antibiotics action on a biofilm follows closely the model in [26]. Since the antibiotics are not produced by the hosting organism, it must be supplied externally. Thus, our equations read

$$\begin{aligned} \dot{A} &= S_A - \mu_A A(t), \\ \dot{B} &= \alpha_{20}\left(1 - \frac{B(t)}{\sigma}\right)B(t) + \alpha_2 B(t)B(t-\tau) - \alpha_3 B(t)A(t). \end{aligned} \tag{3}$$

Here, A denotes the concentration of antibiotics and B the concentration of bacteria. Moreover, the parameters are the logistic reproductive rate of the bacteria α_{20}, the effective carrying capacity of the environment σ, the reproductive rate due to quorum sensing, α_2, the death rate of bacteria due to antibiotics S_A and death rate for the antibiotics μ_A. All these parameters are taken to be positive. The differential equation for B includes a standard logistic term

$$\alpha_{20}\left(1 - \frac{B(t)}{\sigma}\right)B(t), \tag{4}$$

and a quorum sensing (QS) term

$$\alpha_2 B(t)B(t-\tau). \tag{5}$$

It represents the extra increase in population after the QS mechanism has been initiated. The process is not instantaneous. Thus, assuming it takes a certain amount of time τ to produce the signal molecules, the population of bacteria at time t, $B(t)$, will receive these signal molecules from the population of bacteria at time $t-\tau$, *i.e.* $B(t-\tau)$. The coefficient α_2 accounts for the strength of the mechanism. The larger α_2, the more efficient the QS mechanism takes place in the biofilm.

3 Solving stochastic differential equations by Kalman filters

In this section we consider stochastic dynamical systems of the form

$$\dot{\mathbf{x}}(t) = \mathbf{f}\big(\mathbf{x}(t), \mathbf{u}(t), \mathbf{w}(t)\big), \tag{6}$$

$$\mathbf{y}(t) = \mathbf{h}\big(\mathbf{x}(t), \mathbf{v}(t)\big) \tag{7}$$

with $0 \le t \le T_L$,

$$\begin{aligned} &\mathbf{f} : \mathbb{R}^n \times \mathbb{R}^{n_u} \times \mathbb{R}^{n_w} \to \mathbb{R}^n, \\ &\mathbf{h} : \mathbb{R}^n \times \mathbb{R}^{n_v} \to \mathbb{R}^m. \end{aligned}$$

Here, $\mathbf{f}$, $\mathbf{h}$ are linear or nonlinear functions, $\mathbf{x}$ is the state vector, $\mathbf{u}$ is the control vector and $\mathbf{y}$ is the observation vector. We assume the process noise $\mathbf{w}$ and the measurement noise $\mathbf{v}$ to be white, stochastically independent and

$$\mathbf{w} \sim \mathcal{N}\big(0, \mathbb{R}_+^{n_w}\big), \qquad \mathbf{v} \sim \mathcal{N}\big(0, \mathbb{R}_+^{n_v}\big). \tag{8}$$

For real processes, white noise is a somewhat simplified assumption. There are extensions of Kalman filtering to colored noise (*e.g.* [27]) but this will not be picked out here. Equations (6) and (7) describe a stochastic system which is continuous in time. In the following, we restrict ourselves to methods for discrete time systems which are obtained by time discretization of (6). Furthermore we concentrate on non-stiff systems. Of course for the numerical integration of stiff systems semi-implicit or implicit methods are required which means a much higher numerical effort. We begin with linear discrete time systems in Section 3.1. Thereafter, we discuss how nonlinear differential equations can be treated by the extended Kalman filter and the unscented Kalman filter in Sections 3.2 and 3.3. After a comparison of both filters in Section 3.4 we give a reinterpretation of the unscented Kalman filter as a sampling method, see Section 3.5.

3.1 Linear discrete time systems

This section adopts the presentation in [14]. A stochastic linear discrete time system can be written as

$$\mathbf{x}_k = A_{k-1}\mathbf{x}_{k-1} + B_{k-1}\mathbf{u}_{k-1} + W_{k-1}\mathbf{w}_{k-1},$$
$$\mathbf{y}_k = H_k\mathbf{x}_k + V_k\mathbf{v}_k$$

with $\mathbf{w}_{k-1} \sim \mathcal{N}(0, Q_{k-1})$ and $\mathbf{v}_k \sim \mathcal{N}(0, R_k)$ being independent normally distributed random variables. The predictor step of the Kalman filter is given by

$$\hat{\mathbf{x}}_k^- = A_{k-1}\hat{\mathbf{x}}_{k-1} + B_{k-1}\mathbf{u}_{k-1}.$$

For the a priori and a posteriori estimate errors $\mathbf{e}_k^- := \mathbf{x}_k - \hat{\mathbf{x}}_k^-$ and $\mathbf{e}_k := \mathbf{x}_k - \hat{\mathbf{x}}_k$ with corresponding covariances $P_k^- := E(\mathbf{e}_k^- \mathbf{e}_k^{-T})$ and $P_k := E(\mathbf{e}_k \mathbf{e}_k^T)$ holds

$$\mathbf{e}_k^- = A_{k-1}(\mathbf{x}_{k-1} - \hat{\mathbf{x}}_{k-1}) + W_{k-1}\mathbf{w}_{k-1} \quad \text{and} \quad P_k^- = A_{k-1}P_{k-1}A_{k-1}^T + W_{k-1}Q_{k-1}.$$

The corrector of the Kalman filter is given by

$$\hat{\mathbf{x}}_k = \hat{\mathbf{x}}_k^- + K_k\left(\mathbf{y}_k - H_k\hat{\mathbf{x}}_k^-\right),$$

thus for the a posteriori error and its covariances holds

$$\begin{aligned}
\mathbf{e}_k &= \mathbf{x}_k - \hat{\mathbf{x}}_k \\
&= \mathbf{x}_k - \hat{\mathbf{x}}_k^- - K_k\left(\mathbf{y}_k - H_k\hat{\mathbf{x}}_k^-\right) \\
&= \mathbf{x}_k - \hat{\mathbf{x}}_k^- - K_k\left(H_k\mathbf{x}_k + V_k\mathbf{v}_k - H_k\hat{\mathbf{x}}_k^-\right) \\
&= (I - K_kH_k)\left(\mathbf{x}_k - \hat{\mathbf{x}}_k^-\right) - K_kV_k\mathbf{v}_k
\end{aligned}$$

and $P_k = (I - K_kH_k)P_k^-(I - K_kH_k)^T + K_kV_kR_kV_k^TK_k^T$. Minimization of the trace of P_K leads to

$$K_k = P_k^-H_k^T\left[H_kP_k^-H_k^T + V_kR_kV_k^T\right]^{-1}. \tag{9}$$

With this expression for K_k, we derive a simpler formula for P_k,

$$P_k = (I - K_kH_k)P_k^-. \tag{10}$$

The algorithm is summarized in Table 1.

Table 1 Kalman filter

Model	$\mathbf{x}_k = A_{k-1}\mathbf{x}_{k-1} + B_{k-1}\mathbf{u}_{k-1} + W_{k-1}\mathbf{w}_{k-1}$, $\mathbf{w}_{k-1} \sim \mathcal{N}(0, Q_{k-1})$, $\mathbf{y}_k = H_k\mathbf{x}_k + V_k\mathbf{v}_k$, $\mathbf{v}_k \sim \mathcal{N}(0, R_k)$
Initialization	$\hat{\mathbf{x}}_0$, $P_0 = E[\hat{\mathbf{x}}_0\hat{\mathbf{x}}_0^T]$
Predictor	$\hat{\mathbf{x}}_k^- = A_{k-1}\hat{\mathbf{x}}_{k-1} + B_{k-1}\mathbf{u}_{k-1}$, $P_k^- = A_{k-1}P_{k-1}A_{k-1}^T + W_{k-1}Q_{k-1}W_{k-1}^T$
Kalman gain	$K_k = P_k^- H_k^T [H_k P_k^- H_k^T + V_k R_k V_k^T]^{-1}$
Corrector	$\hat{\mathbf{x}}_k = \hat{\mathbf{x}}_k^- + K_k(\mathbf{y}_k - H_k\hat{\mathbf{x}}_k^-)$, $P_k = (I - K_k H_k)P_k^-$

3.2 Extended Kalman Filter (EKF)

The stochastic nonlinear discrete time system can be written as

$$\begin{aligned} \mathbf{x}_k &= \mathbf{f}(\mathbf{x}_{k-1}, \mathbf{u}_{k-1}, \mathbf{w}_{k-1}), \\ \mathbf{y}_k &= \mathbf{h}(\mathbf{x}_k, \mathbf{v}_k). \end{aligned}$$

The predictor of the extended Kalman filter is given by

$$\hat{\mathbf{x}}_k^- = \mathbf{f}(\hat{\mathbf{x}}_{k-1}, \mathbf{u}_{k-1}, 0)$$

with the a priori estimation error and its covariance being

$$\begin{aligned} \mathbf{e}_k^- &= \mathbf{x}_k - \hat{\mathbf{x}}_k^- \doteq A_{k-1}(\mathbf{x}_{k-1} - \hat{\mathbf{x}}_{k-1}) + W_{k-1}\mathbf{w}_{k-1}, \\ P_k^- &= A_{k-1}P_{k-1}A_{k-1}^T + W_k Q_{k-1} W_k^T. \end{aligned}$$

Here we use the abbreviations

$$A_{k-1} := \left.\frac{\partial \mathbf{f}}{\partial \mathbf{x}}\right|_{(\hat{\mathbf{x}}_{k-1}, \mathbf{u}_{k-1}, 0)} \quad \text{and} \quad W_{k-1} := \left.\frac{\partial \mathbf{f}}{\partial \mathbf{w}}\right|_{(\hat{\mathbf{x}}_{k-1}, \mathbf{u}_{k-1}, 0)}.$$

The corrector of the extended Kalman filter has the form

$$\hat{\mathbf{x}}_k = \hat{\mathbf{x}}_k^- + K_k\left(\mathbf{y}_k - \mathbf{h}\left(\hat{\mathbf{x}}_k^-, 0\right)\right)$$

and we write the a posteriori estimation error as

$$\begin{aligned} \mathbf{e}_k &= \mathbf{x}_k - \hat{\mathbf{x}}_k \\ &= \mathbf{x}_k - \hat{\mathbf{x}}_k^- - K_k\left(\mathbf{y}_k - \mathbf{h}\left(\hat{\mathbf{x}}_k^-, 0\right)\right) \\ &= \mathbf{x}_k - \hat{\mathbf{x}}_k^- - K_k\left(\mathbf{h}(\mathbf{x}_k, \mathbf{v}_k) - \mathbf{h}\left(\hat{\mathbf{x}}_k^-, 0\right)\right) \\ &\doteq \mathbf{x}_k - \hat{\mathbf{x}}_k^- - K_k\left(H_k\left(\mathbf{x}_k - \hat{\mathbf{x}}_k^-\right) + V_k v_k\right) \\ &= (I - K_k H_k)\left(\mathbf{x}_k - \hat{\mathbf{x}}_k^-\right) - K_k V_k v_k \end{aligned}$$

with

$$H_k := \left.\frac{\partial \mathbf{h}}{\partial \mathbf{x}}\right|_{(\hat{\mathbf{x}}_k, 0)} \quad \text{and} \quad V_k := \left.\frac{\partial \mathbf{h}}{\partial \mathbf{v}}\right|_{(\hat{\mathbf{x}}_k, 0)}.$$

Table 2 Extended Kalman filter

Model	$\mathbf{x}_k = \mathbf{f}(\mathbf{x}_{k-1}, \mathbf{u}_{k-1}, \mathbf{w}_{k-1}),\ \mathbf{w}_{k-1} \sim \mathcal{N}(0, Q_{k-1}),$ $\mathbf{y}_k = \mathbf{h}(\mathbf{x}_k, \mathbf{v}_k),\ \mathbf{v}_k \sim \mathcal{N}(0, R_k)$
Initialization	$\hat{\mathbf{x}}_0,\ P_0 = E[\hat{\mathbf{x}}_0 \hat{\mathbf{x}}_0^T]$
Predictor	$\hat{\mathbf{x}}_k^- = \mathbf{f}(\hat{\mathbf{x}}_{k-1}, \mathbf{u}_{k-1}, 0),$ $P_k^- = A_{k-1} P_{k-1} A_{k-1}^T + W_{k-1} Q_{k-1} W_{k-1}^T$
Kalman gain	$K_k = P_k^- H_k^T [H_k P_k^- H_k^T + V_k R_k V_k^T]^{-1}$
Corrector	$\hat{\mathbf{x}}_k = \hat{\mathbf{x}}_k^- + K_k(\mathbf{y}_k - \mathbf{h}(\hat{\mathbf{x}}_k^-, 0)),$ $P_k = (I - K_k H_k) P_k^-$

For the Kalman gain and the a posteriori error covariance the same formulae (9), (10) as in the linear case apply. The corresponding algorithm of the extended Kalman filter is summarized in Table 2.

3.3 Unscented Kalman Filter (UKF)

For the extended Kalman filter, derivatives of f and h have to be provided which is not possible in many practical applications. So in [15] a nonlinear filter, called unscented filter, without this drawback was introduced. The unscented filter can be used to estimate the mean and covariance of a nonlinear stochastic process $f(\mathbf{w})$ where $\mathbf{w} \in \mathbb{R}^{n_w}$ is a normally distributed random vector with mean $E(\mathbf{w})$ and covariance $P^{ww} \in \mathbb{R}^{n_w \times n_w}$. So-called sigma points $\mathcal{X}^{(i)}$, together with weights W_i^{mean} and W_i^{cov}, are constructed and mapped to $\mathcal{Z}^{(i)} = f(\mathcal{X}^{(i)})$ for $i = 0, \ldots, p$. The unscented filter then yields an approximation of the mean μ and covariance P^{zz} of the nonlinear function by

$$E(z) \approx \sum_{i=0}^{p} W_i^{\text{mean}} \mathcal{Z}^{(i)},$$

$$P^{zz} \approx \sum_{i=0}^{p} W_i^{\text{cov}} \left(\mathcal{Z}^{(i)} - \mathbf{y}\right)\left(\mathcal{Z}^{(i)} - E(z)\right)^T.$$

By Taylor expansion one can show second order accuracy of mean and covariance, [15, 28].

Example A set of $p + 1 = 2n + 1$ weights and sigma points with respect to mean $E(x) \in \mathbb{R}^n$ and covariance $P^{xx} \in \mathbb{R}^{n \times n}$ is given by [29]

$$\mathcal{X}^{(i)} = \mathbf{x} + \left(\sqrt{nP^{xx}}\right)_i,$$

$$W_i = 1/(2n),$$

$$\mathcal{X}^{(i+n)} = \mathbf{x} - \left(\sqrt{nP^{xx}}\right)_i,$$

$$W_{i+n} = 1/(2n)$$

for $i = 1, \ldots, n$ respectively, where $(\sqrt{P^{xx}})_i$ denotes the i-th column of the matrix root L of P given by $P^{xx} = LL^T$. Thus, the corresponding weights are $W_i^{\text{mean}} = W_i^{\text{cov}} = W_i$ for $i = 1, \ldots, 2n$.

Subsequently, we apply the unscented Kalman filter to the recursive estimation of dynamical processes.

3.3.1 Additive process and measurement noise

For the sake of simplicity we begin with the special case of additive process and measurement noise

$$\mathbf{x}_k = \mathbf{f}(\mathbf{x}_{k-1}, \mathbf{u}_{k-1}, \mathbf{w}_{k-1}) = \tilde{\mathbf{f}}(\mathbf{x}_{k-1}, \mathbf{u}_{k-1}) + \mathbf{w}_{k-1},$$
$$\mathbf{y}_k = \mathbf{h}(\mathbf{x}_k, \mathbf{v}_k) = \tilde{\mathbf{h}}(\mathbf{x}_k) + \mathbf{v}_k.$$

At first, we construct a set of sigma points $\mathcal{X}_{k-1}^{(i)}$ with mean $\hat{\mathbf{x}}_{k-1}$ and covariance P_{k-1}. The dependence of the sigma points on $\hat{\mathbf{x}}_{k-1}$ and P_{k-1} is denoted by $\mathcal{X}_{k-1}^{(i)} = \mathcal{X}_{k-1}^{(i)}(\hat{\mathbf{x}}_{k-1}, P_{k-1})$. The predictor of the unscented Kalman filter and the error covariance are given by

$$\mathcal{X}_k^{-(i)} = \mathbf{f}\left(\mathcal{X}_{k-1}^{(i)}, \mathbf{u}_{k-1}, 0\right), \quad \text{for } i = 0, \ldots, p, \tag{11}$$

$$\hat{\mathbf{x}}_k^- = \sum_{i=0}^{p} W_i^{\text{mean}} \mathcal{X}_k^{-(i)},$$

$$P_k^- = \sum_{i=0}^{p} W_i^{\text{cov}} \left(\mathcal{X}_k^{-(i)} - \hat{\mathbf{x}}_k^-\right)\left(\mathcal{X}_k^{-(i)} - \hat{\mathbf{x}}_k^-\right)^T + Q_{k-1}. \tag{12}$$

The predicted observation is

$$\mathcal{Y}_k^{(i)} = \mathbf{h}\left(\mathcal{X}_k^{-(i)}, 0\right), \quad \text{for } i = 0, \ldots, p, \tag{13}$$

$$\hat{\mathbf{y}}_k^- = \sum_{i=0}^{p} W_i^{\text{mean}} \mathcal{Y}_k^{(i)} \tag{14}$$

with innovation covariance and cross covariance

$$\begin{aligned} P_k^{yy} &= \sum_{i=0}^{p} W_i^{\text{cov}} \left(\mathcal{Y}_k^{(i)} - \hat{\mathbf{y}}_k^-\right)\left(\mathcal{Y}_k^{(i)} - \hat{\mathbf{y}}_k^-\right)^T + R_k, \\ P_k^{xy} &= \sum_{i=0}^{p} W_i^{\text{cov}} \left(\mathcal{X}_k^{-(i)} - \hat{\mathbf{x}}_k^-\right)\left(\mathcal{Y}_k^{(i)} - \hat{\mathbf{y}}_k^-\right)^T. \end{aligned} \tag{15}$$

The Kalman gain has the form

$$K_k = P_k^{xy}\left(P_k^{yy}\right)^{-1}. \tag{16}$$

The corrector of the unscented Kalman filter and the a posteriori covariance are given by

$$\begin{aligned} \hat{\mathbf{x}}_k &= \hat{\mathbf{x}}_k^- + K_k\left(\mathbf{y}_k - \hat{\mathbf{y}}_k^-\right), \\ P_k &= P_k^- - K_k P_k^{yy} K_k^T. \end{aligned} \tag{17}$$

The relationship between the Kalman gains K_k of the extended and unscented filters (9), (16) is given by

$$P_k^{xy} = P_k^- H_k^T \quad \text{and} \quad P_k^{yy} = H_k P_k^- H_k^T + R_k.$$

Table 3 Unscented Kalman filter

Model	$\mathbf{x}_k = \mathbf{f}(\mathbf{x}_{k-1}, \mathbf{u}_{k-1}, \mathbf{w}_{k-1}) = \tilde{\mathbf{f}}(\mathbf{x}_{k-1}, \mathbf{u}_{k-1}) + \mathbf{w}_{k-1}, \mathbf{w}_{k-1} \sim \mathcal{N}(0, Q_{k-1})$, $\mathbf{y}_k = \mathbf{h}(\mathbf{x}_k, \mathbf{v}_k) = \tilde{\mathbf{h}}(\mathbf{x}_k) + \mathbf{v}_k, \mathbf{v}_k \sim \mathcal{N}(0, R_k)$
Initialization	$\hat{\mathbf{x}}_0, P_0 = E[\hat{\mathbf{x}}_0\hat{\mathbf{x}}_0^T]$
Weights	$W_i^{\text{mean}}, W_i^{\text{cov}}, i = 0, \ldots, p$
Sigma points	$\mathcal{X}_{k-1}^{(i)} = \mathcal{X}_{k-1}^{(i)}(\hat{\mathbf{x}}_{k-1}, P_{k-1}), i = 0, \ldots, p$
Predictor	$\mathcal{X}_k^{-(i)} = \mathbf{f}(\mathcal{X}_{k-1}^{(i)}, \mathbf{u}_{k-1}, 0), i = 0, \ldots, p$, $\hat{\mathbf{x}}_k^- = \sum_{i=0}^p W_i^{\text{mean}} \mathcal{X}_k^{-(i)}$, $P_k^- = \sum_{i=0}^p W_i^{\text{cov}}(\mathcal{X}_k^{-(i)} - \hat{\mathbf{x}}_k^-)(\mathcal{X}_k^{-(i)} - \hat{\mathbf{x}}_k^-)^T + Q_{k-1}$
Predicted observation	$\mathcal{Y}_k^{(i)} = \mathbf{h}(\mathcal{X}_k^{-(i)}, 0)$, $\hat{\mathbf{y}}_k^- = \sum_{i=0}^p W_i^{\text{mean}} \mathcal{Y}_k^{(i)}, i = 0, \ldots, p$, $P_k^{yy} = \sum_{i=0}^p W_i^{\text{cov}}(\mathcal{Y}_k^{(i)} - \hat{\mathbf{y}}_k^-)(\mathcal{Y}_k^{(i)} - \hat{\mathbf{y}}_k^-)^T + R_k$, $P_k^{xy} = \sum_{i=0}^p W_i^{\text{cov}}(\mathcal{X}_k^{-(i)} - \hat{\mathbf{x}}_k^-)(\mathcal{Y}_k^{(i)} - \hat{\mathbf{y}}_k^-)^T$
Kalman gain	$K_k = P_k^{xy}(P_k^{yy})^{-1}$
Corrector	$\hat{\mathbf{x}}_k = \hat{\mathbf{x}}_k^- + K_k(\mathbf{y}_k - \hat{\mathbf{y}}_k^-), P_k = P_k^- - K_k P_k^{yy} K_k^T$

Also P_k in (17) corresponds to (10) because substitution of (16) in (17) yields

$$\begin{aligned} P_k &= P_k^- - K_k P_k^{yy} K_k^T \\ &= P_k^- - P_k^{xy}(P_k^{yy})^{-1}(P_k^{xy})^T \\ &= P_k^- - K_k H_k P_k^- \\ &= (I - K_k H_k) P_k^-. \end{aligned}$$

The algorithm is summarized in Table 3.

3.3.2 Nonlinear noise

For the general case of nonlinear noise augmented sets of sigma points for prediction and corrector steps have to be defined [29]. Let

$$\bar{\mathbf{x}} = (\mathbf{x}^T, \mathbf{w}^T)^T, \tag{18}$$

$$\bar{\bar{\mathbf{x}}} = (\mathbf{x}^T, \mathbf{v}^T)^T, \tag{19}$$

$$\bar{\mathbf{f}}(\bar{\mathbf{x}}, \mathbf{u}) = \mathbf{f}(\mathbf{x}, \mathbf{u}, \mathbf{w}), \tag{20}$$

$$\bar{\mathbf{h}}(\bar{\bar{\mathbf{x}}}, \mathbf{u}) = \mathbf{h}(\mathbf{x}, \mathbf{v}), \tag{21}$$

For the predictor step, define sigma points $\bar{\mathcal{X}}_{k-1}^{(i)}$ with respect to mean

$$\bar{\mathbf{x}}_{k-1} := \begin{pmatrix} \hat{\mathbf{x}}_{k-1} \\ 0 \end{pmatrix} \tag{22}$$

and covariance

$$\bar{P}_{k-1} = \begin{pmatrix} P_{k-1} & 0 \\ 0 & Q_{k-1} \end{pmatrix}. \tag{23}$$

Table 4 General unscented Kalman Filter

Model	$\mathbf{x}_k = \mathbf{f}(\mathbf{x}_{k-1}, \mathbf{u}_{k-1}, \mathbf{w}_{k-1}),\ \mathbf{w}_{k-1} \sim \mathcal{N}(0, Q_{k-1}),$ $\mathbf{y}_k = \mathbf{h}(\mathbf{x}_k, \mathbf{v}_k),\ \mathbf{v}_k \sim \mathcal{N}(0, R_k)$
Initialization	$\hat{\mathbf{x}}_0,\ P_0 = E[\hat{\mathbf{x}}_0 \hat{\mathbf{x}}_0^T]$
Weights	$W_i^{\text{mean}},\ W_i^{\text{cov}},\ i = 0, \ldots, p$
Sigma points	$\bar{\mathcal{X}}_{k-1}^{(i)} = \bar{\mathcal{X}}_{k-1}^{(i)}(\bar{\mathbf{x}}_{k-1}, \bar{P}_{k-1}),\ i = 0, \ldots, p$
Predictor	$\mathcal{X}_k^{-(i)} = \bar{\mathbf{f}}(\bar{\mathcal{X}}_{k-1}^{(i)}, \mathbf{u}_{k-1}),\ i = 0, \ldots, p,$ $\hat{\mathbf{x}}_k^- = \sum_{i=0}^p W_i^{\text{mean}} \mathcal{X}_k^{-(i)},$ $P_k^- = \sum_{i=0}^p W_i^{\text{cov}} (\mathcal{X}_k^{-(i)} - \hat{\mathbf{x}}_k^-)(\mathcal{X}_k^{-(i)} - \hat{\mathbf{x}}_k^-)^T$
Sigma points	$\bar{\bar{\mathcal{X}}}_k^{(i)} = \bar{\bar{\mathcal{X}}}_k^{(i)}(\bar{\bar{\mathbf{x}}}_k, \bar{\bar{P}}_k),\ i = 0, \ldots, p$
Predicted observation	$\mathcal{Y}_k^{(i)} = \bar{\mathbf{h}}(\bar{\bar{\mathcal{X}}}_k^{(i)}),$ $\hat{\mathbf{y}}_k^- = \sum_{i=0}^p W_i^{\text{mean}} \mathcal{Y}_k^{(i)},\ i = 0, \ldots, p,$ $P_k^{yy} = \sum_{i=0}^p W_i^{\text{cov}} (\mathcal{Y}_k^{(i)} - \hat{\mathbf{y}}_k^-)(\mathcal{Y}_k^{(i)} - \hat{\mathbf{y}}_k^-)^T,$ $P_k^{xy} = \sum_{i=0}^p W_i^{\text{cov}} (\mathcal{X}_k^{-(i)} - \hat{\mathbf{x}}_k^-)(\mathcal{Y}_k^{(i)} - \hat{\mathbf{y}}_k^-)^T$
Kalman gain	$K_k = P_k^{xy} (P_k^{yy})^{-1}$
Corrector	$\hat{\mathbf{x}}_k = \hat{\mathbf{x}}_k^- + K_k(\mathbf{y}_k - \hat{\mathbf{y}}_k^-),\ P_k = P_k^- - K_k P_k^{yy} K_k^T$

Then (11) is substituted by

$$\mathcal{X}_k^{-(i)} = \bar{\mathbf{f}}\left(\bar{\mathcal{X}}_{k-1}^{(i)}, \mathbf{u}_{k-1}\right), \quad i = 0, \ldots, p \tag{24}$$

and the Q-term in (12) vanishes. Accordingly define sigma points $\bar{\bar{\mathcal{X}}}_k^{(i)}$ for the corrector step with mean

$$\bar{\bar{\mathbf{x}}}_k := \begin{pmatrix} \hat{\mathbf{x}}_k^- \\ 0 \end{pmatrix} \tag{25}$$

and covariance

$$\bar{\bar{P}}_k = \begin{pmatrix} P_k^- & 0 \\ 0 & R_k \end{pmatrix}. \tag{26}$$

Then (13) is substituted by

$$\mathcal{Y}_k^{(i)} = \bar{\mathbf{h}}\left(\bar{\bar{\mathcal{X}}}_k^{(i)}\right), \quad i = 0, \ldots, p \tag{27}$$

and the R-term in (15) vanishes. The algorithm is summarized in Table 4.

3.4 Comparison of EKF and UKF in crack growth example

Figure 1 shows a comparison of the extended Kalman filter, the unscented Kalman filter and a Monte Carlo simulation for the estimation of mean and covariance of the time dependent crack depth. In this test, we only assume the initial crack size a_0 to be uncertain, whereas we set the other parameters in (1) to deterministic values. It can be seen from the figures that both extended and unscented filter are fairly accurate in the linear regime of the crack growth, whereas in the nonlinear regime only the unscented filter shows sufficient accuracy.

Figure 1 Comparison of estimated mean values (Upper) and covariances (Lower) of crack depth.

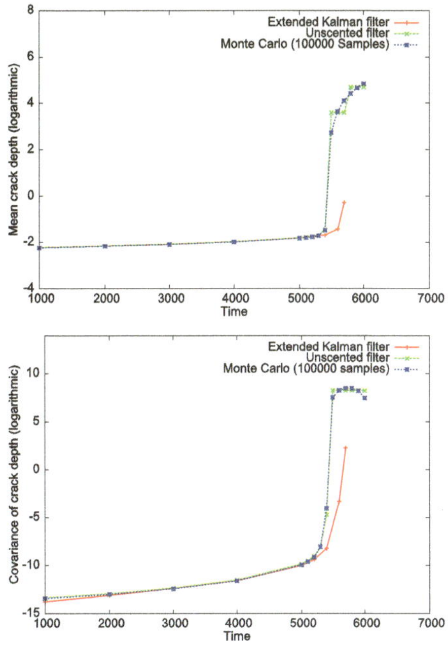

3.5 Reinterpretation of UKF

The extended Kalman filter does not treat the nonlinearities of the above example and is therefore no longer considered. The Kalman filter techniques are developed to deal with uncertainties in the state and the measurements. In order to allow a comparison to the methods based on Polynomial Chaos expansions, *cf.* Sections 4 and 5, we omit the treatment of measurements and only consider the state estimation. Thus, the resulting equations that we are considering within the following are ordinary differential equations with random parameters, *cf.* (30) and (33). In this setting, we interpret the construction of sigma points as special stochastic collocation method which is quadratic in each dimension (28). We reconstruct the solution between the sampled points by the polynomial

$$\mathbf{x}(T_L, \theta_1, \dots, \theta_{n_w}) = \sum_{i=1}^{n_w} \left(q_{0,i} + q_{1,i}\theta_i + q_{2,i}\left(\theta_i^2 - 1\right) \right). \tag{28}$$

In a post-processing step, we will apply a Monte-Carlo simulation based on this polynomial approximation, to obtain approximations of any quantity of interest, *i.e.* the moments or density of the solution.

4 Stochastic Collocation method and random delay differential equations

We will consider the delay differential equation (DDE) with one single, time-independent delay, defined as

$$\dot{\mathbf{x}}(t) = \mathbf{f}\left(\mathbf{x}(t), \mathbf{x}(t - \tau)\right), \quad t > 0,$$

with $\mathbf{f} : \mathbb{R}^n \times \mathbb{R}^n \to \mathbb{R}^n$ and $\tau : \mathbb{R} \to [0, +\infty]$. The parameter τ is called the delay. Given the initial data set (also called history function) $\mathbf{x}(t) = \phi(t)$, $t \in [-\tau, 0]$ we obtain the initial value problem (IVP)

$$IVP := \begin{cases} \dot{\mathbf{x}}(t) = \mathbf{f}(\mathbf{x}(t), \mathbf{x}(t-\tau)), & t > 0, \\ \mathbf{x}(t) = \phi(t), & t \in [-\tau, 0]. \end{cases} \tag{29}$$

When one or more parameters in (29) are affected by uncertainty we will be speaking of a random delay differential equation (RDDE)

$$\dot{\mathbf{x}}(t, \mathbf{w}) = \mathbf{f}\big(\mathbf{x}(t, \mathbf{w}), \mathbf{x}(t-\tau, \mathbf{w}), \mathbf{w}\big), \quad t > 0, \tag{30}$$

if $\mathbf{w} \in L_2$. Eventually, the initial history function can also exhibit a random behavior, $x(t, \mathbf{w}) = \phi(t, \mathbf{w})$, $t \in [-\tau, 0]$.

In order to account for the uncertainties the stochastic collocation schema of the Polynomial Chaos Expansion is used. The schema is explained below, along with the numerical method applied to solve for the underlying deterministic equations.

4.1 Wiener expansions

We will denote a probability space by $(\Omega, \mathcal{F}, P)$ and the space of square integrable random variables on it by $L_2(\Omega, \mathcal{F}, P)$. We will abbreviate it by L_2 in the forthcoming. This space is equipped with the inner product

$$(w, v) := \int_\Omega w(\theta) v(\theta)\, dP(\theta)$$

and the induced norm $\|w\| := \sqrt{(w, w)}$. Let ξ be a standard Gaussian distributed random variable on $(\Omega, \mathcal{F}, P)$, where $\mathcal{F}$ is generated by ξ. N. Wiener [16] and later R. Ghanem and P. Spanos [10] showed, that every random variable $\mathbf{x} \in L_2$ can be expanded into a series

$$\mathbf{x}\big(\mathbf{w}(\cdot)\big) = \sum_{i=0}^{\infty} \mathbf{q}_i \Psi_i\big(\mathbf{w}(\cdot)\big)$$

with basis polynomials $\{\Psi_i\}_{i=0}^{\infty}$. A detailed discussion on this expansions can be found in [1, 4, 30]. In [11] D. Xiu and G. Karniadakis generalized this expansion to polynomials of the Askey scheme. For convergence results of these generalized expansions we refer to [31].

The Wiener expansions can be used to solve operator equations with square integrable random parameters. Assume the operator $T : X \to \mathbb{R}^n$, mapping from the Banach space X into $\mathbb{R}^n$, given by

$$T(x; \mathbf{w}) = 0, \tag{31}$$

with $x \in X$ and parameters $\mathbf{w} \in L_2$. If the solution x is in L_2, it can be expanded by a generalized Wiener expansion. A truncation of this expansion after $\hat{p} + 1$ terms results in an approximation for x in the finite dimensional subspace linspan$\{\Psi_i : i = 0, \ldots, \hat{p}\} \subset L_2$

$$\mathbf{x}(\mathbf{w}) \approx \sum_{i=0}^{\hat{p}} \mathbf{q}_i \Psi_i(\mathbf{w}). \tag{32}$$

Thus the solution $\mathbf{x}$ can be approximated by computing the coefficients $\{\mathbf{q}_i\}_{i=0}^{\hat{p}}$. Different methods can be used to calculate the coefficients. We opt here for the stochastic collocation.

4.2 Stochastic collocation

Stochastic collocation methods make use of the fact that the Wiener expansions are generalized Fourier series. Therefore, the coefficients in (32) are given by

$$\mathbf{q}_i = \frac{(\mathbf{x}, \Psi_i)}{(\Psi_i, \Psi_i)}, \quad i = 0, \ldots, \hat{p}.$$

For the computation of the coefficients $\{\mathbf{q}_i\}_{i=0}^{\hat{p}}$, an estimation of the inner products is necessary. This can be accomplished with cubature rules, which approximate the multidimensional integral as a discrete sum

$$(\mathbf{x}, \Psi_i) \approx \sum_{j=1}^{N_Q} \mathbf{x}\left(\mathbf{w}^{(j)}\right) \Psi_i\left(\mathbf{w}^{(j)}\right) W^{(j)},$$

where $\mathbf{w}^{(j)}$, meant as $\mathbf{w}(\theta^{(j)})$, and $W^{(j)}$ are the nodes and weights of the N_Q-node cubature and are given for each type of cubature, see [4] for more details.

The procedure for the stochastic collocation scheme would be then as follows: For each of the given nodes calculate the solution of (31) and sum them up with the corresponding weights. The products (Ψ_i, Ψ_i) are independent from the operator equations and can be conveniently pre-computed and stored for efficient use in different calculations.

This approach offers some advantages. Contrary to MC techniques, the sampling is not random, but obey the cubature rules to minimize the needed number of nodes for a given precision. Thus, the amount of realizations is significantly reduced. Usually, 10 nodes per dimension are accurate enough. Nevertheless, for increasing number of random variables in the parametrization, the numbers of nodes grows exponentially (curse of dimensionality), and the dimension-independent MC becomes eventually more efficient. Sparse grid cubatures have been proposed to mitigate the curse of dimensionality [4].

Another important feature that makes the stochastic collocation schema attractive within the Wiener expansion methodology is the splitting of the uncertainty quantification and the system governing equations. Since the values of the random variables are chosen in advance, we only need to solve the underlying deterministic equations for the given value of the parameters. Hence, no modification of our existing deterministic software is needed in order to account for uncertainties. This is specially valuable in the case of commercial programs, which cannot be modified, or complex software, whose adaptation for uncertainty quantification can be error-prone. A further advantage is the modularity, making implementation of refinements in the Wiener expansion algorithms and the underlying governing equations independent.

For solving the RDDEs (30) we use the stochastic collocation schema combined with the program RADAR for the computation of the DDEs.

4.3 Numerical solution of the DDEs

The software RADAR is used as the deterministic solver for DDEs. It has been developed by Ernst Hairer and Nicola Guglielmi. For more details a manual with examples is available [32].

In a simplified form for presentation purposes, the DDE

$$\dot{\mathbf{x}}(t) = \mathbf{f}\big(\mathbf{x}(t), \mathbf{x}\big(\alpha\big(t, \mathbf{x}(t)\big)\big)\big), \quad \alpha\big(t, \mathbf{x}(t)\big) < t,$$
$$\mathbf{x}(t) = \mathbf{g}(t), \quad t < t_0,$$

with $\alpha(t, \mathbf{x}(t))$ the (eventually state-dependent) time-lag function, is solved using an implicit Runge-Kutta method called RADAU IIA [33]. If we denote the stepsize by $h_k = t_{k+1} - t_k$ and the coefficient matrix by $A = (a_{ij})_{i,j=1}^s$, $c_i = \sum_{j=1}^s a_{ij}$, weights $\{b_i\}_{i=1}^s$ the implicit schema reads

$$\mathbf{Y}_i^{(k)} - \mathbf{x}_k = h_k \sum_{j=1}^{s} a_{ij} \mathbf{f}\big(\mathbf{Y}_j^{(k)}, \mathbf{Z}_j^{(k)}\big),$$

where $\mathbf{x}_k$ is the approximation of $\mathbf{x}(t_k)$ and $\mathbf{Y}_s = \mathbf{x}_{k+1}$, if we use the weights $b_i = a_{s,i}$, $i = 1, \ldots, s$. $\mathbf{Z}_j^{(k)}$ is defined by

$$\mathbf{Z}_j^{(k)} = \begin{cases} \mathbf{g}(\alpha_j^{(k)}), & \alpha_j^{(k)} < t_0, \\ \mathbf{u}_m(\alpha_j^{(k)}), & \alpha_j^{(k)} \in [t_m, t_{m+1}], \end{cases}$$

where $\alpha_j^{(k)} = \alpha(t_k + c_j h_k, \mathbf{Y}_j^{(k)})$ and $\mathbf{u}_m$ is a polynomial approximation of the solution $\mathbf{x}(t)$ on the interval $[t_m, t_{m+1}]$. Here, the Lagrange interpolation formula is used to build the approximation.

RADAR allows steps larger and smaller than the delay and performs stepsize control by estimating the local error at grid points and the error in the continuous numerical approximation to the solution.

5 Stochastic Galerkin method and random ordinary differential equations

Within Section 4 we discussed the treatment of RDDEs by the collocation method based on generalized Wiener expansions. In this section we will present the stochastic Galerkin method (SGM) for the approximation of the solution of differential equations with uncertain parameters, *cf.* [1]. We again consider the differential equation (30), $\dot{\mathbf{x}}(t, \mathbf{w}) = \mathbf{f}(\mathbf{x}(t, \mathbf{w}), \mathbf{x}(t - \tau, \mathbf{w}), \mathbf{w})$, and omit the delay within the forthcoming discussion to keep the notation simple:

$$\dot{\mathbf{x}}(t, \mathbf{w}) = \mathbf{f}\big(\mathbf{x}(t, \mathbf{w}), \mathbf{w}\big) \tag{33}$$

with initial condition $\mathbf{x}(0, \mathbf{w}) \in L_2$. The extension of the SGM including the treatment of delays is straightforward using interpolation for the approximation of the solution at past times. In Section 6 we will apply the SGM to the test example of the crack growth (an RODE) and the biofilm model (an RDDE) as stated in the introductory Section 1. Subsequently, we intend to present the main idea of the SGM and restate the main results of its convergence analysis from [21, 34].

For the numerical approximation of the solution of (33) we search for an approximation of $\mathbf{x}(t,\mathbf{w})$ within the finite dimensional subspace

$$\mathcal{S} = \text{linspan}\{\Psi_i : i = 0,\dots,\hat{p}\} \subset L_2,$$

the approximation space. Without loss of generality we assume that the orthogonal polynomials $\{\Psi_i\}_{i=0}^{\infty}$ are normalized, *i.e.* $(\Psi_i, \Psi_i) = 1$.

Besides RODE (33) we introduce the perturbed RODE

$$\frac{\partial}{\partial t}\tilde{\mathbf{x}}(t,\mathbf{w}) = \mathbf{f}\big(\pi_{\hat{p}} \circ \tilde{\mathbf{x}}(t,\mathbf{w}),\mathbf{w}\big) =: \tilde{\mathbf{f}}\big(\tilde{\mathbf{x}}(t,\mathbf{w}),\mathbf{w}\big), \tag{34}$$

where $\pi_{\hat{p}} : L_2 \to \mathcal{S}$ denotes the orthogonal projection from L_2 onto $\mathcal{S}$. We define the restricted stochastic weak form of (34) by

$$\left(\frac{\partial}{\partial t}\tilde{\mathbf{x}}(t,\mathbf{w}), \Psi(\mathbf{w})\right) = \left(\tilde{\mathbf{f}}\big(\tilde{\mathbf{x}}(t,\mathbf{w}),\mathbf{w}\big), \Psi(\mathbf{w})\right) \tag{35}$$

for all $\Psi \in \mathcal{S}$. Equation (35) can be written as a system of ordinary differential equations (ODEs) for the coefficients $\{\tilde{\mathbf{q}}_i\}_{i\leq\hat{p}}$:

$$\frac{d}{dt}\tilde{\mathbf{q}}_i(t) = (\tilde{\mathbf{f}}, \Psi_i) \quad \text{for } i \leq \hat{p}. \tag{36}$$

This system of ODEs is deterministic and can be solved by standard integrators. The projected system (36) can be solved by means of a Runge-Kutta method (RKM) of order $p \geq 1$ and stepsize $h > 0$. We obtain the approximation $\hat{\mathbf{x}}$ of the solution of (34) in terms of the numerically computed coefficients $\{\hat{\mathbf{q}}_i\}_{i\leq\hat{p}}$ using the SGM and a RKM,

$$\tilde{\mathbf{x}}(t,\mathbf{w}) \approx \hat{\mathbf{x}}(t,\mathbf{w}) := \sum_{i\leq\hat{p}} \hat{\mathbf{q}}_i(t)\Psi_i(\mathbf{w}).$$

The following scheme depicts the stochastic Galerkin Runge-Kutta method (SGRKM):

$$\begin{aligned} \frac{\partial}{\partial t}\tilde{\mathbf{x}} = \tilde{\mathbf{f}}(\tilde{\mathbf{x}},\mathbf{w}) \quad &\overset{\text{SGM}}{\longrightarrow} \quad \frac{d}{dt}\tilde{\mathbf{q}} = \mathbf{F}(\tilde{\mathbf{q}}) \\ &\overset{\text{RKM}}{\longrightarrow} \quad \hat{\mathbf{x}} = \sum_{i\leq\hat{p}} \hat{\mathbf{q}}_i\Psi_i(\mathbf{w}). \end{aligned} \tag{37}$$

In [34] it is shown that the local discretization error of the SGRKM does not vanish for decreasing stepsize $h \to 0$. Thus, this global approach of the stochastic Galerkin method fails to converge to the exact solution $\mathbf{x}$. To overcome this problem a generalized approach has to be applied. X. Wan and G.E. Karniadakis proposed the Multi-Element generalized Polynomial Chaos (MEgPC) method to attain convergence of the SGM. We recall that this method is based on a partitioning of the parameter space Ω into subsets, so called random elements. The SGM is then applied locally on the partition of the parameter space, *i.e.* the problem is conditionally restricted to each random element. We refer to the literature for a detailed description [18].

Using the MEgPC approach it can be shown that the local discretization error of the SGRKM obeys

$$\left\|\mathbf{x}(t+h,\mathbf{w})-\hat{\mathbf{x}}(t+h,\mathbf{w})\right\|_2 \le \mathcal{O}\left(h\Delta^{\hat{p}+1}\right)+\mathcal{O}\left(h^{p+1}\right), \tag{38}$$

where we denote the maximal diameter of the stochastic elements by Δ, see [34].

5.1 The adaptive algorithm for the solution of RODEs

We use (38) to control the stepsize h in time and the maximal size Δ of the stochastic elements $\{\Omega_i\}_{i=1}^{\nu}$. In a first step, we compute the conditionally restricted solution $\tilde{\mathbf{x}}(t+h,\mathbf{w})$ on every element Ω_i and check the local truncation error of the projected restricted system (36). See [35] for details on adaptive stepsize control of RKMs. This procedure results in the term $\mathcal{O}(h^{p+1})$ being negligible up to a prescribed tolerance. In a second step, we check the error $\mathtt{ERR}$, which is introduced by the truncation of $\tilde{\mathbf{x}}(t+h,\mathbf{w})$ to $\mathcal{S}$. Therefore, we split Ω_i into two equal subsets and compute the solution of (36) conditionally restricted to those subsets. We take the L_2-norm of the difference of the refined solution and the solution $\tilde{\mathbf{x}}(t+h,\mathbf{w})$ on Ω_i as substitute for the actual truncation error $\mathtt{ERR}_i$. We apply this procedure in every random element at every timestep, so we omit the corresponding index i of the subsets in the following. According to (38) it holds

$$\mathtt{ERR} = ch\Delta^{\hat{p}+1}, \tag{39}$$

where we neglect the error term $\mathcal{O}(h^{p+1})$ of the RKM. Furthermore, we require the optimal size Δ_{opt} of the respective element to satisfy

$$\varepsilon = c\Delta_{\text{opt}}^{\hat{p}+1} \tag{40}$$

for prescribed tolerance $\varepsilon > 0$. Division of (40) and (39) results in

$$\Delta_{\text{opt}} = \sqrt[\hat{p}+1]{\frac{\varepsilon h}{\text{ERR}}}\,\Delta =: \mathcal{I}(\text{ERR})\Delta.$$

We use the estimator $\mathcal{I}(\text{ERR})$ to decide about the refinement of the elements $\{\Omega_i\}_{i=1}^{\nu}$. If $\mathcal{I}(\text{ERR}) < \rho_1$, then the element will be refined. If $\mathcal{I}(\text{ERR})$ exceeds a coarsening threshold $\rho_2 > \rho_1$, then the element will be coarsened. For a more detailed description of this procedure see [34].

6 Computational aspects and numerical results

In this section, we apply the three methods, the UKF approach, the stochastic collocation and the SGRKM, to the benchmark problems stated in the introduction. From the solution we compute the densities and the first two moments and compare the results. For the numerical time integration in the SGM we use an embedded explicit RKM of order 4(3). After a discussion of the efficiencies of the methods, we first consider the crack growth example (1) with the initial crack size being uncertain. Thereafter, we study the setting of two uncertain parameters in the model of crack growth. We finish this section by the comparison of the three methods applied to the biofilm model (3).

Table 5 Number of function evaluations for UKF, SCM and MEgPC approach based on a Gaussian cubature rule of order *s*

UKF	$2n+1$
SCM	$(s+1)^n$
MEgPC	$(s+1)^n$

Numbers given per timestep of the time integration method.

6.1 Computational aspects

For UKF and SCM, the efficiency in terms of function evaluations can directly be given as function of the stochastic dimension n (Table 5). Here the number of sigma points for UKF has been chosen as in the example Section 3.3. For SCM a standard Gaussian cubature rule of order s has been used. SCM and MEgPC show an exponential growth in the number of function evaluations (suitable for 2D problems). Sparse cubature rules which are necessary to scope with high dimensional random spaces are considered in [22]. The MEgPC approach results in ν projected systems of ordinary differential equations (36) of dimension

$$\frac{(n+\hat{p})!}{n!\hat{p}!}$$

if a polynomial basis of total order $\hat{p}$ is applied. The number of needed function evaluations depends on the cubature rule applied to compute the projection integrals, for example using a Gaussian cubature tensor product rule we need $(s+1)^n$ function evaluations per timestep of the time integration method. However, the MEgPC approach is most efficient, if the projection integrals can be computed analytically in a preprocessing step before the time-integration starts. If this so called intrusive approach is not feasible, we have to resort to cubature rules to compute the projection. In the case of multiple random dimensions, $n \geq 2$, sparse integration methods have to be applied in order to reduce the curse of dimensionality.

6.2 Comparison of UKF, Stochastic Collocation and SGM in crack growth example

In this section, we compare the presented methods and Monte Carlo Sampling (10^6 samples) in the crack example (1).

6.2.1 One dimensional random space

At first, we assume only the initial crack size a_0 to be a random parameter following a uniform distribution with expected value 10^{-1} and variance 10^{-6}. The other parameters are set to $K_{I_c} = 10^{-12}$, $\Delta_\sigma = 16$ and $C_F = 0$. We remark that although the right-hand side function f in (1) is not polynomial in the initial value a_0, the convergence results from Section 5 apply, see [21, 34] for a detailed discussion.

To compute the numerical approximation $\hat{a}(t, a_0)$ of the crack size $a(t, a_0)$ by means of the MEgPC method, we use the parameters stated in Table 6. The adaptive MEgPC approach uses a truncated generalized Wiener expansion of degree $\hat{p} = 3$ for the approximation of the solution. The set of outcomes Ω is refined where the local error exceeds the prescribed tolerance $\varepsilon = 10^{-8}$. The approximated response surfaces at $t \in \{5{,}000, 5{,}300, 5{,}400, 5{,}420\}$ are shown in Figure 2. The respective elements of the approximation are indicated by dots on the axis of the outcomes Ω.

Table 6 Setting of parameters of the adaptive MEgPC approach for the solution of RODE (1)

Degree of approximation	$\hat{p}$	3
Order of approximation	p	3
Tolerance (time)	TOL	10^{-7}
Tolerance (elements)	ε	10^{-8}
Refinement threshold	ρ_1	1
Coarsening threshold	ρ_2	4

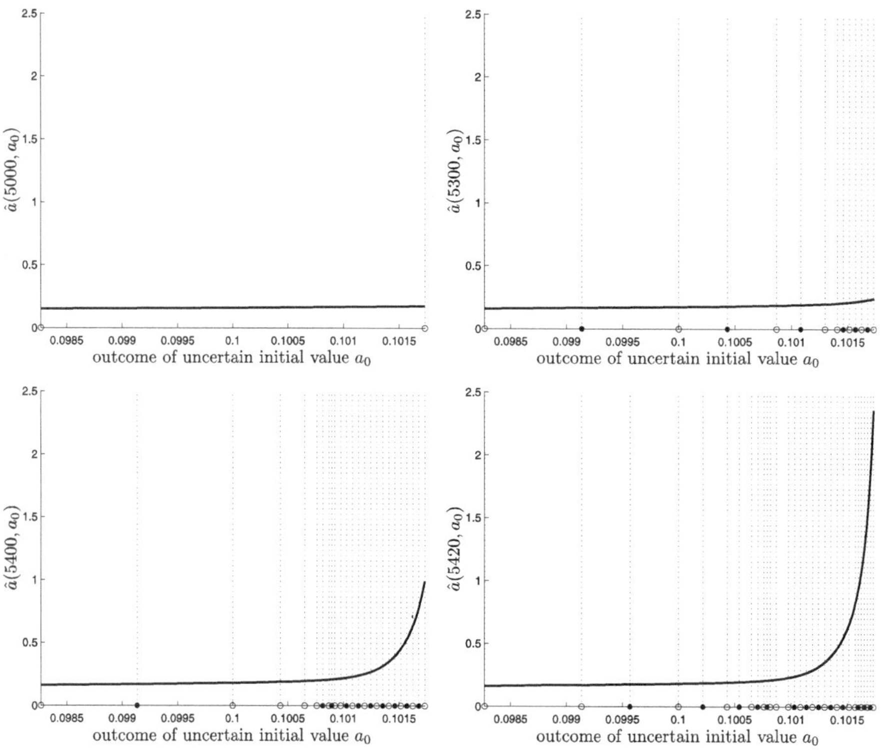

Figure 2 Response surfaces of the solution $\hat{a}(t, a_0)$ at $t \in \{5{,}000, 5{,}300, 5{,}400, 5{,}420\}$.

Although the initial crack size varies only by a small amount, the impact on the outcome is not negligible. A large variation in the crack size occurs at the final time $t = 5{,}420$. The dependence of the solution $a(t, a_0)$ on the uncertain initial crack size a_0 is highly non-linear. This is recognized by the MEgPC algorithm, which refines the parameter space where the response surface can not be represented by a polynomial of degree $\hat{p} = 3$ up to the prescribed tolerance.

In Figure 3 we compare the densities of the approximations $\hat{a}(t, a_0)$, $t \in \{5{,}000, 5{,}420\}$, computed by the stochastic collocation of order 5 (SCM5), MEgPC and the UKF approach respectively. We also plot the density computed by Monte Carlo simulation as a reference for the approximatively exact density. On the one hand, we observe that the UKF is not able to compute a good approximation of the density of the crack size a. This can be explained by the interpretation of the UKF as a stochastic collocation method of polynomial order 2, see Section 3.5. On the other hand, the MEgPC method, as we expect from our considerations in Section 5, yields a good approximation of the density of the crack size a in dependence on an uncertain initial crack size.

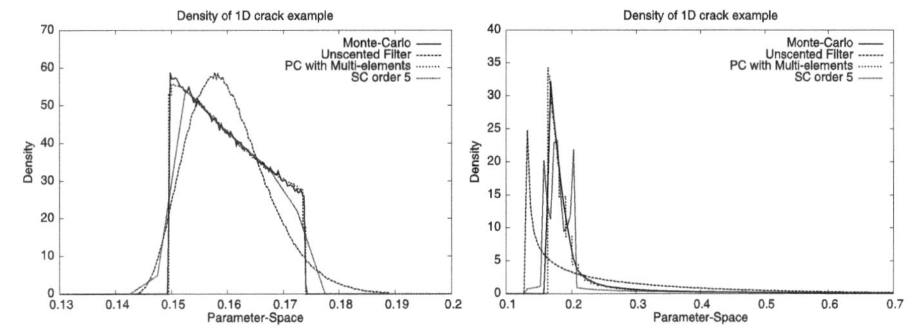

Figure 3 Density of the crack size at two final times $t \in \{5{,}000, 5{,}420\}$, assuming the initial crack size to be an uncertain parameter. Comparison of four different methods: Monte Carlo simulation, UKF, SCM5 and MEgPC.

Table 7 The expected values and variances of $a(t, a_0)$ at times $t \in \{5{,}000, 5{,}420\}$ computed by SCM5, UKF, MEgPC and Monte Carlo simulation

Time	Expected value				Variance			
	SCM5	MEgPC	UKF	MCS	SCM5	MEgPC	UKF	MCS
5,000	0.1.601	0.1.601	0.1.602	0.1601	$4.727 \cdot 10^{-5}$	$4.729 \cdot 10^{-5}$	$5.113 \cdot 10^{-5}$	$4.728 \cdot 10^{-5}$
5,420	0.2.681	0.2.726	0.2.671	0.2749	$5.655 \cdot 10^{-2}$	$7.791 \cdot 10^{-2}$	$2.449 \cdot 10^{-2}$	$8.627 \cdot 10^{-2}$

Table 8 Setting of parameters of the adaptive MEgPC approach for the solution of RODE (1)

Degree of approximation	$\hat{p}$	$(3, 3)$
Order of approximation	p	3
Tolerance (time)	TOL	10^{-8}
Tolerance (elements)	ε	10^{-8}
Refinement threshold	ρ_1	1
Coarsening threshold	ρ_2	4

In addition to the comparison of the densities we also compare the first two moments, see Table 7. We remark that although the UKF approach does not capture the overall density, it is able to approximate the first two moments in good accuracy. Moreover, the first two moments, which we computed by means of the MEgPC method and the Monte Carlo simulation match very good.

6.2.2 Two dimensional random space

In this section we consider the crack example (1) with two uncertain parameters. In addition to Section 6.2.1 we assume the parameter K_{Ic} to be uniformly distributed with expected value 0.25 and variance 0.02.

To compute the numerical approximation $\hat{a}(t, a_0, K_{Ic})$ of the crack size $a(t, a_0, K_{Ic})$ by means of the MEgPC method, we use the parameters as stated in Table 8. The notation $\hat{p} = (3, 3)$ means that we use a polynomial approximation of degree 3 in both random variables a_0 and K_{Ic}.

In Figure 4 we plot the densities computed by SCM5, UKF and MEgPC at times $t \in \{3{,}000, 3{,}500, 3{,}550\}$. Additionally, we plot the density of the reference solution computed by Monte Carlo simulation. We see that the difference between the density computed by the UKF and the reference density is negligible at $t = 3{,}000$ and becomes larger as time evolves. However, the overall structure of the density can be captured by the UKF ap-

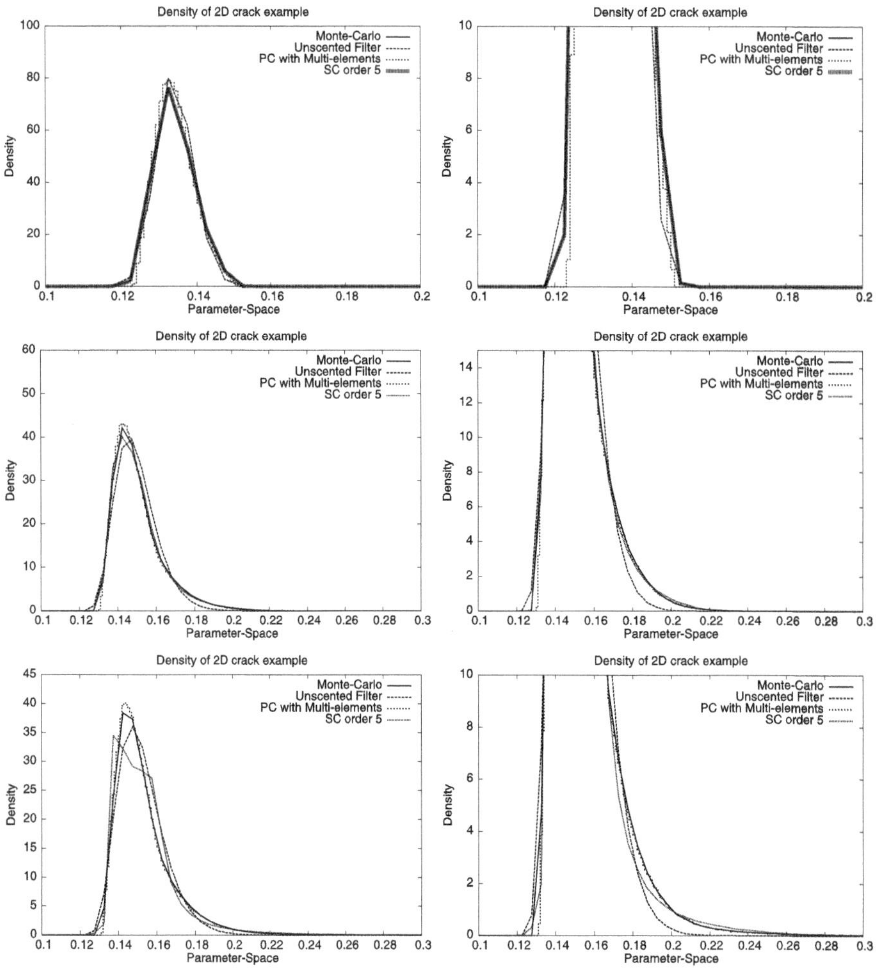

Figure 4 Density of the crack size at 4 final times resulting from two uncertain parameters. Comparison of 3 different methods: Monte-Carlo simulation, UKF, SCM5 and MEgPC. Left: complete density, right: zoom into the density.

Table 9 The expected values and variances of $a(t, a_0, K_{Ic})$ at times $t \in \{3{,}000, 3{,}500, 3{,}550\}$ computed by SCM5, UKF, MEgPC and Monte Carlo simulation based on 10^6 samples

Time	Expected value				Variance			
	SCM5	MEgPC	UKF	MCS	SCM5	MEgPC	UKF	MCS
3,000	0.1343	0.1343	0.1342	0.1343	$2.557 \cdot 10^{-5}$	$2.562 \cdot 10^{-5}$	$2.142 \cdot 10^{-5}$	$2.566 \cdot 10^{-5}$
3,500	0.1504	0.1504	0.1497	0.1504	$1.663 \cdot 10^{-4}$	$1.674 \cdot 10^{-4}$	$1.098 \cdot 10^{-4}$	$1.667 \cdot 10^{-4}$
3,550	0.1534	0.1534	0.1521	0.1534	$2.661 \cdot 10^{-4}$	$2.718 \cdot 10^{-4}$	$1.404 \cdot 10^{-4}$	$2.718 \cdot 10^{-4}$

proach. This is also reflected by the good approximation of the first two moments, see Table 9. Nevertheless, due to the increasing non-linear dependence on the crack size to the uncertain parameters the accuracy of the UKF approach becomes worse with increasing time. Again, this can be explained by the interpretation of the UKF as a polynomial approximation of degree 2 of the solution, see Section 3.5. Similar to the case of only one random parameter the densities computed by the MEgPC method and Monte Carlo simulation show perfect agreement. Moreover, the first two moments of the solutions agree very well, see Table 9.

Table 10 Setting of parameters of the adaptive MEgPC approach for the biofilm model

Degree of approximation	$\hat{p}$	5
Order of approximation	p	3
Tolerance (time)	TOL	$5 \cdot 10^{-6}$
Tolerance (elements)	ε	×
Refinement threshold	ρ_1	×
Coarsening threshold	ρ_2	×

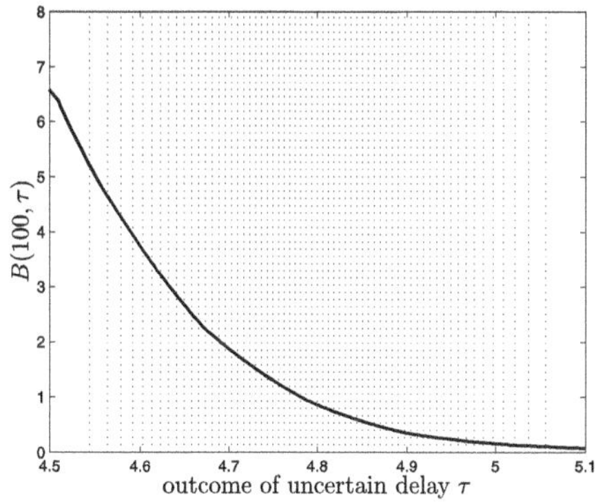

Figure 5 Response surface of the bacteria concentration $B(100, \tau)$.

6.3 Comparison of UKF, Stochastic Collocation and SGM in biofilm example

The introductory model for the biofilm (3) with delay Quorum Sensing has been studied in [36] in great detail. Here we compare the numerical approximations of the solution by means of its densities and first two moments. Therefore, we compute them by the UKF approach, the stochastic collocation and the MEgPC method at the time $t = 100$. The uncertain parameter is assumed to be the delay τ, a beta(2, 2) distributed random variable with expected value 4.8. Moreover, we use the choice of parameters $\alpha_{20} = 0.25$, $\sigma = \alpha_2 = \alpha_3 = 0.6$, $S_A = 0.4$ and $\mu_A = 0.5$ as well as the initial populations $B(0) = 1$ and $A(0) = 0$.

For the MEgPC method we use the choice of parameters as stated in Table 10.

As indicated by the crosses we do not use the adaptive mesh refinement strategy in this example. The reason is that we have to evaluate the solution at past timesteps in order to get an evaluation of the right-hand side function. In order to evaluate the solution at a time in the past the interpolation of the solution between two timesteps is needed. Therefore, we choose to have the solution defined on the same mesh of elements at every timestep. In order to avoid different meshes at different timesteps we switch off the mesh adaptivity and prescribe a sufficiently fine mesh, see the dotted lines in Figure 5, from the beginning of the computations. The response surface of the solution at time $t = 100$, computed by the MEgPC approach, is shown in Figure 5.

From the solution of the 3 methods we compute the density at time $t = 100$, see Figure 6 (left). The right plot shows a magnification of the density to the tail of the density, *i.e.* to the interval [3, 7]. On the one hand, we see that the density computed by the stochastic collocation of order 5 simulation and the MEgPC method match very good (therefore we here omit the Monte-Carlo solution) . On the other hand, we again observe a difference to the density computed by the UKF approach. The UKF density shows a much smaller weight in [0, 0.5] than the two other methods. This can be explained by the polynomial approximation of degree 2, where more mass of the response surface lies within this range

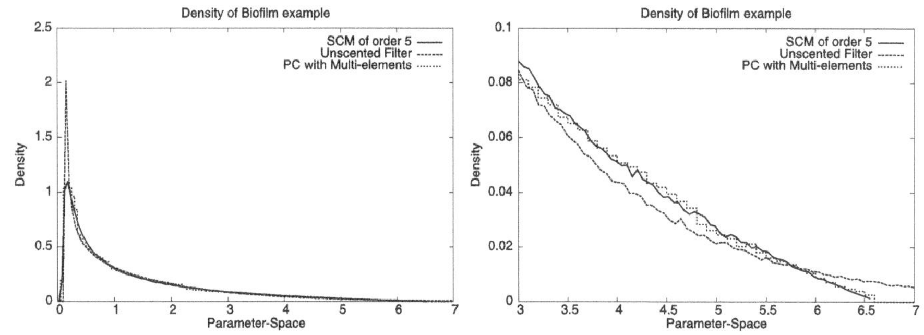

Figure 6 Density (left) of the biofilm model at the final time $t = 100$ resulting from one uncertain parameter. Comparison of 3 different methods: SCM, UKF and MEgPC approach. Left: complete density, right: zoom into the density.

Table 11 The expected values and variances of the biofilm model at time $t = 100$ computed by the MEgPC approach, the UKF and the stochastic collocation method

Method	Expected value	Variance
MEgPC	1.349	1.707
UKF	1.388	1.801
SCM order 5	1.352	1.706
SCM order 9	1.352	1.706

$[0, 0.5]$ of outcomes. Nevertheless, the overall shape of the density can be captured well by the UKF. In Table 11 we state the first two moments, respectively computed by the three methods. We see good agreement of the first two moments.

7 Conclusions

In the present article, we discussed the numerical treatment of problems with uncertain parameters. We gave a detailed introduction to the unscented Kalman filter and its application to random ordinary differential and random delay differential equations, see Section 3. We dropped the measurements in the equations and model the uncertainty by random variables. In this setting, the unscented Kalman filter can be interpreted as a stochastic collocation method of order 2, see Section 3.5. Sampling of the resulting surrogate polynomial model yields an approximation of the moments of the solution. Due to the fact that the sampling is restricted to the directions given by the columns of the root of the covariance matrix P^{xx}, only $2n+1$ realizations of the underlying random differential equation (RDE), resp. an RDDE or an RODE, are needed in the UKF approach. Here, n denotes the number of uncertain parameters. Thus, it is a cheap, in terms of computational costs, method to approximate the random behavior of the solution.

We compare this simplified UKF approach with the SCM and the MEgPC method. In the SCM more realizations, as compared to the UKF, of the solution of the RDE have to be computed. The number of realizations depends on the order of the approximation and the underlying cubature method. For most cubature rules, even in sparse grid approximations, the number of function evaluations is considerably greater than $2n+1$ and therefore the SCM is computationally more expensive. The SGM, as opposed to the SCM and the UKF, is not based on sampling, but on a spectral expansion of the solution of the RDE into orthogonal polynomials. An approximate solution is found by a Galerkin approach which results in the stochastic weak form (34). If we are able to compute it in a pre-processing step, only a system of deterministic differential equations for the coefficients of the truncated generalized Wiener expansion has to be solved. The dimension of the projected RDE is small, because we only need a low order approximation within the respective stochastic elements. The overall computational costs scale linearly with the number of stochastic elements. In the present article we did not take advantage of the pre-processing of the projection in the stochastic weak form. On the one hand, the projection in the examples from Section 6 can be efficiently performed by an adaptive Simpson cubature rule. On the other hand, pre-processing the stochastic weak form would result in an projection error, which can not be easily estimated a priori.

In Section 6 we applied the three methods to the benchmark problems of the crack growth model (1) and the biofilm model (3). In the case of the crack growth, which is an RODE, we discussed the prescription of up to two random parameters. It revealed that the UKF approach results in a non-sufficient representation of the overall density, but gives a good approximation of the first two moments. Opposed to that, the MEgPC method computes a good approximation of the density and of the first two moments. This is in agreement of the convergence results in Section 5. In the second benchmark problem, the example of the biofilm (3), we again compared the densities of the solutions computed by the three methods. The SCM, as well as the SGM method showed perfect agreement of the results - in the density and in the first two moments. Again, opposed to that the UKF approach yielded only a rough approximation of the overall density. Nevertheless, it is able to compute a good approximation of the first two moments.

From the numerical results in Section 6 we see that the UKF approach is a cheap alternative to get a rough impression of the random behavior of the solution of the RDE. If detailed information about the solution process is needed, more sophisticated methods like SCM and multi-element SGM have to be applied. This is for example in failure detection the case, where the tail of the probability distribution is of great interest.

References

1. Augustin F, Gilg A, Paffrath M, Rentrop P, Wever U: **Polynomial chaos for the approximation of uncertainties: chances and limits.** *Eur. J. Appl. Math.* 2008 **19**:149-190.
2. Xiu D: *Numerical Methods for Stochastic Computations: A Spectral Method Approach.* Princeton: Princeton University Press; 2010.
3. Karniadakis GE, Su CH, Xiu D, Lucor D, Schwab C, Todor RA: **Generalized polynomial chaos solution for differential equations with random inputs.** Report Nr. 2005-01, ETH Zürich. Seminar für Angewandte Mathematik; 2005.
4. Le Maître OP, Knio OM: *Spectral Methods for Uncertainty Quantification. Scientific Computation.* New York: Springer; 2010.
5. Fox BL: *Strategies for Quasi-Monte Carlo.* Berlin: Springer; 1999.
6. Babuska I, Tempone R, Zouraris G: **Galerkin finite element approximation of stochastic elliptic partial differential equations.** *SIAM J. Numer. Anal.* 2004, **42**(2):800-825.
7. Pulch R: **Polynomial chaos for multirate partial differential algebraic equations with random parameters.** *Appl. Numer. Math.* 2009, **59**(10):2610-2624.
8. Xiu D: **Fast numerical methods for stochastic computations.** *Commun. Comput. Phys.* 2008, **5**:242-272.
9. Maître OPL, Knio OM, Najm HN, Ghanem RG: **Uncertainty propagation using Wiener-Haar expansions.** *J. Comput. Phys.* 2004, **197**:28-57.
10. Ghanem R, Spanos P: *Stochastic Finite Elements: A Spectral Approach.* New York: Springer; 1991.
11. Xiu D, Karniadakis G: **The Wiener-Askey polynomial chaos for stochastic differential equations.** *SIAM J. Sci. Comput.* 2002, **24**(2):619-644.
12. Schevenels M, Lombaerts G, Degrande G: **Application of the stochastic finite element method for Gaussian and non-Gaussian systems.** In *ISMA2004 International Conference on Noise and Vibration Engineering.* Leuven, Belgien; 2004.
13. Kalman R: **A new approach to linear filtering and prediction problems.** *Trans. ASME, J. Basic Eng.* 1960, **82**:35-45.
14. Welch G, Bishop G: **An introduction to the Kalman filter.** Tr 95-041, UNC-Chapel Hill; 2006.
15. Julier SJ, Uhlmann JK, Durrant-Whyte HF: **A new method for the nonlinear transformation of means and covariances in filters and estimators.** *IEEE Trans. Autom. Control* 2000, **45**(3):477-482.

16. Wiener N: **The homogeneous chaos.** *Am. J. Math.* 1938, **60**:897-936.
17. Cameron R, Martin W: **The orthogonal development of nonlinear functionals in series of Fourier-Hermite functionals.** *Ann. Math.* 1947, **48**(2):385-392.
18. Wan X, Karniadakis G: **An adaptive multi-element generalized polynomial chaos method for stochastic differential equations.** *J. Comput. Phys.* 2005, **209**:617-642.
19. Maître OPL, Njam HN, Ghanem RG, Knio OM: **Multi-resolution analysis of Wiener-type uncertainty propagation schemes.** *J. Comput. Phys.* 2004, **197**(2):502-531.
20. Wan X, Karniadakis G: **Multi-element generalized polynomial chaos for arbitrary probability measures.** *SIAM J. Sci. Comput.* 2006, **26**(3):901-928.
21. Augustin F, Rentrop P: **Stochastic Galerkin techniques for random ordinary differential equations.** *Numer. Math.* 2012, **122**:399-419.
22. Paffrath M, Wever U: **Stochastic integration methods and their application to reliability analysis.** In *ASME Turbo Expo*. Copenhagen, Denmark; 2012.
23. Villegas M: **Random delay differential equations: application to biofilm modeling.** *Phd thesis*. Techn. Universität München (Fachbereich Mathematik, Prof. Rentrop); 2011.
24. Harris DO: **Probabilistic crack growth and modeling.** In *Reliability-Based Mechanical Design*. Edited by Cruse T. New York: Marcel Dekker; 1997.
25. Forman RG: **Fatigue crack growth computer program, NASA/FLAGRO version 2.0.** Johnson Space Center Report JSC-22267A, NASA; 1993.
26. Juan Z, Zhongua Z, Yaohong S: **Hopf bifurcation of a bacteria-immunity system with delayed quorum sensing.** *Appl. Math. Comput.* 2010, **215**:3936-3949.
27. Kuhlmann H: **Kalman-filtering with coloured measurement noise for deformation analysis.** In *Proceedings, 11th FIG Symposium on Deformation Measurements*. Santorini, Greece; 2003.
28. Weber B: **Numerik nichtlinearer Systeme mit Unsicherheiten.** *Master thesis*. Technische Universität München, Prof. Rentrop; 2008.
29. Julier S, Uhlmann J: **Unscented filtering and nonlinear estimation.** *Proc. IEEE* 2004, **92**(3):401-422.
30. Janson S: *Gaussian Hilbert Spaces*. Cambridge: Cambridge University Press; 1997.
31. Ernst OG, Mugler A, Starkloff HJ, Ullmann E: **On the convergence of generalized polynomial chaos expansions.** *ESAIM Math. Model. Numer. Anal.* 2012, **46**:317-339.
32. Guglielmi N, Hairer E: **Users' guide for the code RADAR5 - version 2.1.** *Technical Report - available at http://www.unige.ch/~hairer/software.html* 2005.
33. Hairer E, Wanner G: *Solving Ordinary Differential Equations II. Stiff and Differential-Algebraic Problems*. Berlin: Springer; 1996.
34. Augustin F: *Generalized Wiener Expansions for the Numerical Solution of Random Ordinary Differential Equations*. München: Dr. Hut; 2012.
35. Hairer E, Nørsett SP, Wanner G: *Solving Ordinary Differential Equations I*. Berlin: Springer; 1993.
36. Villegas M: **Biofilms with quorum sensing: A numerical approach for delay differential equations with uncertainties.** *Berichte aus den numerischen Arbeitsgruppen*. Technische Universität München; 2011, 8/2011.

6

3D finite element modeling and simulation of industrial semiconductor devices including impact ionization

Aurelio Mauri[1*], Andrea Bortolossi[2], Giovanni Novielli[3] and Riccardo Sacco[4]

*Correspondence:
aureliogianc@micron.com
[1]Process R & D, Micron Technology, via Olivetti 2, Agrate Brianza, MB 20864, Italy
Full list of author information is available at the end of the article

Abstract

In this article we address the numerical study of 3D semiconductor devices for applications in electronics industry including charge generation phenomena due to impact ionization. With this aim, we propose two novel 3D finite element (FE) models (methods A and B), for electron and hole Drift-Diffusion (DD) current densities. Method A is based on a primal-mixed formulation of the DD model as a function of the quasi-Fermi potential gradient, while method B is a modification of the standard DD formula based on the introduction of an artificial diffusion matrix. Method A is a Galerkin FE approximation of the density current (written in generalized ohmic form) where the harmonic average of the electrical conductivity is used instead of the standard average while method B is a genuine 3D extension of the classic 1D Scharfetter-Gummel difference formula interpreted as a stabilized Galerkin FE approximation with the use of an 'optimal' artificial diffusion. The proposed methods are compared in the 3D simulation of a *p*-*n* junction diode and of a *p*-MOS transistor in the on-state regime. Results show that method A outperforms method B in physical accuracy and numerical stability. Method A is then used in the 3D simulation of a *n*-MOS transistor in the off-state regime including impact ionization. Results demonstrate that the model is able to accurately compute the I-V characteristic of the device until drain-to-bulk junction breakdown.

PACS Codes: 85.30
MSC: 35Jxx; 65N30

Keywords: Drift-Diffusion model; finite element method; numerical simulation; impact ionization; semiconductor device design

1 Introduction and motivation

Semiconductor technology is undergoing a continuously increasing advancement in the aggressive scaling of device length [1]. In this scenario, three-dimensional (3D) device modeling and numerical simulation techniques play a critical role in the prediction of the electrical performance of the system under investigation. In the case of novel memory devices, due to the different undergoing physical phenomena, a self-consistent multiphysics approach is preferred with respect to the simulation of independent phenomena such as chemical reactions, electrical conduction and material properties modification. To respond to this need the software FEMOS (Finite Element Modeling Oriented Simulator) has been designed: FEMOS is a general-purpose modular numerical code based on the use of the Galerkin Finite Element Method (GFEM) implemented in a fully 3D framework through shared libraries using an objected-oriented programming language (C++). The first results of this new approach have been recently presented in [2, 3] and [4]. In the present work we extend the FEMOS computational platform in the study of the *Drift Diffusion model* (DD) [5, 6] and focus on the issue of endowing the simulation tool of a consistent, stable and accurate procedure for the approximation of electron and hole current densities in the device. This is of utmost importance in: (i) visualization and post-processing; (ii) evaluation of conduction currents at device terminals; and (iii) inclusion in the DD model of generation phenomena due to Impact Ionization (II). Here, our attention is devoted to (iii), because of the critical role of II in the convergence and numerical stability of the iterative algorithm used to solve the DD system (see [7], Chapter 3 and [8]), although the methods we propose for the treatment of (iii) can also be profitably employed for (i) and (ii).

To allow a consistent treatment of the generation term due to II within the FE procedure, we propose two novel discrete models for electron and hole current densities over the computational grid. The two methods provide a constant approximation of the current density inside each mesh element and for this reason they can be easily implemented in any simulation environment not necessarily employing the GFEM but utilizing, instead, the Box Integration Method (BIM) that is widely employed in contemporary device simulation tools. We refer to [9, 10] for an introduction to the mathematical structure of the BIM; we refer also to [5], Chapter 6, for a detailed discussion of the implementation of the BIM in the context of a finite box geometrical discretization, and to [11] for a general overview of the BIM in the solution of a general transport model including electro-thermal effects.

The first scheme (method A) is based on the use of a primal-mixed formulation of the DD model written as a function of the quasi-Fermi potential gradient (see [12] and [13]). The second scheme (method B) is a modification of the standard DD formula based on the introduction of an artificial diffusion matrix (see also [14]). Method A can be classified as a Galerkin FE approximation of the density current (written in generalized ohmic form) where the harmonic average of the electrical conductivity is used instead of the standard average (cf. [15, 16]), while method B is a genuine 3D extension of the classic 1D Scharfetter-Gummel (SG) difference formula for the computation of the current density over a 1D element [17], having interpreted this latter formula as the result of a stabilized Galerkin FE approximation with the use of an 'optimal' artificial diffusion (cf. [18]). The proposed finite element models are validated in the numerical study of 3D device structures (*p-n* junction diode and MOS transistors) under on-state and off-state regimes. Results clearly indicate that method A provides the best performance in terms of physical accuracy and numerical stability, and demonstrate the ability of the simulation model in accurately computing the I-V characteristic of the device until the onset of drain-to-bulk junction breakdown.

The outline of the article is as follows. In Section 2 we review the DD model and the algorithms for its finite element discretization. In Section 3 we illustrate the novel methods proposed to calculate the current densities in the device. In Section 4 we carry out their extensive validation by comparing the obtained results with a reference simulation suite in the study of 3D *p-n* and MOS structures including the II generation term. Concluding remarks and perspectives are addressed in Section 5.

2 Physical models, solution map and finite element discretization

In this section we present the mathematical model of the problem, the functional iteration for its decoupled solution and the finite element approximation of the linearized subproblems obtained through decoupling.

2.1 Drift-Diffusion model

The Drift Diffusion model (DD) is described by the Poisson equation coupled with the continuity equations for the mobile carriers, electrons and holes (see [5–7]):

$$\nabla \cdot \mathbf{D} = q(p - n + D), \tag{1a}$$

$$-q\frac{\partial n}{\partial t} + \nabla \cdot \mathbf{J}_n = q(R - G), \tag{1b}$$

$$q\frac{\partial p}{\partial t} + \nabla \cdot \mathbf{J}_p = -q(R - G), \tag{1c}$$

$$\mathbf{D} = \epsilon \mathbf{E}, \tag{1d}$$

$$\mathbf{E} = -\nabla \varphi, \tag{1e}$$

$$\mathbf{J}_n = q\mu_n n\mathbf{E} + qD_n \nabla n, \tag{1f}$$

$$\mathbf{J}_p = q\mu_p p\mathbf{E} - qD_p \nabla p, \tag{1g}$$

where $\mathbf{D}$, $\mathbf{E}$, $\mathbf{J}_n$ and $\mathbf{J}_p$ are the electric displacement, electric field and electron and hole current densities, while ϵ is the dielectric constant, φ the electrostatic potential, p and n the hole and electron concentrations, D the total net doping, q the elementary charge, μ_n and μ_p the electron and hole mobilities, D_n and D_p the electron and hole diffusivities, while R and G are the net recombination and generation rates. Adopted models for μ_n and μ_p account for an appropriate description of the scattering mechanism from lattice and impurities and the degradation due to high electric fields. Adopted models for R account for two-particle (Shockley Read and Hall model) and three-particle net recombination mechanisms (Auger model) while the function G is a pure generation term and accounts for particle rate per unit volume due to Impact Ionization (Van Overstraeten-De Man model) with the projection of electric field along current density direction as a driving force. We refer to [5], Chapter 4 for an exhaustive discussion of recombination/generation mechanisms in semiconductors and associated model parameters used in numerical simulations.

2.2 Geometrical notation and boundary conditions

The DD equation system (1a)-(1g) is solved in a 3D computational domain Ω that represents the geometrical model of the semiconductor device under investigation. In our analysis stationary conditions are always assumed (i.e., $\partial n/\partial t = \partial p/\partial t = 0$), without compromising the validity of the proposed approximation methods.

The device domain Ω is the union of two open disjoint subsets, Ω_{Si} (doped silicon part), and Ω_{ox} (oxide part) which it is assumed to be a perfect insulator. The device boundary $\partial\Omega$ is divided into three disjoint subsets, Γ_D, Γ_N and $\Gamma_{\text{int}} = \partial\Omega_{\text{Si}} \cap \partial\Omega_{\text{ox}}$. The subset Γ_D includes the ideal ohmic contacts of the device, i.e., equipotential surfaces where Dirichlet boundary conditions for the dependent variables φ, n and p are applied. On the subset

Γ_N homogeneous Neumann boundary conditions are applied to prevent any exchange of electric and current flux with the surrounding environment. On the interface subset Γ_{int} continuity of electric potential and of the normal component of the electric displacement is assumed, while tangentially flowing electron and hole current fluxes are enforced. This latter condition reflects the fact that the oxide region is assumed to behave as a perfect insulator and prevents from simulating current injection from the semiconductor into the oxide region.

The complete set of boundary conditions is:

$$\begin{aligned} &\varphi = \varphi_D, & &\mathbf{D}\cdot\mathbf{n} = 0, & &[\![\mathbf{D}\cdot\mathbf{n}]\!] = 0, \\ &n = n_D \quad \text{on } \Gamma_D, & &\mathbf{J}_n\cdot\mathbf{n} = 0 \quad \text{on } \Gamma_N, & &\mathbf{J}_n\cdot\mathbf{n} = 0 \quad \text{on } \Gamma_{\text{int}}, \\ &p = p_D, & &\mathbf{J}_p\cdot\mathbf{n} = 0, & &\mathbf{J}_p\cdot\mathbf{n} = 0, \end{aligned} \tag{2}$$

where φ_D, p_D and n_D are boundary data for the dependent variables (computed as described in [5], Chapter 5), $\mathbf{n}$ is the outward unit normal vector over $\partial\Omega = \Gamma_D \cup \Gamma_N$ and over Γ_{int}, while $[\![\cdot]\!]$ is the jump operator across the interface Γ_{int}.

2.3 The Gummel solution map

The highly nonlinear coupled nature of the DD system makes its analytical treatment very difficult, if not impossible. For this reason, numerical schemes must be used to compute an approximate solution. To this purpose, we introduce the Maxwell-Boltzmann (MB) statistics for electrons and holes:

$$n = n_i \exp\big((\varphi - \varphi_n)/V_{\text{th}}\big), \tag{3a}$$

$$p = n_i \exp\big((\varphi_p - \varphi)/V_{\text{th}}\big), \tag{3b}$$

where n_i is the intrinsic concentration in the semiconductor material, φ_n and φ_p are the quasi-Fermi potentials for electrons and holes while V_{th} is the thermal voltage. The Gummel's decoupled algorithm for the iterative solution of the DD system (2.1) consists of solving first the Poisson equation (1a) with respect to the dependent variable φ, having expressed n and p with the Maxwell-Boltzmann statistics (3a) and (3b). This makes (1a) exponentially nonlinear with respect to the electric potential so that to update this latter dependent variable the Newton method is employed [19]. The continuity equations (1b)-(1c) are then successively solved in linear form with respect to the dependent variables n and p inside the adopted decoupled Gummel map algorithm [20]. The lagging procedure introduced and analyzed in [6] is used to treat the recombination/generation terms in such a way that each decoupled continuity equation enjoys a continuous maximum principle. This ensures the well-posedness of each subproblem and the strict positivity of the carrier densities as physically required. Then, once a step of the Gummel iteration is completed, the MB relations (3a) and (3b) are inverted to update the quasi-Fermi potentials using the computed iterates for φ, n and p, until convergence. Linear convergence of the Gummel algorithm is theoretically proved in [6], although practical computational experience often shows superlinear convergence behavior.

2.4 Finite element approximation

The simulation domain Ω is divided in a discrete partition $\mathcal{T}_h$ made by elements K, each element K being a tetrahedron such that $\overline{\Omega} = \bigcup_{K\in\mathcal{T}_h} \overline{K}$. Then, each differential problem in the Gummel decoupled algorithm of Section 2.3 is written in weak form (see [21], Chapter 5) and numerically approximated using the displacement-based Galerkin Finite Element Method (GFEM) with piecewise linear conforming elements for potential and carrier densities (see [21], Chapter 6). In the particular case of the linearized continuity equations, to avoid numerical instabilities due to possible dominance of the drift term, the variant of the GFEM denoted Edge Averaged Finite Element (EAFE) method proposed and analyzed in [22–25] is used in the numerical approximation. The EAFE scheme enjoys several good properties (numerical stability, robustness and convergence with respect to the mesh discretization parameter). In particular, in the 2D case if $\mathcal{T}_h$ is a Delaunay triangulation, then the stiffness matrix associated with (1c) and (1b) is a M-matrix [24]. As a consequence, the computed solution (electron and hole density) satisfies the Discrete Maximum Principle (DMP) and is strictly positive over the computational domain. Unfortunately this property is no longer valid in a 3D framework because the Delaunay condition on the mesh is not enough to guarantee the M-matrix property, so that a more restrictive condition, sufficient to ensure the DMP, is presented in [22] (cf. Eq. (2.5), Lemma 2.1). In the numerical experiments shown in Section 4 we have used this latter condition to check the geometrical quality of the triangulation $\mathcal{T}_h$ in the critical case where the generation term due to impact ionization phenomena becomes the dominating driving force in the simulation of charge transport in the device.

3 Finite element models for the current density

The construction of a stable and accurate approximate current density field in a primal-based FE formulation is not a trivial task because of possible numerical problems arising from differentiation and cancellation in the DD transport relations (1f)-(1g). Throughout the remainder of this section we assume that the solution of the DD system (1a)-(1g) is carried out using the Gummel decoupled algorithm described in Section 2.3. We first introduce a simple finite element formula for the representation of the current density over each mesh element which is a straightforward implementation of (1f)-(1g). Then, we propose two novel finite element methodologies for current density discretization. The novel approaches have a much simpler implementation than other more sophisticated formulations (such as the dual-mixed hybridized FEM with exponential fitting, see [16, 26]) and are completely compatible for use in the classical Box Integration Method (BIM) [9, 10]. Of the three examined methods, one reveals to be particularly robust and accurate, as demonstrated by several tests discussed in Section 4. In the remainder of the article, for a given element K in $\mathcal{T}_h$ we denote the volume of K by $\mathrm{vol}(K)$; moreover, with the subscript K we refer to a quantity defined in the interior of K while the subscript h refers to a quantity defined at the vertices of K. For a given function $f : K \to \mathbb{R}$ we define $\langle f \rangle_K := \int_K f \, dK / \mathrm{vol}(K)$ the mean integral value of f over K. We also assume that carrier mobility (and the associated diffusivity through Einstein's relation) is constant in K, while the electric potential is linear (so that the associated electric field is a constant vector in K).

3.1 The DD formula

The simplest FE model for the current densities (method DDFE) is obtained by substituting into the transport equations (1f)-(1g) the functions n, p and φ with their corresponding FE approximations n_h, p_h and φ_h, and then by computing the integral average of the resulting expressions over the element K. This yields

$$\mathbf{J}_{n,K} = q\mu_n \langle n \rangle_K \mathbf{E}_K + qD_n \nabla n_{h,K}, \tag{4a}$$

$$\mathbf{J}_{p,K} = q\mu_p \langle p \rangle_K \mathbf{E}_K - qD_p \nabla p_{h,K}. \tag{4b}$$

It is immediate to see that the discrete current densities (4a) and (4b) automatically reproduce the limiting cases of pure diffusive flow ($\mathbf{E}_K = \mathbf{0}$) and pure ohmic flow ($n_{h,K}$ = constant and $p_{h,K}$ = constant). In the case of thermal equilibrium ($\mathbf{J}_{n,K} = \mathbf{J}_{p,K} = \mathbf{0}$), we can anticipate computational difficulties with the use of (4a) and (4b) because of exact cancellation of drift and diffusive current contributions. Thus, by a continuity argument, we also see that method DDFE does not seem appropriate in the numerical treatment of the subthreshold current regime, where currents are not exacty equal to zero but are very small. The two following formulations are designed to overcome this limitation.

3.2 Method A

To describe method A we consider the case of electron continuity equation because a completely similar treatment holds for the hole continuity equation. Using (3a) into (1f) yields the equivalent ohmic form of the electron current density

$$\mathbf{J}_n = \sigma_n \mathbf{E}_n, \tag{5a}$$

where:

$$\sigma_n = q\mu_n n = q\mu_n n_i \exp\big((\varphi - \varphi_n)/V_{\text{th}}\big), \tag{5b}$$

$$\mathbf{E}_n = -\nabla \varphi_n \tag{5c}$$

are the electrical conductivity and the effective electric field experienced by electrons, respectively. We notice that the use of the ohmic representation yields a nonlinear gradient form of the DD current density. To construct the finite element model for $\mathbf{J}_n$ as in (5a), we use the primal-mixed (PM) FEM introduced and analyzed in [12] and recently extended to the case of advective-diffusive operators in [13]. In the PM FEM of lowest order, the approximate current density is constant over each $K \in \mathcal{T}_h$ while the approximate quasi-Fermi potential is piecewise linear and continuous over $\mathcal{T}_h$. Let us introduce the finite element spaces of piecewise constant and piecewise linear continuous functions over $\mathcal{T}_h$:

$$Q_h = \big\{ w \in L^2(\Omega) : w|_K \in \mathbb{P}_0(K)\ \forall K \in \mathcal{T}_h \big\}, \tag{5d}$$

$$V_h = \big\{ u \in C^0(\overline{\Omega}) : u|_K \in \mathbb{P}_1(K)\ \forall K \in \mathcal{T}_h \big\} \tag{5e}$$

and the approximate electrical conductivity $\sigma_{n,h} := q\mu_n n_i \exp((\varphi_h - \varphi_{n,h})/V_{\text{th}})$ where $\varphi_{n,h} \in V_h$ and $\varphi_h \in V_h$ are the finite element discrete analogues of φ_n and φ, respectively. Then, the PM-FE approximation of (5a) reads: find $\mathbf{J}_{n,h} \in [Q_h]^3$ such that

$$\int_\Omega \sigma_{n,h}^{-1} \mathbf{J}_{n,h} \cdot \mathbf{q}_h \, d\Omega + \int_\Omega \nabla \varphi_{n,h} \cdot \mathbf{q}_h \, d\Omega = 0 \quad \forall \mathbf{q}_h \in [Q_h]^3, \tag{5f}$$

where $\mathbf{J}_{n,h} \in [Q_h]^3$ is the finite element discrete analogue of $\mathbf{J}_n$. Since functions in $[Q_h]^3$ are discontinuous, relation (5f) amounts to

$$\int_K \sigma_{n,h}^{-1} \mathbf{J}_{n,h} \cdot \mathbf{q}_h \, dK + \int_K \nabla \varphi_{n,h} \cdot \mathbf{q}_h \, dK = 0 \quad \forall \mathbf{q}_h \in \left[\mathbb{P}_0(K)\right]^3. \tag{5g}$$

Using in (5g) the standard basis functions for $[\mathbb{P}_0(K)]^3$ we obtain

$$[\mathbf{J_{n,h}}]_i = -\mathcal{H}_K(\sigma_{n,h}) \frac{\partial \varphi_{n,h}}{\partial x_i}, \quad i = 1, 2, 3, \forall K \in \mathcal{T}_h, \tag{5h}$$

where $\mathcal{H}_K(\sigma_{n,h}) := (\langle \sigma_{n,h}^{-1} \rangle_K)^{-1}$ is the harmonic average of $\sigma_{n,h}$ over the element K.

We notice that the approximation of the current density with a constant vector may appear inadequate because of the exponential dependence of the electron conductivity on the quasi-Fermi potential (cf. (5b)), except in the case of a constant electron concentration. It can be proved (see [12]) that $\|\mathbf{J}_n - \mathbf{J}_{n,h}\|_{L^2(\Omega)} = \mathcal{O}(h)$, where h denotes the mesh parameter, so that an appropriate reduction of the grid size may be required to achieve a desired fixed tolerance.

To numerically compute in a simple and accurate manner the harmonic average of the electrical conductivity, we use the following quadrature rule (see also [27, 28])

$$\left(\frac{\int_K \sigma_{n,h}^{-1} \, dK}{\mathrm{vol}(K)} \right)^{-1} \simeq \left(\frac{\int_{e*} \sigma_{n,h}^{-1} \, de}{|e^*|} \right)^{-1} \tag{5i}$$

e^* being the edge of ∂K where the maximum drop of $\sigma_{n,h}$ occurs and $|e^*|$ its Euclidean length. Upon introducing the linear dimensionless potential $\Phi_n := (\varphi_h - \varphi_{n,h})/V_{\mathrm{th}}$ and the two vertices of K: $\mathbf{x}_m$ s.t. $\Phi_n(\mathbf{x}_m) = \Phi_{n,m} := \min_K(\Phi_n)$ and $\mathbf{x}_M$ s.t. $\Phi_n(\mathbf{x}_M) = \Phi_{n,M} := \max_K(\Phi_n)$, we define e^* to be the edge of ∂K which connects $\mathbf{x}_m$ and $\mathbf{x}_M$, so that (5i) yields

$$\int_K \sigma_{n,h}^{-1} \, dK \simeq q\mu_n n_i \exp(\Phi_{n,m}) \mathcal{B}(\Phi_{n,m} - \Phi_{n,M}), \tag{5j}$$

where $\mathcal{B}(Z) := Z/(\exp(Z) - 1)$ is the inverse of the Bernoulli function, such that $\mathcal{B}(0) = 1$. Using MB statistics (3a), relation (5j) can be written as

$$\int_K \sigma_{n,h}^{-1} \, dK \simeq q\mu_n n_m \mathcal{B}(\Phi_{n,m} - \Phi_{n,M}), \tag{5k}$$

where $n_m := n_i e^{\Phi_{n,m}}$. Expression (5j) can be also written as

$$\int_K \sigma_{n,h}^{-1} \, dK \simeq q\mu_n n_i \exp(\Phi_{n,m}) \exp(\Phi_{n,M}) \exp(-\Phi_{n,M}) \mathcal{B}(\Phi_{n,m} - \Phi_{n,M})$$

which, using again (3a) and the property $\exp(Z)\mathcal{B}(Z) = \mathcal{B}(-Z)$, becomes

$$\int_K \sigma_{n,h}^{-1} \, dK \simeq q\mu_n n_M \mathcal{B}(\Phi_{n,M} - \Phi_{n,m}), \tag{5l}$$

where $n_M := n_i e^{\Phi_{n,M}}$. Combining (5k) and (5l), we find

$$\mathbf{J}_{n,K} = -q\mu_n \left[\frac{n_m \mathcal{B}(-\Delta\Phi_{n,\max}) + n_M \mathcal{B}(\Delta\Phi_{n,\max})}{2} \right] \nabla \varphi_{n,h}, \tag{5m}$$

where $\Delta\Phi_{n,\max} := \Phi_{n,M} - \Phi_{n,m}$. In the case where $\Delta\Phi_{n,\max}$ is sufficiently large, we see that the 1D harmonic average of n_h along the edge e^* has the effect of 'favoring' the smallest nodal concentration in the element K rather than the largest one, so that we approximately have

$$\mathbf{J}_{n,K} \simeq -q\mu_n n_m \max_{\partial K} \frac{(\varphi_h - \varphi_{n,h})}{2V_{\text{th}}} \nabla\varphi_{n,h}. \tag{5n}$$

If, instead, $\Delta\Phi_{n,\max}$ is small then the 1D harmonic average behaves appreciably as the standard average and we obtain

$$\mathbf{J}_{n,K} \simeq -q\mu_n \frac{n_m + n_M}{2} \nabla\varphi_{n,h}. \tag{5o}$$

Following the same procedure as above also for the hole current density we have

$$\mathbf{J}_{p,K} = -q\mu_p \left[\frac{p_m \mathcal{B}(-\Delta\Phi_{p,\max}) + p_M \mathcal{B}(\Delta\Phi_{p,\max})}{2} \right] \nabla\varphi_{p,h}, \tag{5p}$$

where $\Phi_p := (\varphi_{p,h} - \varphi_h)/V_{\text{th}}$ and $\Delta\Phi_{p,\max} := \max_K \Phi_p - \min_K \Phi_p$.

3.3 Method B

In the previous section the discrete form of the current density is constructed by starting from the equivalent 'ohmic' representation in terms of the quasi Fermi potential, and then by performing a suitable approximation of the electrical conductivity over the finite element K. Here, we continue along the same line of thought, but starting from the classic DD format (1f)-(1g), with the intent of using the method of Streamline Upwind artificial diffusion proposed in [18] for the advective-diffusive model to stabilize the computation in the presence of a high electric field.

3.3.1 The 1D SG method as an artificial diffusion scheme

In the 1D setting the artificial diffusion technique consists of replacing the electron diffusion coefficient D_n with the modified quantity

$$D_{n,h} = D_n + D_n \Phi(\mathbb{P}e|_K), \tag{6a}$$

where Φ is a suitable nonnegative stabilization function of the 1D local Pèclet number $\mathbb{P}e|_K = (h|\partial_x \varphi_h|)/(2V_{\text{th}}) = |\Delta\varphi|/(2V_{\text{th}})$, h and $\Delta\varphi$ being the length of the 1D interval and the potential drop over the interval, respectively. The local Pèclet number gives a measure of how much the drift term dominates over the diffusion term in the transport mechanism. If $\mathbb{P}e|_K > 1$ the problem is locally drift (advection)-dominated and in such a case we need introduce an extra amount of diffusion in (6a) (given by $D_n\Phi(\mathbb{P}e|_K)$) to prevent the occurrence of unphysical spurious oscillations in the computed solution, which might even lead to a negative electron concentration. If $\mathbb{P}e|_K < 1$ the problem is locally diffusion-dominated and there is no need of adding an extra diffusion, so that the standard GFEM is enough for obtaining an accurate and numerically stable solution. Based on the last observation, the function Φ has to satisfy the property of consistency

$$\lim_{\mathbb{P}e|_K \to 0} \Phi(\mathbb{P}e|_K) = 0 \quad \forall K \in \mathcal{T}_h. \tag{6b}$$

The 1D approximation of the electron current density to be used in a stabilized GFEM is thus given by the following relation

$$J_{n,h}(n_h)|_K = q\mu_n \langle n\rangle_K E_{h,K} + qD_n\big(1 + \Phi(\mathbb{P}e|_K)\big)\partial_x n_h \quad \forall K \in \mathcal{T}_h. \tag{6c}$$

To design in a physically sound and consistent manner the optimal stabilization function Φ, we assume the modified method to exactly satisfy some limiting cases that often occur in practical important electronic applications. Using a 3D notation, for sake of generality and because this will be used in later extension, the considered cases are:

C1. Constant carrier concentrations (only drift contribution): $\mathbf{J}_n = q\mu_n n\mathbf{E}$.

C2. Constant potential ($\mathbf{E} = 0$, only diffusive contribution): $\mathbf{J}_n = qD_n\nabla n$.

C3. Constant quasi Fermi potential (no current flow): $\mathbf{J}_n = -q\mu_n n\nabla\varphi_n = \mathbf{0}$.

Notice that case C3 implies that

$$n = Ce^{\varphi/V_{\rm th}}, \tag{6d}$$

where C is an arbitrary constant such that $C = \exp(-\overline{\varphi}_n/V_{\rm th})$, $\overline{\varphi}_n$ being a given constant value. Thanks to assumption (6b) the stabilized current (6c) automatically satisfies cases C1 and C2. Case C3 is recovered by imposing (in the 1D setting)

$$J_{n,h}\big(\Pi_1^K(Ce^{\varphi/V_{\rm th}})\big) = 0, \tag{6e}$$

where Π_1^K is the $\mathbb{P}_1$-interpolant over the element K. Using (6e) in (6c), noting that $E_K = -\partial_x\varphi_h$ and using Einstein's relation, we obtain the following relation for the stabilization function

$$\Phi(\mathbb{P}e|_K) = \frac{\langle n\rangle_K}{V_{\rm th}}\frac{\partial_x\varphi_h}{\partial_x\Pi_1^K(e^{\varphi/V_{\rm th}})} - 1. \tag{6f}$$

Enforcing relation (6d) at the two vertices x_i of the interval, $i = 1, 2$, yields

$$\Phi(\mathbb{P}e|_K) = \sigma\,\mathbb{P}e|_K\frac{e^{\varphi_1/V_{\rm th}} + e^{\varphi_2/V_{\rm th}}}{e^{\varphi_2/V_{\rm th}} - e^{\varphi_1/V_{\rm th}}} - 1 = \sigma\,\mathbb{P}e|_K\frac{e^{2\sigma\mathbb{P}e|_K} + 1}{e^{2\sigma\mathbb{P}e|_K} - 1} - 1, \tag{6g}$$

where n_1 and n_2 are the two nodal values of n_h while $\sigma := \mathrm{sign}(\Delta\varphi)$. Setting for brevity $X := 2\sigma\mathbb{P}e|_K$, relation (6g) becomes

$$\begin{aligned}\Phi(X) &= \frac{X}{2}\left(\frac{e^X}{e^X - 1} + \frac{1}{e^X - 1}\right) - 1 = \frac{1}{2}\big(\mathcal{B}(-X) + \mathcal{B}(X)\big) - 1\\ &= \frac{1}{2}\big(X + \mathcal{B}(X) + \mathcal{B}(X)\big) - 1 = \mathcal{B}(X) + \frac{X}{2} - 1,\end{aligned}$$

and replacing the definition of X we obtain for both $\Delta\varphi > 0$ and $\Delta\varphi < 0$

$$\Phi(\mathbb{P}e|_K) = \mathcal{B}(2\mathbb{P}e|_K) + \mathbb{P}e|_K - 1 \tag{6h}$$

which, upon substitution into (6c), recovers the well known 1D Scharfetter-Gummel scheme [17]. This latter method was originally proposed as an 'exponential fitting' difference scheme by Allen and Southwell in [29] and subsequently referred to as an artificial diffusion stabilized scheme with 'optimal viscosity' by Brooks and Hughes in [18].

3.3.2 The 3D SG Artificial Diffusion method

In a 3D framework a straightforward extension of the 1D Scharfetter-Gummel stabilization (6g)-(6h) can be obtained by introducing a 3×3 diagonal stabilizing tensor $\underline{\underline{\mathbf{\Phi}}}^K$ defined on each element $K \in \mathcal{T}_h$ as follows

$$\underline{\underline{\mathbf{\Phi}}}^K_{ii} = \frac{\langle \Pi_1^K (e^{(\varphi_h - \varphi_M)/V_{\text{th}}}) \rangle_K \partial_{x_i} \varphi}{\partial_{x_i} \Pi_1^K (e^{(\varphi_h - \varphi_M)/V_{\text{th}}}) V_{\text{th}}} - 1, \quad i = 1,2,3, \tag{6i}$$

where we have used, as a reference value for the potential reference to avoid overflow exceptions in the machine evaluation of (6i), the maximum φ_M of φ_h in K.

The 3D approximate electron current density is then

$$\mathbf{J}_{n,K} = q\mu_n \langle n \rangle_K \mathbf{E}_K + qD_n(\underline{\underline{\mathcal{I}}} + \underline{\underline{\mathbf{\Phi}}}_K)\nabla n_h, \tag{6j}$$

where $\underline{\underline{\mathcal{I}}}$ is the 3×3 identity tensor. In a completely similar manner we have the relation for the hole current density

$$\mathbf{J}_{p,K} = q\mu_p \langle p \rangle_K \mathbf{E}_K - qD_p(\underline{\underline{\mathcal{I}}} + \underline{\underline{\mathbf{\Phi}}}_K)\nabla p_h, \tag{6k}$$

where

$$\underline{\underline{\mathbf{\Phi}}}^K_{ii} = \frac{\langle \Pi_1^K (e^{(\varphi_m - \varphi_h)/V_{\text{th}}}) \rangle_K \partial_{x_i} \varphi}{\partial_{x_i} \Pi_1^K (e^{(\varphi_m - \varphi_h)/V_{\text{th}}}) V_{\text{th}}} - 1, \quad i = 1,2,3, \tag{6l}$$

φ_m being the minimum of φ over K. It is immediate to check that the two proposed novel approximations (6j) and (6k) satisfy all cases C1, C2 and C3 in Section 3.3.1.

4 Numerical experiments

This section is devoted to the illustration of several numerical experiments. Specifically, Section 4.1 is aimed to check the correctness of the implementation of the 3D-EAFE method within the FEMOS code and to compare the performance of this finite element approach with a reference-commercial tool based on the BIM [30]. Section 4.2 is aimed to compare the performance of the various methods for current calculation of Section 3 against physical expectation. The conclusion of these tests is that method A turns out to be the most stable and accurate, so that Section 4.3 employs such method in a self-consistent simulation of semiconductor devices in the presence of II generation terms. 3D ideal and realistic devices commonly employed in semiconductor industry are taken into consideration: (1) a *p-n* ideal diode; (2) a *p* and *n*-MOS ideal devices; and (3) a *p*-MOS device resulting from a realistic 2D-process simulation accounting for non ideal doping profiles. Device (3) represents a severe benchmark to highlight the accuracy and the stability of the different current density calculation methods.

4.1 Accuracy of the 3D-EAFE discretization

Several tests have been performed in order to assess the correctness of the EAFE approach in the simulation of semiconductor devices in 3D framework. For reading benefits we just report a benchmark with a reference commercial tool based on *n-p*-channel MOSFET structure of Figure 1(a) (*GateLength* : 100 nm; *GateOx* : 30 nm and contacts S = *Source*,

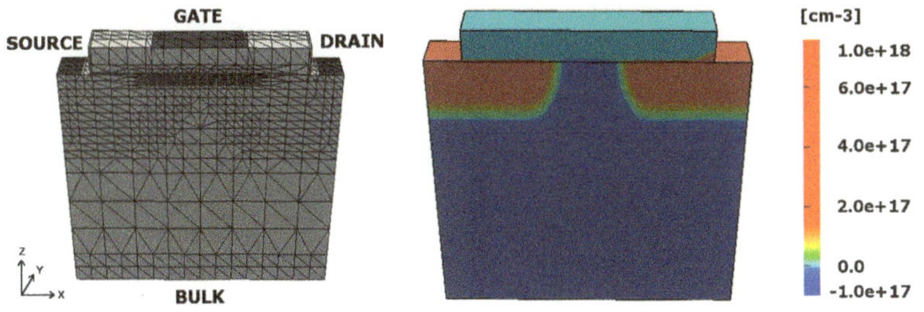

(a) Finite element mesh (b) Doping profile

Figure 1 MOS structure: Left: 3D tetrahedral mesh. Right: Doping profile in the case of the *n*-channel device.

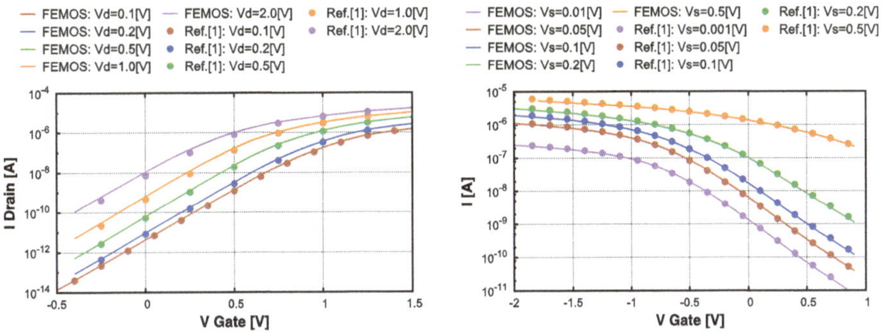

(a) *n*-MOSFET (b) *p*-MOSFET

Figure 2 MOSFET forward bias: comparison between FEMOS (line) and Ref. [30] (symbol). Left: *n*-channel Id-Vg. Right: *p*-channel Id-Vg.

Gate, D = Drain, Body) with analytical doping profiles: Figure 1(b) reports the doping profile for the *n*-channel device (the *p*-channel device is obtained just by the inverting the doping type). The agreement with the reference simulation tool is very good as reported in Figure 2(a) for the *n*-channel device and in Figure 2(b) for the *p*-channel device for several *Drain* bias conditions (electron and hole mobilities have been calculated using the Canali model [31]). We point out that the calculation of the current at the contacts is carried out by extending to the 3D case the approach proposed in [32] for the 2D case: this novel formulation, known as *residual method,* is based on the approach proposed in [33]. We also notice that the 3D-profiles of carriers are indistinguishable between FEMOS and Ref. [30] as reported in Figure 3 in the case of electrons in the *n*-channel device at $V_{\text{Gate}} = 2$ V and $V_{\text{Drain}} = 0.1$ V.

4.2 Evaluation of the performance of current calculation methods

To evaluate the simulation performance of the methods for current calculation introduced in Section 3 we have firstly considered a diode structure with a *p*-*n* junction geometrically represented by the cubic region $\Omega = (0, 0.3)^3$ μm^3 (see Figure 4(a)). A Gaussian implantation of donors with a peak of 10^{19} cm^{-3} and a depth of 0.15 μm (see Figure 4(b)) is made over a *p*-type region with a constant acceptor profile of 10^{18} cm^{-3} magnitude. One contact is defined for each of the doped regions: for the *n*-type part a rounded-shape contact is used (Top), while the *p*-type part is contacted at the bottom face (Body). As mentioned in Section 2.2, contacts are considered of pure Ohmic-type with appropriate Dirichlet

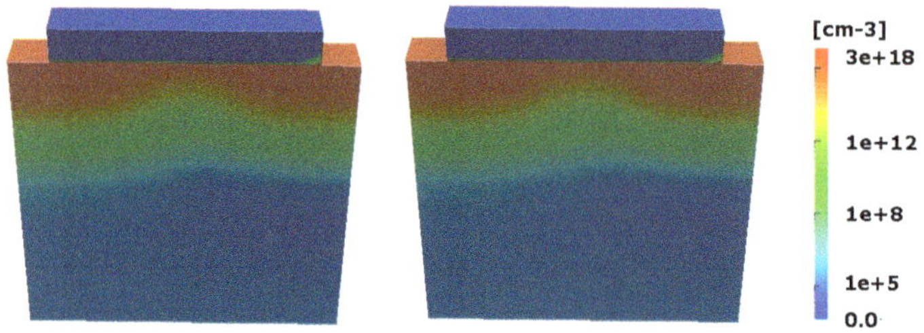

(a) FEMOS (b) Ref. [30]

Figure 3 ***n*-MOSFET electron concentration at $V_{\text{Gate}} = 2$ V and $V_{\text{Drain}} = 0.1$ V.** Left: FEMOS calculation. Right: Ref. [30] calculation.

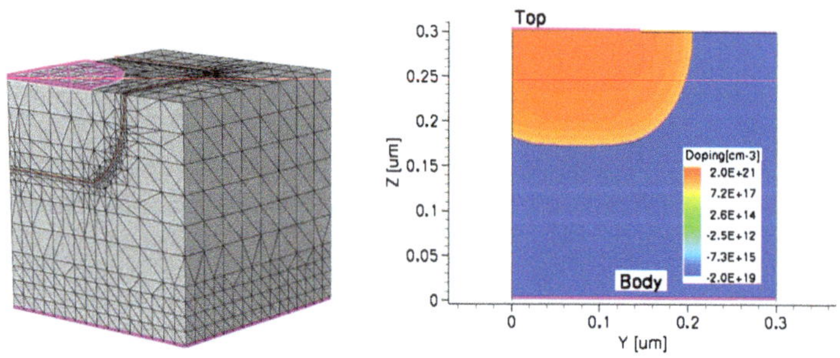

(a) Mesh (b) 2D cut of Mesh for doping profile visualization

Figure 4 Diode structure: Left: mesh. Right: 2D cut of the doping profile.

boundary conditions: the Top is maintained at ground while the Body is ramped at 0.8 V. In Figure 5 the results of the calculation of the current density vector field for electrons and holes in the semiconductor bulk are represented through streamlines connecting the Body with the Top contact. As expected the current calculation obtained with method DDFE (eq. (4a) and (4b)) is affected by a critical behavior, in particular close to and inside the *n*-junction as shown in Figure 5(a) and Figure 5(b) where instability has to be ascribed to numerical cancellation of the drift and diffusion contributions. Results get definitely better by employing method B using Eqs. (6j) and (6k) where the improvement can be appreciated in Figure 5(e) and Figure 5(f). However, a careful inspection of the hole current density reveals that some numerical instability is still evident inside the *n*-junctions. The extension of the 1D SG scheme to 3D provided by method A in Eq. (5h) and (5p) results in the streamlines presented in Figure 5(c) and Figure 5(d): no spurious instability can be observed anymore and our calculations are in excellent agreement with the results of a commercial code (not shown here). The comparison between the different current computational methods has also been carried out on a *p*-channel MOSFET. The doping profiles have been obtained by using a 2-D process simulator with implantation and diffusion steps [30] with the purpose to have a realistic doping as reported in Figure 6(a) with a $GateLength: 180$ nm and a $GateOx: 4.5$ nm. The presence of floating non-compensated *p*-type regions in the channel body increases the computational difficulties. The 2D doping profile has been then extruded in three spatial dimensions, and the generated mesh is shown in Figure 6(b) where the device contact has been highlighted with purple color (the body contact is not shown in the picture but is located in the bottom face of the sili-

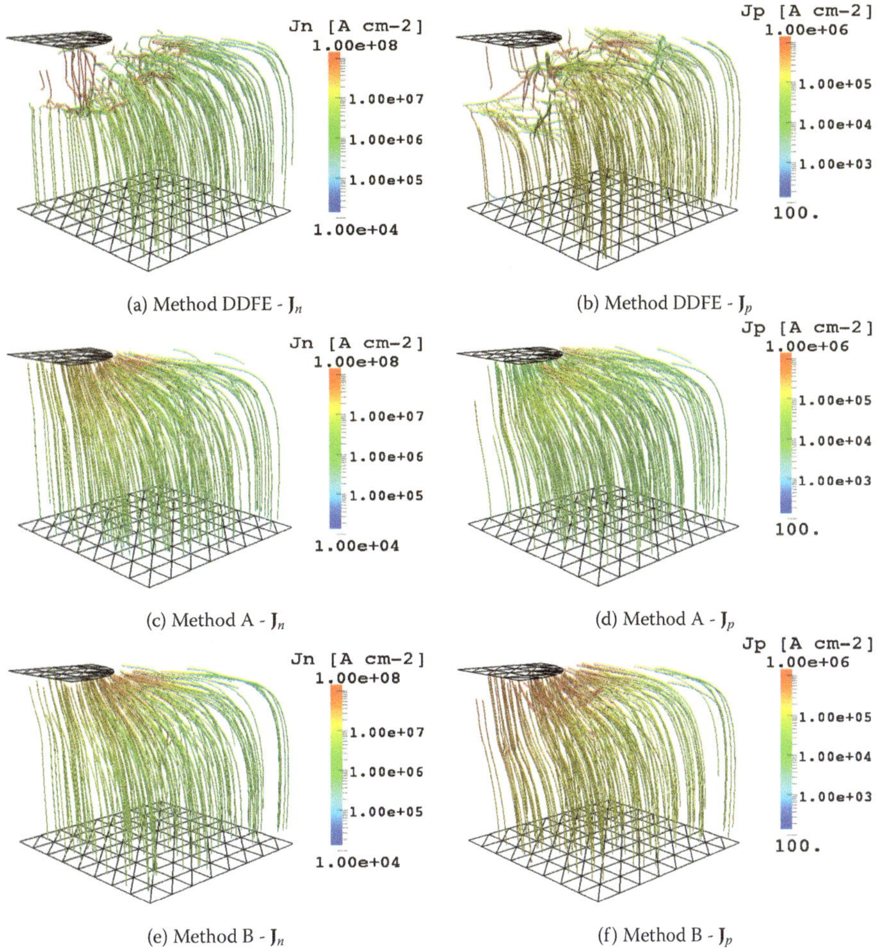

(a) Method DDFE - $\mathbf{J}_n$ (b) Method DDFE - $\mathbf{J}_p$

(c) Method A - $\mathbf{J}_n$ (d) Method A - $\mathbf{J}_p$

(e) Method B - $\mathbf{J}_n$ (f) Method B - $\mathbf{J}_p$

Figure 5 Electron (left column) and hole (right column) current densities of Diode test case at $V_{\mathbf{body}}$ = 0.8 V. Top: Method DDFE. Middle: Method A. Bottom: Method B.

con region). Regarding the calculation of $\mathbf{J}_p$, the numerical difficulties found with method DDFE are still confirmed as clearly depicted in Figure 7(a) not only inside the p-type region but also around the floating regions present in the body (the visualization are referred to the bias conditions $V_G = -1$ V and $V_D = -0.1$ V). The marginality found using formula (6k) is increased as reported in Figure 7(c): this comes by the fact that the evaluation of the coefficient in (6i) is again undergoing numerical problems related to roundoff error. However method B is giving a much better current density evaluation with respect to the pure application of the Drift-Diffusion approach at the element level. We conclude this section by noting that, again, the best description of the expected physical behavior of the device is obtained by adopting method A, that turns out to provide an accurate and stable 3D extension of the 1D SG formula, as clearly demonstrated by Figure 7(b).

4.3 Application of method A to the study of a reverse-biased MOS with II

In this section we implement method A for current calculation inside the Gummel map algorithm to evaluate the II generation term by projecting the electric field along the obtained current streamlines. Since we are using the Van Overstraeten-de Man model for II we have run the simulation not beyond the limits of validity of this model, i.e. when the

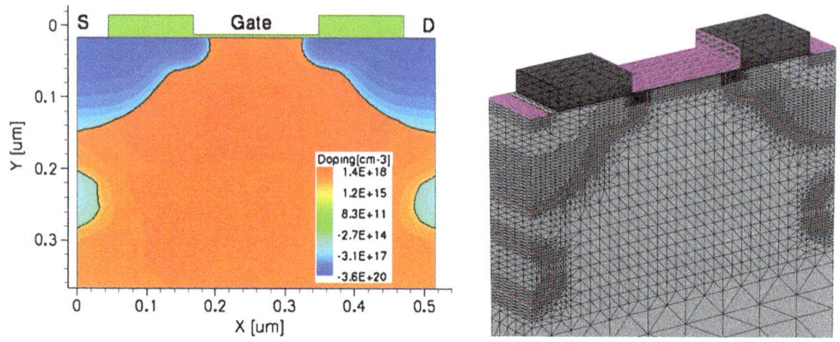

(a) Doping (b) Mesh

Figure 6 ***p*-MOSFET structure:** Left: 2D cut of the doping profile. Right: Mesh.

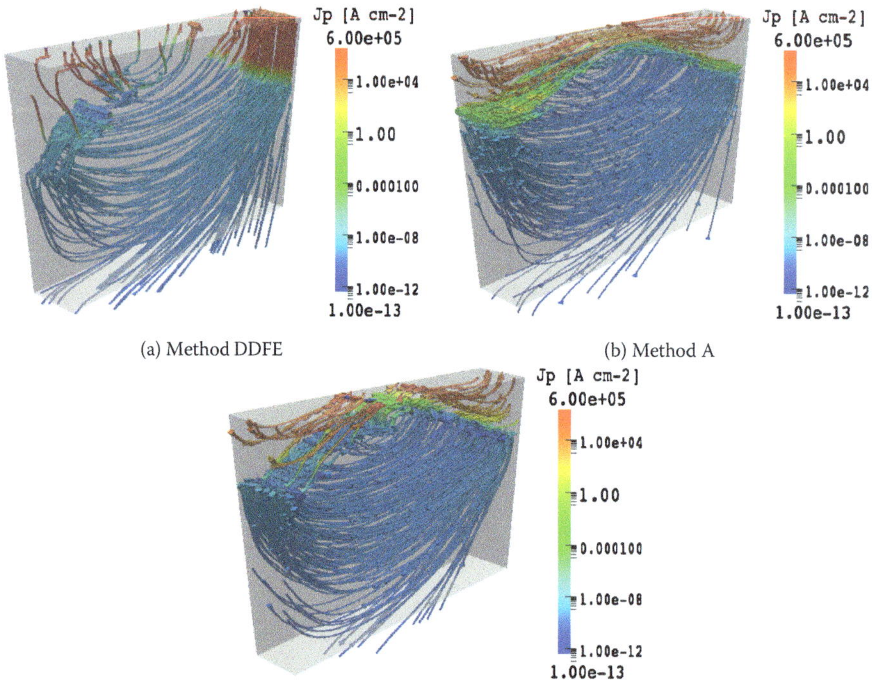

(a) Method DDFE (b) Method A

(c) Method B

Figure 7 **Hole current density calculation of *p*-MOSFET at $V_G = -1.0$ V and $V_D = -0.1$ V:** Top left: Method DDFE. Top right: Method A. Bottom: Method B.

avalanche field is in the range of $6 \cdot 10^5$ Vcm^{-1}. Computations are performed by keeping all the contacts at ground and applying a reverse bias to the *Drain* contact. Figure 8(a) and Figure 8(b) compare the FEMOS simulations with those obtained using [30] for both *n* and *p* channel device under condition of reverse bias. The current calculated at *Drain* and *Source* contact are quite aligned with the reference tool while the *Bulk* component is lower due to the different generation value of II terms (not shown here). Figure 9 and Figure 10 visualize the increase of the II generation term with the increasing of the reverse bias in the case of the *n*-MOS and *p*-MOS device. The drain-to-bulk breakdown is starting in the *Drain-Gate* overlap region and proceed until it reaches the drain contact where the current streamlines are more dense as shown in Figure 11(b). A final insight into

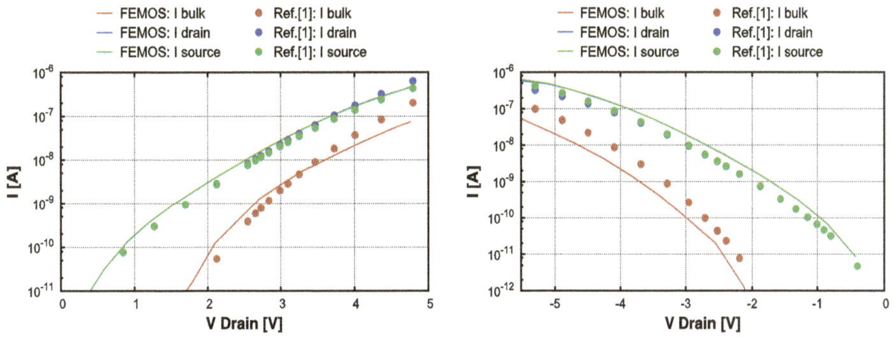

(a) n-MOSFET (b) p-MOSFET

Figure 8 MOSFET in reverse bias (all contacts are grounded while Drain is reverse biased): comparison between FEMOS (line) and Ref. [30] (symbol). Left: n-channel device. Right: p-channel device.

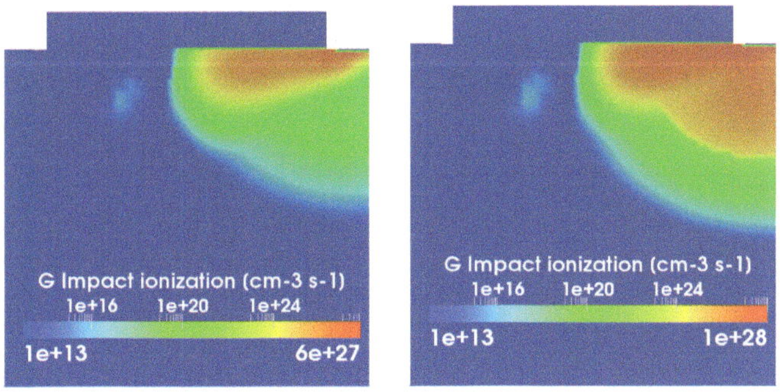

(a) II, $V_D = 3.4$ V (b) II, $V_D = 4.8$ V

Figure 9 n-MOSFET: evolution of the II generation term (all contacts are grounded except drain). Left: $V_D = 3.4$ V. Right: $V_D = 4.8$ V.

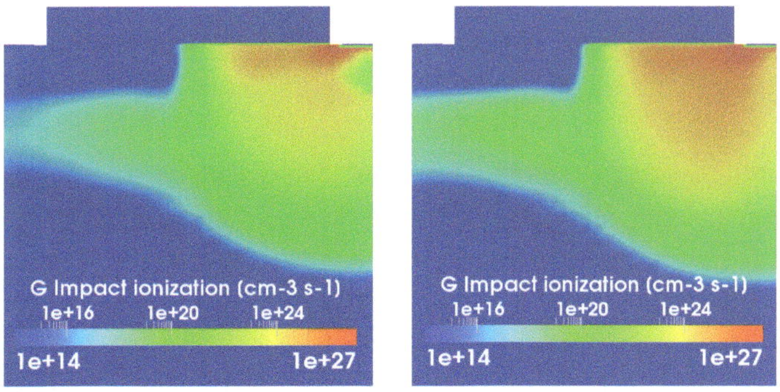

(a) II, $V_D = -3.4$ V (b) II, $V_D = -4.8$ V

Figure 10 p-MOSFET: evolution of the II generation term (all contacts are grounded except drain). Left: $V_D = -3.4$ V. Right: $V_D = -4.8$ V.

the simulation of the n-MOS device using method A is offered by Figure 11 which reports the visualization of $\mathbf{J}_n$ in the off and on states: in accordance to physical expectation, in the on-state the electron current density is confined under the gate from drain-to-source while in the off-state it is flowing from source-to-drain and from source-to-body.

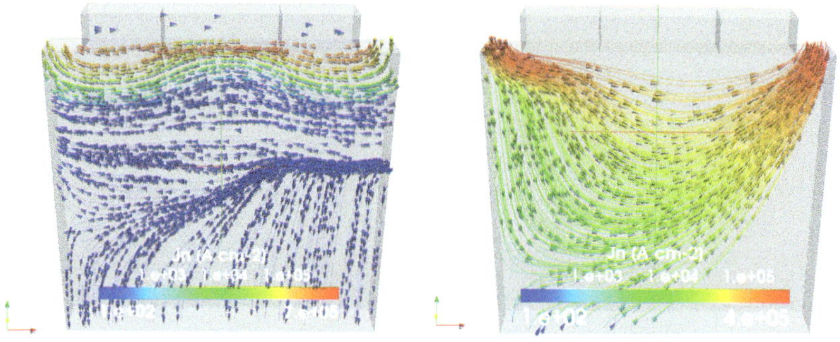

(a) $\mathbf{J}_n$ on-state (b) $\mathbf{J}_n$ off-state

Figure 11 ***n*-MOSFET: electron current density streamlines obtained with method A.** Left: On-state at $V_G = 1$ V and $V_D = 0.1$ V. Right: Off-state at $V_G = 0$ V and $V_D = 0.85$ V.

5 Conclusions

In this article we have addressed the problem of representing in 3D the carrier current density by extending the beneficial properties emanating from the classic 1D Scharfetter-Gummel difference formula. To this purpose, we have adopted the Galerkin Finite Element Method for the numerical simulation of the Drift-Diffusion model in the 3D case, and we have proposed two novel methods for current density evaluation, denoted method A and B, to which, for comparison, we have added also the basic method DDFE using the DD formula. The three schemes compute a piecewise constant approximation of the current density over a tetrahedral partition of the device domain.

Method DDFE turns out to provide the worst results in the test experiments. Method B is a 3D extension of the method of optimal artificial diffusion and gives reasonably accurate results. Method A is based on a primal-mixed formulation endowed with a suitable quadrature formula for the approximate evaluation of the harmonic average of the electrical conductivity. It is by far the best of the three considered approaches, providing simulation results in excellent agreement with a commercial software. The three FE formulations are numerically verified in the study of realistic 3D device structures, also including the presence of Impact Ionization phenomena. Even in this latter case, method A is able to correctly describe the complex carrier flow patterns inside the device bulk and to track the I-V curve of the device up to the avalanche breakdown region.

Despite the fact that the proposed formulations are illustrated and validated in the study of the classic DD transport model in semiconductors, they can be applied in a straightforward manner to the numerical treatment of general conservation laws for advective-diffusive fluxes where the advective term is in gradient form, as is the case for ion electrodiffusion in electrochemistry and biology with the Poisson-Nernst-Planck model [34] and hydrodynamic and quantum-corrected charge transport in semiconductors [35–37].

Competing interests

The authors declare that they have no competing interests.

Authors' contributions

All authors contributed equally to the writing of this paper. All authors read and approved the final manuscript.

Author details

[1]Process R & D, Micron Technology, via Olivetti 2, Agrate Brianza, MB 20864, Italy. [2]IBM Italia, Via Circonvallazione Idroscalo, Segrate, MI 20090, Italy. [3]STMicroelectronics, via Olivetti 2, Agrate Brianza, MB 20864, Italy. [4]Dipartimento di Matematica, Politecnico di Milano, Piazza L. da Vinci 32, Milano, 20133, Italy.

Acknowledgements
Aurelio Mauri and Riccardo Sacco gratefully thank Drs. Claudio Lombardi and Chandra Mouli of Micron Technology, for supporting Andrea Bortolossi and Giovanni Novielli during their MSc thesis in the development of the program FEMOS.

References

1. Cogez P, Graef M, Huizing B, Mahnkopf R, Ishiuchi H, Ikumi N, Choi S, Choi JH, Diaz CH, See YC, Gargini P, Kingscott T, Steff I. The International Technology Roadmap for Semiconductors: 2013. Technical report. Jointly sponsored by ESIA, JEITA, KSIA, TSIA and SIA. 2013. http://www.itrs.net/Links/2013ITRS/Summary2013.htm.
2. Novielli G, Ghetti A, Varesi E, Mauri A, Sacco R. Atomic migration in phase change materials. In: Electron devices meeting (IEDM), 2013 IEEE international. 2013. p. 22.3.1-22.3.4. doi:10.1109/IEDM.2013.6724683.
3. Mauri A, Sacco R, Verri M. Electro-thermo-chemical computational models for 3D heterogeneous semiconductor device simulation. Appl Math Model. 2014 in press, doi:10.1016/j.apm.2014.12.008.
4. Benvenuti A, Ghetti A, Mauri A, Liu H, Mouli C. Current and future prospects of non-volatile memory modeling. In: Simulation and Semiconductor Process and Devices (SISPAD), 2014 IEEE international. 2014. p. 249-52.
5. Selberherr S. Analysis and simulation of semiconductor devices. Berlin: Springer; 1984.
6. Jerome JW. Analysis of charge transport. Berlin: Springer; 1996.
7. Markowich PA. The stationary semiconductor device equations. Computational microelectronics. Berlin: Springer; 1986. ISBN 9783211818923.
8. Micheletti S, Quarteroni A, Sacco R. Current-voltage characteristics simulation of semiconductor devices using domain decomposition. J Comput Phys. 1995;119(1):46-61. doi:10.1006/jcph.1995.1115.
9. Bank RE, Rose D, Fichtner W. Numerical methods for semiconductor device simulation. SIAM J Sci Stat Comput. 1983;4(3):416-35. doi:10.1137/0904032.
10. Bank RE, Rose D, Fichtner W. Numerical methods for semiconductor device simulation. IEEE Trans Electron Devices. 1983;30(9):1031-41. doi:10.1109/T-ED.1983.21257.
11. Ciampolini P, Pierantoni A, Liuzzo A, Baccarani G. 3-D simulations of silicon devices: physical models and numerical algorithms. In: Baccarani G, editor. Process and device modeling for microelectronics. Amsterdam: Elsevier; 1993. p. 53-107.
12. Roberts J, Thomas JM. Mixed and hybrid methods. Handb Numer Anal. 1991;2:523-633.
13. Sacco R, Carichino L, de Falco C, Verri M, Agostini F, Gradinger T. A multiscale thermo-fluid computational model for a two-phase cooling system. Comput Methods Appl Mech Eng. 2014;282(1):239-68. doi:10.1016/j.cma.2014.08.003.
14. Bank RE, Burgler JF, Fichtner W, Smith RK. Some upwinding techniques for finite element approximations of convection-diffusion equations. Numer Math. 1990;58:185-202.
15. Babuška I, Osborn JE. Generalized finite element methods: their performance and their relation to mixed methods. SIAM J Numer Anal. 1983;20:510-36.
16. Brezzi F, Marini L, Pietra P. Two-dimensional exponential fitting and applications to Drift-Diffusion models. SIAM J Numer Anal. 1989;26(6):1342-55. doi:10.1137/0726078.
17. Gummel HK, Scharfetter D. Large-signal analysis of a silicon Read diode oscillator. IEEE Trans Electron Devices. 1969;16:64-77.
18. Brooks AN, Hughes TJR. Streamline Upwind/Petrov-Galerkin formulations for convection dominated flows with particular emphasis on the incompressible Navier-Stokes equations. Comput Methods Appl Mech Eng. 1982;32:199-259. doi:10.1016/0045-7825(82)90071-8.
19. Ortega JM, Rheinboldt WC. Iterative solution of nonlinear equations in several variables. New York: Academic Press; 1970.
20. Gummel HK. A self-consistent iterative scheme for one-dimensional steady state transistor calculations. IEEE Trans Electron Devices. 1964;11:455-65.
21. Quarteroni A, Valli A. Numerical approximation of partial differential equations. Berlin: Springer; 2008.
22. Xu J, Zikatanov L. A monotone finite element scheme for convection-diffusion equations. Math Comput. 1999;68:1429-46.
23. Zikatanov LT, Lazarov RD. An exponential fitting scheme for general convection-diffusion equations on tetrahedral meshes. Comput Appl Math. 2012;1:60-9.
24. Bank RE, Coughran W, Cowsar LC. The Finite Volume Scharfetter-Gummel method for steady convection diffusion equations. Comput Vis Sci. 1998;1:123-36.
25. Gatti E, Micheletti S, Sacco R. A new Galerkin framework for the Drift-Diffusion equation in semiconductors. East-West J Numer Math. 1998;6:101-36.
26. Degond P, Juengel A, Pietra P. Numerical discretization of energy-transport models for semiconductors with nonparabolic band structure. SIAM J Sci Comput. 2000;22(3):986-1007. doi:10.1137/S1064827599360972.
27. Brezzi F, Marini L, Pietra P. Numerical simulation of semiconductor devices. Comput Methods Appl Mech Eng. 1989;75:493-514.
28. Brezzi F, Marini LD, Markowich PA, Pietra P. On some numerical problems in semiconductor device simulation. In: Toscani G, Boffi V, Rionero S, editors. Mathematical aspects of fluid and plasma dynamics. Lecture notes in mathematics. vol. 1460. 1991. p. 31-42.
29. Allen DNG, Southwell RV. Relaxation methods applied to determine the motion, in two dimensions, of a viscous fluid past a fixed cylinder. Q J Mech Appl Math. 1955;8:129-45.
30. Sentaurus device user guide. Synopsis Inc. 2013.
31. Canali C. Electron and hole drift velocity measurements in silicon and their empirical relation to electric field and temperature. IEEE Trans Electron Devices. 1975;22:1045-7.
32. Gusmeroli R, Spinelli AS. Accurate boundary integrals calculation in semiconductor device simulation. IEEE Trans Electron Devices. 2006;53:1730-3.
33. Hughes TJR, Engel G, Mazzei L, Larson MG. The continuous Galerkin method is locally conservative. J Comput Phys. 2000;163:467-88.

34. Rubinstein I. Electrodiffusion of ions. Philadelphia: SIAM; 1990.
35. Forghieri A, Guerrieri R, Ciampolini P, Gnudi A, Rudan M, Baccarani G. A new discretization strategy of the semiconductor equations comprising momentum and energy balance. IEEE Trans Comput-Aided Des Integr Circuits Syst. 1988;7(2):231-42. doi:10.1109/43.3153.
36. de Falco C, Lacaita AL, Gatti E, Sacco R. Quantum-corrected Drift-Diffusion models for transport in semiconductor devices. J Comput Phys. 2005;204:533-61.
37. de Falco C, Jerome JW, Sacco R. Quantum-corrected drift-diffusion models: solution fixed point map and finite element approximation. J Comput Phys. 2009;228(5):1770-89. doi:10.1016/j.jcp.2008.11.010.

Modelling, asymptotic analysis and simulation of the gas dynamics in a chimney

Elisabetta Felaco and Ingenuin Gasser*

*Correspondence:
gasser@math.uni-hamburg.de
Department Mathematik,
Universität Hamburg, Bundesstraße
55, Hamburg, 20146, Germany

Abstract

This paper presents a one dimensional model for the gas dynamics in a chimney. This is a prototype example of a small Mach number flow with strong heat sources. Due to the small Mach number of the gas flow an asymptotic model is derived from a fully compressible model, which then is compared to the original model and to the often used Boussinesq approximation. The Boussinesq approximation is shown to be inappropriate for this application, whereas the small Mach number asymptotic model we propose is shown to be a very good approximation. In particular it allows very fast numerical simulations. Finally, all this is underlined by numerical simulations where we validate the various models.

MSC: 35C20; 35Q35; 76M45; 76N15; 80A20

Keywords: chimney; gas dynamics; Low Mach number; Boussinesq approximation

1 Introduction

Chimneys are typical representatives of the beginning of the industrial era in the 18th century. However there is a much longer story behind. Starting from the 13th century chimneys were build and used to channel away smoke and hot air. Nowadays chimneys are common presences in our everyday life, from industrial applications to the heating systems in our houses. Just to give some examples, we can think to modern heating systems, whit different kinds of fuel, and also to Controlled Ventilation Systems, where the basic principle of chimney is used, often with the help of a fan, to keep an healthy level of oxygen in modern buildings [1].

Here by chimney we mean a vertical structure for venting into the outside atmosphere hot flue gases produced *e.g.* in a combustion process like in a boiler, in a stove, in a furnace or in a fireplace. The underlying main physical principles are well understood. Therefore it is considered only a technical issue to give precise and quantitative predictions of the gas dynamics in a chimney. On one hand there is huge technological know how (for example see [2] and [3]) with empirical formulas for very specific situations based on restrictive (often unrealistic) assumptions. On the other hand one can find sophisticated CFD gas-dynamics simulations based on time expensive numerical simulations. There seems to be a peculiar lack of models in between these two extreme areas, *i.e.* models which are physically reasonably realistic and numerically cheap at the same time. This paper goes exactly in this direction.

When describing the gas dynamics in a chimney a few crucial facts can be identified.

- The flow is driven by density gradients (buoyancy force).

- The flow velocities are slow compared to the speed of sound.
- The temperature differences are big and decisive for the flow behaviour.

Small flow velocities indicate an incompressible model. In combination with the buoyancy as driving force a model with variable density is necessary. In such a context a widely used approach is the Boussinesq approximation going back originally to Boussinesq [4] with later justifications by Bois, Zeytounian *etc.* (see *e.g.* [5–7] and citations therein). However we will show that here a Boussinesq type approach is not appropriate. As an alternative we use a variable density low Mach number model (see [8–12] for the non isentropic or the non isothermal cases) adapted to this approximation which results to be realistic and numerically cheap at the same time.

The aim of the paper is to study the flow dynamics both qualitatively and quantitatively. A common measure for the quality of the full process in a chimney is the draft in the chimney. And the draft in the chimney depends *e.g.* on the heat source, on the chimney geometry, on the chimney friction and heat loss parameters, on the outside meteorological situation *etc.*

In *Section* 2 we describe the underlying equations. In *Section* 3 we will scale the equation and identify various asymptotic regions. In *Section* 4 we discuss the non validity of a Boussinesq type model and propose a variable density low Mach number model. Finally in *Section* 5 we will provide several numerical examples, first using simulations on a full system of equations as a reference to underline the validity of our approach, and then analysing in more details the dependence on the parameters.

2 The model

We describe the motion of smoke and flue gases produced by combustion trough a chimney. An ideal example that we can have in mind is that of a home fireplace (Figure 1), but the model applies to a great variety of other cases (modern heating and ventilation systems, industrial chimneys *etc.*).

The fresh cold air enters at the base of the structure and, after flowing through a small section of the chimney, it reaches the fireplace. There a combustion process takes place and the air gains some heat and mass (in the combustion process the air mixture typically looses oxygen and gains carbon dioxide and water vapor, the total variation resulting in an net increase of the air mass). As a result the density decreases and the velocity increases. The hot air smoke mixture enters the main part of the chimney, where the so called chimney effect acts as the main driving force. Clearly, in that part there are also other effects like friction and heat losses. Finally the hot air smoke mixture exits at the top of the structure.

We assume that the main dynamics acts in vertical direction, as we do not expect significant flow phenomena in other directions. Then almost all gas particles experience the

Figure 1 Basic setting of the chimney and the fireplace.

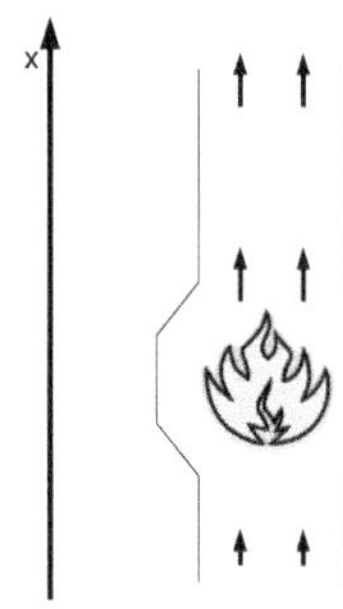

same when passing through the structure. Therefore we describe the gas dynamics by a one dimensional model, where the single dimension x starts at the base of the structure and follows the air flow towards the top of the chimney. All quantities under consideration have to be interpreted as mean values over the cross sectional area. However, this approach still allows to consider variable cross sections *e.g.* a larger cross section in the fire place area.

Keeping in mind the above considerations and simplifications we start with a one dimensional fully compressible model. Let $\tilde{\rho}$, $\tilde{u}$, $\tilde{p}$ and $\tilde{T}$ be respectively the density, the velocity, the pressure and the temperature of the air flow, all averaged on the cross section $\tilde{A}$ and functions of space $\tilde{x}$ and time $\tilde{t}$. The space variable $\tilde{x}$ is zero at the bottom of the structure and its positive direction points toward the chimney top. We consider an unscaled version of the Euler equations of gas dynamics with a balance equation for mass

$$(\tilde{A}\tilde{\rho})_{\tilde{t}} + (\tilde{A}\tilde{\rho}\tilde{u})_{\tilde{x}} = \tilde{A}\tilde{m}, \tag{1}$$

a balance equation for momentum

$$(\tilde{A}\tilde{\rho}\tilde{u})_{\tilde{t}} + \left(\tilde{A}\left(\tilde{\rho}\tilde{u}^2 + \tilde{p}\right)\right)_{\tilde{x}} = \tilde{A}_{\tilde{x}}\tilde{p} - \tilde{A}g\tilde{\rho} - \frac{\lambda}{2D}\tilde{A}\tilde{\rho}\tilde{u}|\tilde{u}| + \tilde{\Delta}p_{fan}\frac{\tilde{A}}{L}, \tag{2}$$

and a balance equation for energy

$$\begin{aligned}&\left(\tilde{A}\left(c_V\frac{\tilde{p}}{R} + \frac{\tilde{\rho}\tilde{u}^2}{2} + g\tilde{\rho}\tilde{h}\right)\right)_{\tilde{t}} + \left(\tilde{A}\tilde{u}\left(c_V\frac{\tilde{p}}{R} + \frac{\tilde{\rho}\tilde{u}^2}{2} + g\tilde{\rho}\tilde{h}\right) + \tilde{A}\tilde{p}\tilde{u}\right)_{\tilde{x}}\\ &= \tilde{A}\tilde{q} - \frac{k}{2}(\tilde{T} - \tilde{T}_w) - \tilde{A}\sigma(\tilde{T} - \tilde{T}_w)^4.\end{aligned} \tag{3}$$

This system is completed by the ideal gas law

$$\tilde{p} = R\tilde{\rho}\tilde{T}.$$

In (1) the term on the right hand side $\tilde{m}$ is due to the additionally produced gas in the combustion process as a consequence of the oxygen consumption. We will propose a model for this mass variation in the next section.

The term $-\tilde{A}g\tilde{\rho}$ on the right hand side of equation (2) describes the gravitational force which is volume dependent and so linearly dependent on the cross section. The friction on the surface of the chimney walls is described by $-\frac{\lambda}{2D}\tilde{A}\tilde{\rho}\tilde{u}|\tilde{u}|$ where λ is the friction factor and

$$D = 2\sqrt{\tilde{A}/\pi}$$

is the so called hydraulic diameter. Realistic values for λ can be found in literature, see, for example [13].

In case of an additional fan (for example in low temperature chimneys) a pressure gain term $\tilde{\Delta}p_{fan}/L$ is introduced, where L is the reference value for the length of the chimney.

In the energy equation (3) $h = h(x)$ is the height of the structure and c_v is the specific heat of dry air at constant volume. On the right hand side of (3) we have the heat source term due to the combustion process $\tilde{A}\tilde{q}$, the heat loss trough the chimney walls $-\frac{k}{2}(\tilde{T} - \tilde{T}_w)$ and the energy loss due to thermal radiation $-\tilde{A}\sigma(\tilde{T} - \tilde{T}_w)^4$. Here $\tilde{T}_w$ is the wall temperature, k depends on the cross section and on the heat loss coefficient α according the relation

$$k = \alpha 2\sqrt{\tilde{A}\pi}.$$

A typical description for the heat loss coefficient is

$$\alpha = St(c_v + R)\tilde{\rho}\tilde{u},$$

where $St = \lambda/8$ is the Stanton number [14]. Using average density and velocity values we can compute α. Note that in *Section* 5 we will use slightly smaller α values to take into account the presence of insulating materials typically used in chimney walls [2, 3, 15, 16].

Finally $\sigma = 5.670400 \times 10^{-8}$ J s^{-1} m^{-2} K^{-4} is the thermal radiation coefficient (Stefan's constant).

2.1 Modelling the mass loss and the heat source

We assume both terms $\tilde{m}$ and $\tilde{q}$ depending on the air flow $\tilde{\rho}\tilde{u}$. Assuming that in the fireplace there is always enough fuel, the reaction rate depends on the amount of oxygen available for the combustion process. The amount of oxygen (being a fixed percentage in the air) is linearly dependent on the mass flow. Thus we set

$$\tilde{q} = \tilde{c}_q\tilde{\rho}|\tilde{u}|$$

and

$$\tilde{m} = \tilde{c}_m\tilde{\rho}|\tilde{u}|,$$

where $\tilde{c}_m$ and $\tilde{c}_q$ are a mass variation and a heat production coefficient. To determine reasonable values to these coefficients we recall that burning wood produces $\sim 10^6$ J/kg of thermal energy ([17, 18] and [19]), while chimneys used for heating in common handholds have usually a power of 10^2 kW [15, 16]. Considering this and taking into account the reference values for density and velocity, we will use a heat production coefficient $\tilde{c}_q \sim 10^5$–10^6 kg m/s^2.

As for the mass variation in the air, there are different possible approaches [20]. We can consider the burning reaction and compare the wood mass loss with the mass gain in the air, and then use the results from [18] and [19]. Otherwise we can use the experimental results on the oxygen variation in air after combustion. We will use $\tilde{c}_m \sim 10^{-2}$ 1/m.

3 Scaling

Now we scale the problem. Every quantity $\tilde{z}$ is associated by $\tilde{z} = z_r z$ to a dimensionless quantity z, where we use the sub-index r to indicate the reference values. Similar, every function $\tilde{f}$ is related by $f_r f(x,t) = \tilde{f}(\tilde{x},\tilde{t})$ to a dimensionless function f. Realistic reference values are listed in *Table* 1.

In this way we obtain the scaled model

$$\rho_t + (\rho u)_x = -\frac{A_x}{A}\rho u + c_m\rho|u|, \quad (4)$$

Table 1 Scaling table with typical reference values

Quantity	Reference value	Typical reference value
x	$x_r = L$	10 m
t	t_r	20 s
A	A_r	0.07 m^2
h	L	10 m
u	$u_r = L/t_r$	0.5 m/s
p	p_r	101,328 kg/(m s^2)
T	$T_r = T_W$	263 K (= −10°C)
ρ	$\rho_r = p_r/(RT_r)$	1.3 kg/m^3
q	$q_r = \tilde{c}_q \rho_r u_r$	5×10^5 W/m^3
m	$m_r = \tilde{c}_m \rho_r u_r$	5×10^{-2} kg/(s m^3)

Table 2 Constants

Constant	Value
g	9.81 m/s^2
R	287 m^2/(s^2 K)
c_V	718 m^2/(s^2 K)
γ	1.4
σ	5.6704×10^{-8} J s^{-1} m^{-2} K^{-4}

$$\rho u_t + \rho u u_x + \frac{1}{\gamma M^2} p_x = -c_m \rho |u| u - \frac{1}{Fr^2}\rho - \frac{\xi}{\sqrt{A}} \rho u |u| + \Delta p_{fan}, \tag{5}$$

$$\left(p + \gamma M^2(\gamma - 1)\frac{\rho u^2}{2} + \frac{\gamma M^2(\gamma - 1)}{Fr^2}\rho h\right)_t + \frac{1}{A}\left[Au\left(p + \gamma M^2(\gamma - 1)\frac{\rho u^2}{2} + \frac{\gamma M^2(\gamma - 1)}{Fr^2}\rho h\right) + (\gamma - 1)Aup\right]_x = c_q \rho |u| - a\frac{(T - T_w)}{\sqrt{A}} - rT^4, \tag{6}$$

while the state equation becomes

$$p = \rho T. \tag{7}$$

(Here $\rho_r := p_r/RT_r$, see *Table* 1.)

We introduced the dimensionless constants

$$\gamma = \frac{c_V + R}{c_V}, \qquad \frac{1}{\gamma M^2} = \frac{p_r}{\rho_r u_r^2}, \qquad \frac{1}{Fr^2} = \frac{L}{u_r^2} g,$$

where M is the Mach-number and Fr is the Froude-number. Moreover we define

$$\xi = \frac{\lambda \sqrt{\pi} L}{4\sqrt{A_r}}, \qquad \Delta p_{fan} = \frac{1}{\rho_r u_r^2} \tilde{\Delta} p_{fan}, \quad r = \frac{t_r T_r^4 \sigma}{p_r}(\gamma - 1)$$

and

$$c_m = \tilde{c}_m L, \qquad c_q = \frac{\tilde{c}_q(\gamma - 1)\rho_r L}{p_r}, \qquad a = \frac{\alpha T_r \sqrt{\pi} t_r}{\sqrt{A_r} p_r}(\gamma - 1).$$

In *Table* 2 we remember the values of the constants used in the model.

We note that in our setting $M \sim 10^{-3}$ and therefore $\frac{1}{M^2} \sim 10^6$. In addition we have $Fr \sim 10^{-1}$ and therefore $\frac{1}{Fr^2} \sim 10^2$. These facts suggest to perform an asymptotic analysis with the intention to further simplify the model.

3.1 Initial and boundary conditions

Equations (4)-(7) need to be endowed with the appropriate initial and boundary conditions. We assign initial data for the velocity u, the density ρ and the pressure p:

$$u(x,0) = u_0(x), \qquad \rho(x,0) = \rho_0(x), \qquad p(x,0) = p_0(x) \tag{8}$$

and boundary condition for p:

$$p(0,t) = p_l, \qquad p(1,t) = p_r, \tag{9}$$

where p_l and p_r are, respectively, pressure at the bottom and pressure at the top of the structure.

Finally, as the density ρ satisfies a transport equation, we prescribe inflow conditions:

$$\begin{aligned} \rho(0,t) &= \rho_l \quad \text{if } u(0,t) > 0, \\ \rho(1,t) &= \rho_r \quad \text{if } u(1,t) < 0. \end{aligned} \tag{10}$$

We still need to assign reasonable values to p_l, p_r, ρ_l and ρ_r. Thus we study the gasdynamics outside the structure. Assuming a stationary state without sources and chimney effects we have from the equations (4)-(6)

$$\begin{aligned} &(\rho_h u)_x = 0, \\ &\rho_h u u_x + \frac{1}{\varepsilon}(p_h)_x = -\frac{1}{Fr^2}\rho_h, \\ &\left[u\left(p_h + \varepsilon(\gamma-1)\frac{\rho_h u^2}{2} + \frac{\varepsilon(\gamma-1)}{Fr^2}\rho_h h\right) + (\gamma-1)u p_h\right]_x = 0, \end{aligned} \tag{11}$$

where ρ_h and p_h are the outside density and pressure. If in addition we assume no flow (*i.e.* $u = 0$) we obtain the scaled adiabatic formulas for density, pressure and temperature [13, 21]

$$\rho_h = \rho_{h0}\left(1 - (\gamma-1)\frac{M^2}{Fr^2}h(x)\right)^{\frac{1}{\gamma-1}}, \tag{12}$$

$$p_h = p_{h0}\left(1 - (\gamma-1)\frac{M^2}{Fr^2}h(x)\right)^{\frac{\gamma}{\gamma-1}}, \tag{13}$$

$$T_h = T_{h0}\left(1 - (\gamma-1)\frac{M^2}{Fr^2}h(x)\right). \tag{14}$$

Remember that in a vertical setting the height is $h(x) = x$, $x = 0$ at the bottom, x axis oriented toward the top of the chimney. In our scaling the values at the bottom are given by $\rho_{h0} = 1$ and $p_{h0} = 1$.

Now we can complete (9) and (10) imposing:

$$p_l = p_h|_{x=0} = 1, \qquad p_r = p_h|_{x=1} = \left(1 - (\gamma - 1)\frac{M^2}{Fr^2}\right)^{\frac{\gamma}{\gamma-1}}, \tag{15}$$

$$\rho_l = \rho_h|_{x=0} = 1, \qquad \rho_r = \rho_h|_{x=1} = \left(1 - (\gamma - 1)\frac{M^2}{Fr^2}\right)^{\frac{1}{\gamma-1}}. \tag{16}$$

Thus the problem to solve consists of the equations (4)-(7), the initial conditions (8), the boundary condition (9) and (10) specified in (15) and (16).

4 Asymptotics

In the previous section we encountered two small parameters, the Mach number $M \sim 10^{-3}$ and the Froude number $Fr \sim 10^{-1}$. In our application they seem to be of a different order of magnitude. Therefore we do not consider the so called quasistatic case

$$M \approx Fr \ll 1.$$

However, the cases

$$Fr^2 \approx M \ll 1, \tag{17}$$

and

$$Fr \text{ fixed}, \quad M \ll 1 \tag{18}$$

should not be ruled out. The case (17) leads to the so called Boussinesq approximation, the case (18) is known as the small Mach number limit. Both approximations are widely used, see [4–7] and citations therein for the Boussinesq approximation. For the small Mach number approximation in general (non isentropic and non isothermal) see [8–12, 22], for numerical issues see [23–26] and references therein. For a variety of one dimensional applications similar to the one in this paper see [27–30]. In the literature (for example in [31]) the Boussinesq number is defined as $B = M^2/Fr^2$. Both cases (17) and (18) refer to the limit $B \to 0$, whereas the quasistatic case would correspond to a limit with $B \to B_0 \neq 0$.

In the following for sake of simplicity we neglect the terms involving the fan and the radiation energy. Also, we deal with a vertical chimney such that $h(x) = x$ holds. We start with the classical Boussinesq approximation.

4.1 Boussinesq approximation

In this section we assume $M \approx Fr^2$ and set $c = \frac{M}{Fr^2} = O(1)$. The following derivation mimics the asymptotic derivation of the Boussinesq approximation presented by [7] extended to the case with mass losses, heat sources *etc.* (see also Remark 2 at the end of this section). Note that from the state equation we get $T = \frac{p}{\rho}$ so that we consider the equations (4)-(6) for the unknowns ρ, p and u. For $z = \rho, p, u$, we start with the following asymptotic expansion in the small Mach number M

$$z(x,t) = z^{(0)}(x,t) + z^{(1)}(x,t)M + z^{(2)}(x,t)M^2 + O(M^3),$$

to obtain in (4)-(6) a hierarchy of equations.

From (4) in $O(M^0)$ and $O(M^1)$ we obtain

$$\rho_t^{(0)} + \left(\rho^{(0)} u^{(0)}\right)_x = -\frac{A_x}{A}\rho^0 u^0 + c_m \rho^{(0)} \left|u^{(0)}\right|, \tag{19}$$

$$\rho_t^{(1)} + \left(\rho^{(1)} u^{(0)} + \rho^{(0)} u^{(1)}\right)_x = \left(c_m - \frac{A_x}{A}\right)\left(\rho^{(1)}\left|u^{(0)}\right| + \rho^{(0)}\left|u^{(1)}\right|\right). \tag{20}$$

From (5) in $O(M^{-2})$, $O(M^{-1})$, $O(M^0)$ we get

$$p_x^{(0)} = 0, \tag{21}$$

$$p_x^{(1)} = -\gamma c \rho^{(0)}, \tag{22}$$

$$\rho^{(0)} u_t^{(0)} + \rho^{(0)} u^{(0)} u_x^{(0)} + \frac{1}{\gamma} p_x^{(2)} = -c_m \rho^{(0)}\left|u^{(0)}\right| u^{(0)} - c\rho^{(1)} - \frac{\xi}{\sqrt{A}}\rho^{(0)} u^{(0)}\left|u^{(0)}\right|. \tag{23}$$

From (6) in $O(M^0)$ and $O(M^1)$ we obtain

$$\left(p^{(0)}\right)_t + \left(\gamma u^{(0)} p^{(0)} + c_m \rho^{(0)} u^{(0)} T^{(0)}\right)_x = -\frac{A_x}{A}\gamma \rho^{(0)} p^{(0)} c_q \rho^{(0)}\left|u^{(0)}\right| - a\frac{\left(\frac{p^{(0)}}{\rho^{(0)}} - T_w\right)}{\sqrt{A}}, \tag{24}$$

$$\left(p^{(1)} + c(\gamma - 1)\gamma \rho^{(0)} x\right)_t + -\frac{1}{A}\left(A\gamma\left(u^{(0)} p^{(1)} + u^{(1)} p^{(0)}\right)\right)_x$$

$$= c_q\left(\rho^{(0)}\left|u^{(1)}\right| + \rho^{(1)}\left|u^{(0)}\right|\right) - a\frac{\frac{p^{(1)}}{\rho^{(0)}} - \frac{\rho^{(1)}}{(\rho^{(0)})^2}}{\sqrt{A}}. \tag{25}$$

Expanding the boundary conditions (15) we find

$$p(0,t) = 1,$$

$$p(1,t) = \left(1 - (\gamma - 1)cM\right)^{\frac{\gamma}{\gamma - 1}} = 1 - \gamma c M + \gamma c^2 M^2,$$

and we deduce

$$p^{(0)}(0,t) = 1, \qquad p^{(0)}(1,t) = 1,$$

$$p^{(1)}(0,t) = 0, \qquad p^{(1)}(1,t) = -c\gamma,$$

$$p^{(2)}(0,t) = 0, \qquad p^{(2)}(1,t) = +\gamma c^2.$$

In a similar way, the inflow condition for $\rho^{(0)}$ and $\rho^{(1)}$ are given by

$$\rho^{(0)}(0,t) = 1 \quad \text{if } u(0,t) > 0, \qquad \rho^{(0)}(1,t) = 1 \quad \text{if } u(1,t) < 0,$$

$$\rho^{(1)}(0,t) = 0 \quad \text{if } u(0,t) > 0, \qquad \rho^{(1)}(1,t) = -c \quad \text{if } u(1,t) < 0.$$

The main assumption in the classical Boussinesq approximation is a constant leading order density $\rho^{(0)} = 1$, where the value is fixed to 1 due to the (scaled) boundary data. From (21), (22) and the boundary and inflow conditions we conclude

$$p^{(0)}(x,t) \equiv 1, \qquad p^{(1)} = p^{(1)}\left(\rho^{(0)}\right) = -\gamma c \rho^{(0)} x.$$

This leads in (19) and (24) to the contradiction

$$u_x^{(0)} = c_m \rho^{(0)} \left| u^{(0)} \right|,$$

$$u_x^{(0)} = c_q \rho^{(0)} \left| u^{(0)} \right| - a \frac{(\frac{p^{(0)}}{\rho^{(0)}} - T_w)}{\sqrt{A}}.$$

This means that a mass loss or a heat source of order one is not compatible with the Boussinesq assumption of a constant leading order density. This is reasonable since order one mass loss or heat source terms are supposed to induce significant order one changes in the density (and therefore in temperature). We conclude that the Boussinesq approximation is NOT applicable in our case of significant mass losses or heat source. Therefore in *Section* 4.2 we pass to the small Mach number asypmtotics.

Remark 1 Only for small mass losses and heat sources, *i.e.* in the case

$$c_m = O(M), \qquad c_q = O(M), \qquad a = O(M)$$

the Boussinesq approximation would be applicable and we obtain the well known 'incompressibility' condition $u_x^{(0)} = 0$ and a $u^{(0)} = u^{(0)}(t)$. Then (20) and (25) become the two equations for the perturbations $u^{(1)}$ and $\rho^{(1)}$. Finally $\rho^{(1)}$ enters in the buoyancy term on the right hand side of (23).

For simplicity let's assume the very simple case $c_m = c_q = a = 0$ corresponding to the outside case with no sources. Then equation (25) reduces to

$$u_x^{(1)} = c u^{(0)}$$

which gives in (20)

$$\rho_t^{(1)} + u^{(0)} \left(\rho^{(1)} \right)_x = -c u^{(0)}.$$

Assuming positive velocities $u^{(0)}$ and the boundary condition $\rho^{(1)}(0, t) = 0$ gives the long time behaviour

$$\rho^{(1)}(x, t) = -cx.$$

Thus

$$\rho(x, t) = 1 - Mcx + O\left(M^2\right) = 1 - \frac{M^2}{Fr^2} x + O\left(M^2\right)$$

which are exactly the first two terms in the expansion of the adiabatic formula (12) for the density. In *Section* 5.2 we come back to these asymptotics when doing numerical comparisons.

Remark 2 We said that this derivation of the Boussinesq approximation is in the line of [7] which is different from the more classical asymptotic derivations going back to [5, 6]. In fact in [5, 6] a multiple scale analysis is used to obtain the final equations. However, the non applicability of the Boussinesq approximation in our application is independent of the chosen approach. For details on the differences between the approaches see [7].

4.2 Small Mach number approximation

Here we keep the Froude number Fr fixed and consider only the Mach number to be small. This limit is well known (see citations at the beginning of this section). Note again that from the state equation we get $T = p/\rho$ so that we consider only the equations (4)-(6) for the unknowns ρ, p and u.

Since the equations (4)-(6) only contain M^2 (and not M itself) for $z = \rho, p, u$, we make an Ansatz of an asymptotic expansion in the small parameter M^2

$$z(x,t) = z^{(0)}(x,t) + z^{(2)}(x,t)M^2 + O(M^4),$$

which leads again to a hierarchy of equations. From (4) in $O(M^0)$ we have

$$\rho_t^{(0)} + \left(\rho^{(0)}u^{(0)}\right)_x = -\frac{A_x}{A}\rho^{(0)}u^{(0)} + c_m\rho^{(0)}\left|u^{(0)}\right|. \tag{26}$$

From (5) in $O(M^{-2})$ and $O(M^0)$ we obtain

$$p_x^{(0)} = 0, \tag{27}$$

$$\rho^{(0)}u_t^{(0)} + \rho^{(0)}u^{(0)}u_x^{(0)} + \frac{1}{\gamma}p_x^{(2)} = -c_m\rho^{(0)}\left|u^{(0)}\right|u^{(0)} - \frac{1}{Fr^2}\rho^{(0)} - \frac{\xi}{\sqrt{A}}\rho^{(0)}u^{(0)}\left|u^{(0)}\right|. \tag{28}$$

Finally (6) gives in $O(M^0)$

$$(p_0)_t + \frac{1}{A}\left(\gamma A u^{(0)}p^{(0)}\right)_x = c_q\rho^{(0)}\left|u^{(0)}\right| - a\frac{\left(\frac{p^{(0)}}{\rho^{(0)}} - T_w\right)}{\sqrt{A}},$$

where we used $T^{(0)} = p^{(0)}/\rho^{(0)}$.

The boundary conditions for p can be expanded

$$p(0,t) = 1,$$

$$p(1,t) = \left(1 - (\gamma - 1)\frac{M^2}{Fr^2}\right)^{\frac{\gamma}{\gamma-1}} \approx 1 - \gamma\frac{M^2}{Fr^2},$$

from which we obtain

$$p^{(0)}(0,t) = 1, \qquad p^{(0)}(1,t) = 1, \tag{29}$$

$$p^{(2)}(0,t) = 0, \qquad p^{(2)}(1,t) = -\frac{\gamma}{Fr^2}. \tag{30}$$

In a similar way we get the inflow conditions for $\rho^{(0)}$:

$$\rho^{(0)}(0,t) = 1 \quad \text{if } u(0,t) > 0, \qquad \rho^{(0)}(1,t) = 1 \quad \text{if } u(1,t) < 0. \tag{31}$$

From the boundary condition (29) on $p^{(0)}$ and from the $O(M^{-2})$ relation (27) obtained from the momentum equation, we conclude $p^{(0)}(x,t) \equiv 1$. Therefore we are left with three unknowns, $\rho^{(0)}$, $u^{(0)}$, $p^{(2)}$ satisfying a set of three equations

$$\rho_t^{(0)} + \left(\rho^{(0)}u^{(0)}\right)_x = -\frac{A_x}{A}\rho^{(0)}u^{(0)} + c_m\rho^{(0)}\left|u^{(0)}\right|, \tag{32}$$

$$\rho^{(0)}u_t^{(0)} + \rho u^{(0)}u_x^{(0)} + \frac{1}{\gamma}(p^{(2)})_x$$

$$= -c_m\rho^{(0)}|u^{(0)}|u^{(0)} - \frac{\rho^{(0)}}{Fr^2} - \frac{\xi}{\sqrt{A}}\rho^{(0)}u^{(0)}|u^{(0)}|, \quad (33)$$

$$\gamma u_x^{(0)} = -\gamma\frac{A_x}{A}u^{(0)} + c_q\rho^{(0)}|u^{(0)}| - \frac{a}{\sqrt{A}}\left(\frac{1}{\rho^{(0)}} - T_w\right). \quad (34)$$

These equations have to be completed with initial data for $\rho^{(0)}$, $u^{(0)}$ and boundary conditions (30) for $p^{(2)}$ and (31) for $\rho^{(0)}$.

Note that $\rho^{(0)}$ is not constant and describes eventually big density changes which imply big temperature changes. The same holds for $u^{(0)}$. Contrary to the Boussinesq approximation we encounter no contradiction with order one mass loss or heat source terms. Therefore we proceed with this set of equations in the following sections.

As we can see in (32)-(34), the model contains 4 parameters which have to be chosen appropriately, *i.e.* c_m, c_q, ξ and a. The former two parameters depend on the chemistry of the fire, the latter two (friction and heat loss) only on the construction of the chimney. The quality of quantitative results obtained by this model depends on the precision of knowledge of the 4 mentioned parameters. By the way, this problem is in common with all approaches, in particular also with all the sophisticated CFD ones. The model, being 1 dimensional and not fully compressible, allows - as we will see also in the examples - fast simulations. Fast direct simulations are the basis of any optimisation strategy. In the planning stage of a chimney the diameter of the chimney, the height of the chimney, the size of the allowed heat loss *etc.* are typical parameters to be optimised.

5 Numerical simulations

5.1 A reformulation of the small Mach number equations

We are now interested in solving numerically equations (32)-(34) with initial data (8) for $\rho^{(0)}$ and $u^{(0)}$, boundary conditions (30) for $p^{(2)}$ and inflow conditions (31) for $\rho^{(0)}$.

First we notice that it is more convenient to consider the air pressure in relation to the atmospheric pressure. So, if we call $p^{ac} = p^{ac(0)} + M^2p^{ac(2)} + O(M^4)$ the altitude corrected pressure, given by

$$p^{ac} = p - p_h = (p^{(0)} - 1) + M^2\left(p^{(2)} + \frac{\gamma}{Fr^2}x\right) + O(M^4)$$

we obtain

$$\rho_t + (\rho u)_x = -\frac{A_x}{A}\rho u + c_m\rho|u|, \quad (35)$$

$$\rho u_t + \rho u u_x + \frac{1}{\gamma}(p^{ac(2)})_x = -c_m\rho|u|u - \frac{(\rho - 1)}{Fr^2} - \frac{\xi}{\sqrt{A}}\rho u|u| + \Delta p_{fan}, \quad (36)$$

$$\gamma u_x = -\gamma\frac{A_x}{A}u + c_q\rho|u| - \frac{a}{\sqrt{A}}\left(\frac{1}{\rho} - T_w\right) - r\left(\frac{1}{\rho}\right)^4, \quad (37)$$

where we dropped the index $^{(0)}$ for $\rho^{(0)}$ and $u^{(0)}$. The boundary conditions for the altitude corrected pressure are given by $p^{ac}(0,t) = p^{ac}(1,t) = 0$. From now on we drop the superscript *ac*.

Also, here we included again the additional fan and the radiation term, that were previously neglected for sake of simplicity.

We reformulate (35)-(37) in order to simplify and in particular to obtain a system of evolution equations. First we integrate (37) in space on $(0,x)$, and define a new function $v(t) := A(0)u(0,t)$; then we rewrite (35) using (37); finally we integrate (36) in space on $(0,1)$.

In this way we get

$$u(x,t) - \frac{v(t)}{A(x)} = \frac{1}{\gamma A(x)} \int_0^x A(y) c_q \rho(y,t) \left| u(y,t) \right| dy$$

$$- \frac{1}{\gamma A(x)} \int_0^x \left\{ a\sqrt{A(y)} \left(\frac{1}{\rho(y,t)} - T_w \right) + r \left(\frac{1}{\rho(y,t)} \right)^4 \right\} dy, \tag{38}$$

$$\rho_t + u\rho_x = -\frac{1}{\gamma} \rho c_q \rho |u| + c_m \rho |u| + \frac{a}{\gamma\sqrt{A}} \rho \left(\frac{1}{\rho} - T_w \right) + \frac{r}{\gamma} \rho \left(\frac{1}{\rho} \right)^4, \tag{39}$$

$$v_t = -\left(\int_0^1 \rho A^{-1} \right)^{-1} \left\{ \int_0^1 \rho u u_x \, dx + \int_0^1 c_m \rho |u| u \, dx + \frac{1}{Fr^2} \int_0^1 (\rho - 1) \, dx \right.$$

$$+ \int_0^1 \frac{\xi}{\sqrt{A}} \rho u |u| \, dx + \frac{\Delta p^{(2)}}{\gamma} - \int_0^1 \Delta p_{fan} \, dx + \int_0^1 \frac{\rho}{\gamma A} \left(\int_0^x A c_q (\rho |u|)_t \, dy \right.$$

$$\left. \left. + \int_0^x a\sqrt{A} \left(\frac{1}{\rho^2} \rho_t \right) dy + 4 \int_0^x rA \left(\frac{1}{\rho} \right)^3 \left(\frac{1}{\rho^2} \rho_t \right) dy \right) dx \right\}, \tag{40}$$

where $\Delta p^{(2)} = (p^{(2)}(1,t) - p^{(2)}(0,t)) = 0$.

Thus we have to solve a PDE for the density $\rho(x,t)$, an ODE for $v(t)$, both evolution equations, and an integral equation for $u(x,t)$.

Finally we have to discuss how the initial, boundary and inflow conditions translate in terms of these new equations.

The initial data (8) can be transformed in initial data for equations (39)-(40), while the boundary conditions for the pressure are included in equation (40). The inflow conditions (31) for the density stay unaltered.

To solve equations (38), (39) and (40) we use

- an explicit upwind for the PDE for ρ in (39),
- an explicit time step for the ODE for v in (40),
- an explicit time step for the integral equation for u in (38).

In what follows i will be the index for the space discretization and n the one for the time discretization; moreover, for the sake of brevity, we summarize equations (38), (39) and (40) as

$$u = \frac{v}{A} + F_u[\rho, u],$$

$$\rho_t = -u\rho_x + F_\rho[\rho, u],$$

$$v_t = F_v[\rho, \rho_t, u, u_x, u_t],$$

where F_u, F_ρ, F_v are the right hand sides of (38), (39), (40), respectively.

Table 3 Geometrical data

	Value
Chimney length before the fire place	0.5 m
Fire place height	0.5 m
Chimney length after the fire place	7 m
Chimney radius	0.125 m
Fireplace radius	0.25 m

Here the cross section A depends on the radius as $A = \pi r^2$.

With this notation, the explicit upwind for the PDE for ρ in (39) is given by:

$$\frac{\rho_i^{n+1} - \rho_i^n}{\Delta t} + \frac{1}{2}\left(u_i^n - |u_i^n|\right)\frac{\rho_{i+1}^n - \rho_i^n}{\Delta x} + \frac{1}{2}\left(u_i^n + |u_i^n|\right)\frac{\rho_i^n - \rho_{i-1}^n}{\Delta x} = F_\rho\left[\rho_i^n, u_i^n\right].$$

To solve the explicit time step for the ODE for v in (40), we use:

$$\frac{v^{n+1} - v^n}{\Delta t} = F\Bigg[\rho_i^n, -\frac{1}{2}\left(u_i^n - |u_i^n|\right)\frac{\rho_{i+1}^n - \rho_i^n}{\Delta x} - \frac{1}{2}\left(u_i^n + |u_i^n|\right)\frac{\rho_i^n - \rho_{i-1}^n}{\Delta x} + F_\rho\left[\rho_i^n, u_i^n\right],$$
$$u_i^n, -\frac{1}{2}\left(u_i^n - |u_i^n|\right)\frac{u_{i+1}^n - u_i^n}{\Delta x} - \frac{1}{2}\left(u_i^n + |u_i^n|\right)\frac{u_i^n - u_{i-1}^n}{\Delta x}, \frac{u_i^n - u_i^{n-1}}{\Delta t}\Bigg].$$

Finally, we compute

$$u_i^{n+1} = \frac{v^{n+1}}{A_i} + F_u\left[\rho_i^n, u_i^n\right].$$

The integrals involved are solved by summation. To use $\frac{(u_i^n - u_i^{n-1})}{\Delta t}$ at the starting time (level $n = 0$), we prescribe (the same) initial values for u also at a level $n = -1$.

To assure stability of the scheme we only need to check CFL condition for the upwind scheme. This method is the simplest reasonable choice (from a numerical point of view) and promises to be fast. We will see that this simple choice already guarantees numerical robustness. In the future we plan to study more sophisticated numerical approaches to simulate our set of equations.

To be more precise we expect to be (much) faster than the corresponding simulations on the full set of compressible equations (4)-(6). This is due to the fact that the CFL condition in the compressible equations (4)-(6) is very restrictive due to big sound speed (due to the small Mach number). Contrary to that the CFL condition in (40)-(39) is determined by the flow velocity (which is of order one) and allows much bigger time steps. In addition every time step in the asymptotic model consists of an ODE and a single PDE timestep, whereas in the fully compressible model we have a time step in a coupled system of three PDE's. In a similar context a precise comparison of the numerical effort was presented [29].

In the following examples we will use as geometrical data for our structure those given in *Table* 3; we will study the dependence on these data in *Section* 5.4.

We will run our simulations with final time $T = 20$ s since within this time range the stationary state is reached in all our examples.

Table 4 Data for basic examples

	Symbol	Value
Friction factor	λ	1
Heat loss coefficient	α	0
Constant for heat exchange	$\tilde{c}_q$	6×10^5 kg m/s^2
Constant for mass loss	$\tilde{c}_m$	10^{-2} m
Wall temperature	T_w	-10°C

5.2 Example 1: Basic description - comparisons with other models

At first we would like to compare the results obtained from our model, with those obtained from the Boussinesq approximation (19)-(25) and from the full system (4)-(7), which will be used as our reference.

For the Boussinesq approximation, we consider equations (19)-(25) and assume

$$c_m = M\tilde{c}_m, \qquad c_q = M\tilde{c}_q, \qquad a = M\tilde{a},$$

with $\tilde{c}_m$, $\tilde{q}_m$, $\tilde{a}$ of order 1. If we also assume $\rho^{(0)} = 1$, trough computation similar to those used for the small Mach number case in *Section* 5.1, we get:

$$u_t^{(0)} = \frac{1}{\int_0^1 \frac{dx}{A}} \left\{ \frac{\Delta p^{(2)}}{\gamma} - \int_0^1 \left(c\rho^{(1)} + \frac{\xi}{\sqrt{A}} u^{(0)} |u^{(0)}| - \frac{A_x}{A} \left(u^{(0)}\right)^2 \right) dx \right\}, \tag{41}$$

$$\rho_t^{(1)} + u^{(0)} \rho_x^{(1)} = \left(\tilde{c}_m - \frac{\tilde{c}_q}{\gamma} \right) |u^{(0)}| - \gamma c u^{(0)} + \frac{\tilde{a}}{\gamma \sqrt{A}} (1 - T_w). \tag{42}$$

Here $\Delta p^{(2)} = (p^{(2)}(1,t) - p^{(2)}(0,t)) = 0$, where we used the altitude corrected pressure; the boundary conditions are those obtained in *Section* 4.1. To solve these equations we use an explicit upwind scheme for $\rho^{(1)}$ and an explicit time step for $u^{(0)}$.

Solving (41)-(42) we will be able to compute the unknowns in the zero order approximation, *i.e.*

$$\rho \sim \rho^{(0)}, \qquad p \sim p^{(0)}, \qquad T \sim T^{(0)} = \frac{p^{(0)}}{\rho^{(0)}}, \qquad u = u^{(0)}, \tag{43}$$

where only the velocity is not constant. In the first order approximation we have

$$\rho \sim \rho^{(0)} + M\rho^{(1)}, \qquad p \sim p^{(0)} + Mp^{(1)}, \qquad T = T^{(0)} + T^{(1)}M = \frac{p^{(0)} + p^{(1)}M}{\rho^{(0)} + \rho^{(1)}M}. \tag{44}$$

In order to solve the full system (4)-(7), we use the standard technique of splitting the problem into two sub-problems, solving first the hyperbolic conservation law by an upwind scheme, and then solving the ODE for the source term with an explicit Euler scheme. In this case, we choose to use a first order scheme in order to have a reasonable comparison with the result obtained from the other two models.

For simplicity, we consider a basic example with no heat loss through the chimney walls and constant cross section. We consider the values for our parameters as given in *Table* 4.

As for the initial data, we consider $u_0(x) \equiv 0$, $\rho_0(x) = \rho_h(x)$ and, in the case of the full system, $p_0(x) = p_h(x)$.

Figure 2 Temperature. Simulations for data as in Table 4. While the results obtained with the full system and the system of equations reduced via small Mach number approximation, coincide, the results obtained with the Boussinesq approximation are unreasonable.

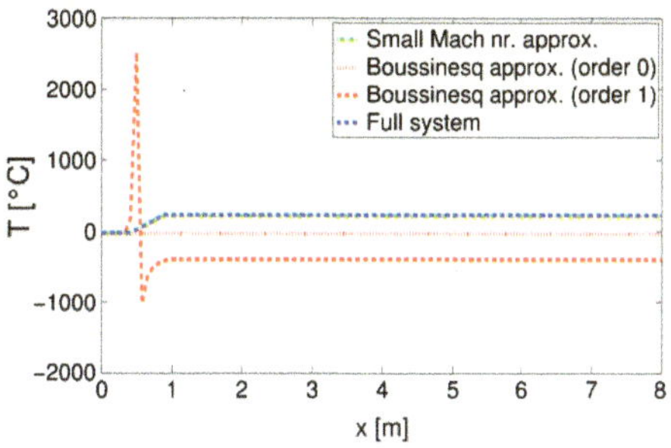

Figure 3 Density. Simulations for data as in Table 4. While the results obtained with the full system and the system of equations reduced via small Mach number approximation, coincide, the results obtained with the Boussinesq approximation are unreasonable.

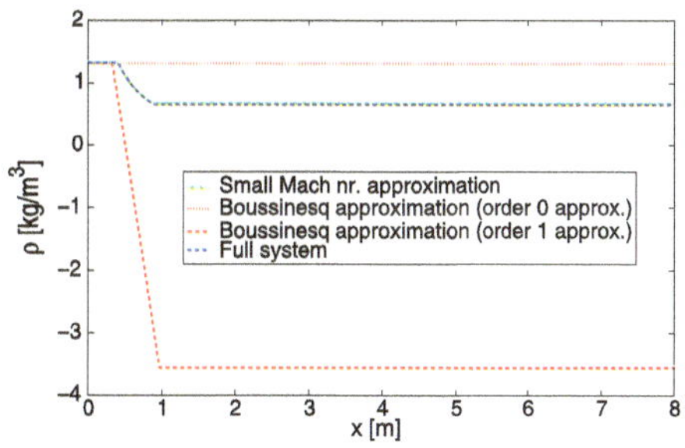

Figure 4 Velocity. Simulations for data as in Table 4. While the results obtained with the full system and the system of equations reduced via small Mach number approximation, coincide, the results obtained with the Boussinesq approximation are unreasonable.

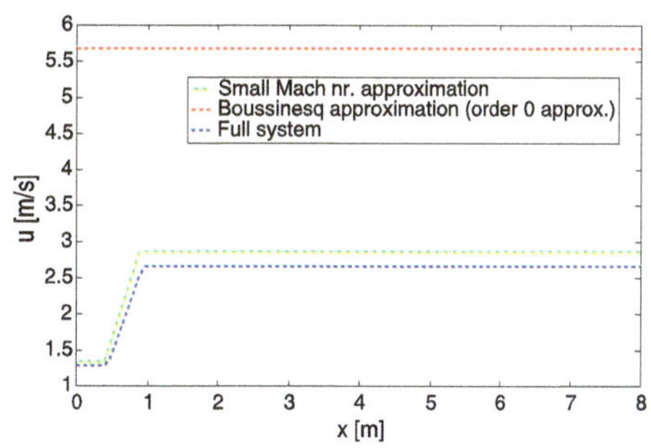

After a simulation time of few seconds, all three models reach stationary values, and we can compare the result for velocity, density and temperature of air through the chimney.

For both the small Mach number and the full model we see that the temperature reaches a value of 250°C (with $\Delta T \sim 10^2$) and a velocity of ~ 3 m/s (Figures 2-4). We have very good agreement although the simulation time of the small Mach approach is about three orders of magnitude lower (and faster than real time).

Meanwhile, the Boussinesq approximation gives unrealistic results, involving negative density (Figure 3).

In Figure 2 after a first period where the density is still positive (and positive temperature), also the temperature becomes negative due to (44). This is not a numerical problem, but a problem of the wrong asymptotics. The reason is that for strong heat sources the Boussinesq approximation is a bad approximation since the higher order terms in the asymptotics are not of higher order but of the same order as the leading order terms.

On the other hand, lowering the heat sources, the Boussinesq approximation becomes reasonable, in the sense that the density and the temperature are no more negative.

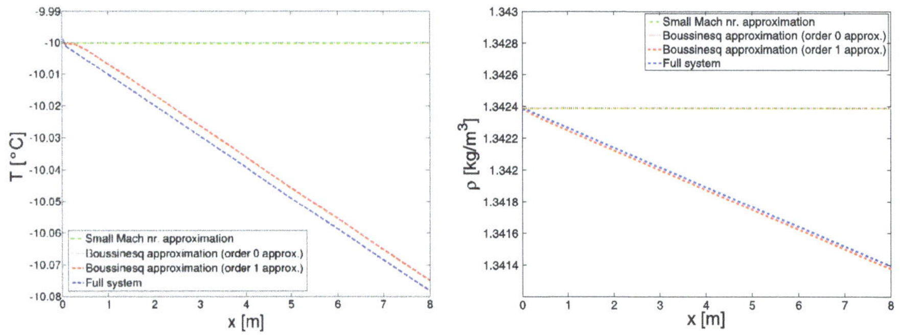

Figure 5 Temperature and density for no heat source. The results obtained with the (first order) Boussinesq approximation coincide with those obtained with the full system, while the small Mach number approximation coincides with the zeroth order Boussinesq approach.

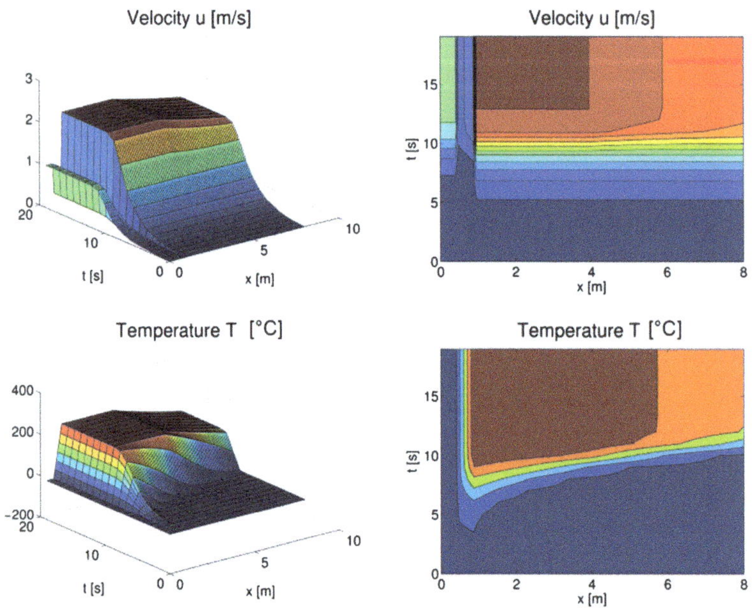

Figure 6 Adding heat loss. Data as in Table 4, but with $\alpha = 20\ \mathrm{W/(m^2K)}$.

When we finally compare the three models for no heat source, the solutions given from Boussinesq approximation, that computes the result up to order one, overlap those given by the full system (Figure 5).

5.3 Example 1.2: adding heat loss

Now we concentrate on the results obtained with our model.

We consider a variable cross section, and add the effect of the heat loss along the last 4 meters of the chimney walls setting the heat loss coefficient $\alpha = 20\ \mathrm{W/(m^2K)}$ (see Figure 6).

This choice is motivated by the fact that many chimneys are not isolated over the last meters (at least in older buildings) since there the chimney passes under the roof (which may be a not heated area) or even outside of the roof. In the heated part of the building chimneys are typically well isolated.

Figure 7 Exit velocity as a function of the chimney radius for three different heat loss coefficients.

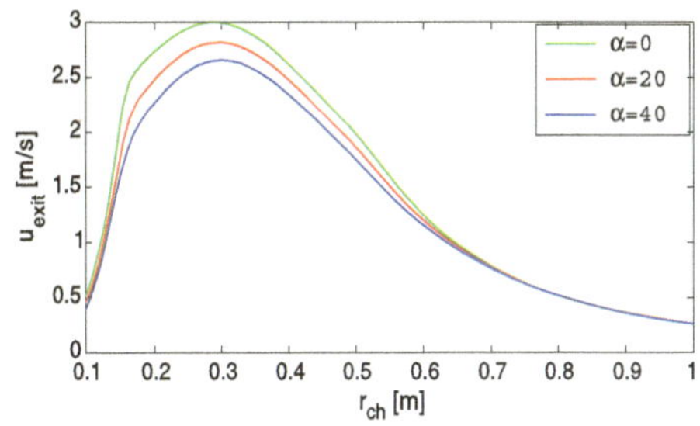

As we can see from Figure 6, when the air flow passes the fireplace, temperature increases (as density decreases). The wider cross section in the fireplace reduces the velocity. The stationary state is reached very rapidly.

The maximal velocity and maximal temperature are, respectively, 2 m/s and 240°C, while at the top of the chimney, the exit velocity is about 1.8 m/s and the exit temperature is about 155°C. All these are reasonable values. In modern chimneys, thanks to the use of special materials, the friction factor and the heat loss coefficient are kept considerably smaller, such that for small heat sources there is still a good draft.

We finally note that adding the radiation term in the energy equation does not affect the previous results since the Stefan constant is too small. Or in other words, the reached temperature is too low to induce significant losses due to radiation.

5.4 Example 2: effects of the geometry

Here we show how the draft velocity depends - for fixed cross sectional area at the fireplace - on the geometry of the structure.

Figure 7 shows the dependence on the radius of the chimney (for three different values of the heat loss coefficient α). For a small diameter the surface effects dominate. Therefore for a small diameter the friction losses are dominant. For bigger diameters the surface effects become less important. Then increasing the diameter the draft is decreasing since the total buoyancy force is distributed over a bigger cross section. As a final result there is an optimal diameter for which a maximal value for the velocity is reached.

On the other side we can vary the effective length of the chimney *i.e.* the length of the chimney above the fire place. In Figure 8 the result for two different values of heat loss coefficient is given. We assume heat loss only in the last part of the chimney (1/4 of the total effective length), all the other data are as in the previous example. In a short chimney the buoyancy force is too low, in a very high chimney the friction losses dominate. Again, there is an optimal effective length for which the velocity reaches a maximum.

5.5 Example 3: dependence on the friction factor

In Examples 1 and 2 we noticed the dependence of the draft on the heat loss coefficient α. Here we discuss the dependence on another surface effect, the friction factor λ. It is not surprising that increasing the friction factor the draft velocity decreases. In Figure 9 we choose the logarithm of the exit velocity, to underline the sensitivity on small variations in the friction factor value.

5.6 Example 4: external influences

Finally we model the presence of strong winds on the chimney top that may interfere with the hot gas flowing out of the structure.

Figure 8 Exit velocity as a function of the effective length for two different heat loss coefficients.

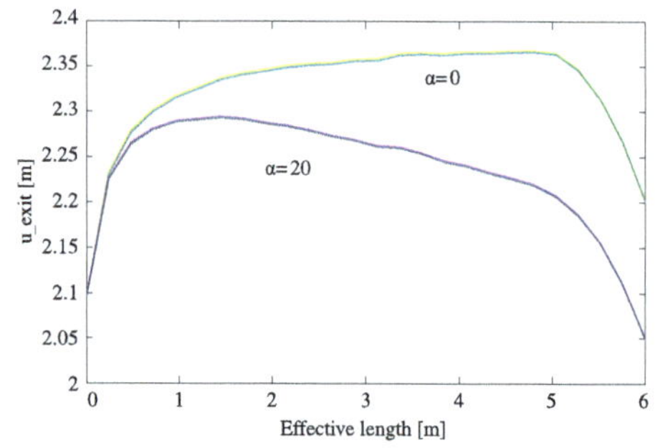

Figure 9 Exit velocity as a function of the friction factor (on a logarithmic scale).

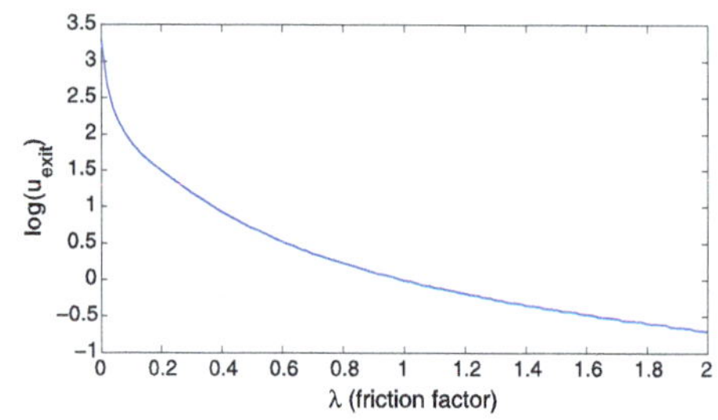

Figure 10 Exit temperature as a function of the external pressure on top of the chimney. In the dark blue area in the lower right part no updraft in the chimney is possible. In the upper left part the chimney is active.

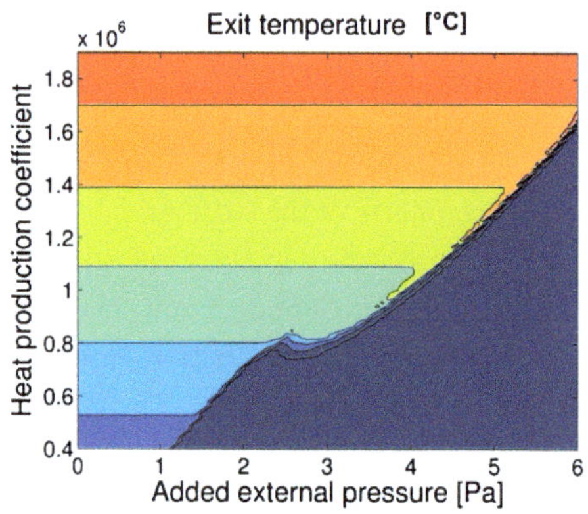

We assume that the pressure variation at the top of the chimney due to wind refers mainly to horizontal wind directions. In case of a vertical wind component there would be an influence on the direction of the vertical part of the velocity but, thanks to rotational symmetry, no influence on the direction of the horizontal part is expected.

So we simulate the outside wind by increasing the external pressure.

The results are shown in Figure 10 where we draw the gas temperature as a function of the fire strength (represented by the heat production coefficient) over the added external pressure. In the dark blue area in the lower right part the outside pressure is too high for the given fire strength and no updraft in the chimney is possible. In the upper left part the chimney is active. There, a higher fire strength induces a higher gas temperature. A higher gas temperature induces more buoyancy. Therefore the chimney works even for stronger winds (*i.e.* higher outside pressures).

Another outside parameter is the external temperature. As we are working with the assumption of constant pressure, when modifying the external temperature we modify the reference values T_r and ρ_r (see *Table* 1). Consequently we modify the heat production co-

Figure 11 Exit velocity as a function of the external temperature for three different heat loss parameters.

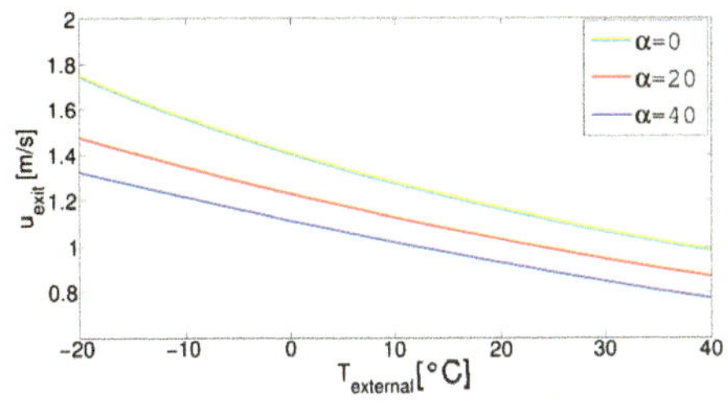

efficient $\tilde{c}_q$ (see definition of $\tilde{c}_q$ in *Section* 3). Thus, increasing the external temperature decreases the air velocity (Figure 11). This result agrees with empirical experience.

6 Conclusions

This paper proposes a small Mach number model to describe the gas dynamics in a chimney. It is shown that in this small Mach number application with strong heat sources the Boussinesq approach is not adequate. However, other small Mach number asymptotic models are shown to be a very good and fast alternative to a fully compressible model. A numerical scheme is proposed and completed with numerical simulations. Being a compromise between highly sophisticated CFD simulations and basic empirical formulas this approach is appropriate to perform parameter studies or to apply parameter studies (*e.g.* diameter, height, heat loss *etc.*) or to apply control procedures.

References

1. McDermott H: *Handbook of Ventilation for Contaminant Control*; 1985.
2. [http://www.eca-europe.org]
3. [http://www.nace.org.uk]
4. Boussinesq J: *Théorie analytique de la chaleur: mise en harmonie avec la thermodynamique et avec la théorie mécanique de la lumière*. Volume 2. Paris: Gauthier-Villars; 1903.
5. Bois PA: **Asymptotic aspects of the Boussinesq approximation for gases and liquids.** *Geophys Astrophys Fluid Dyn* 1991, **58**(1-4):45-55.
6. Zeytounian RK: **Joseph Boussinesq and his approximation: a contemporary view.** *C R, Méc* 2003, **331**(8):575-586.
7. Principe J, Codina R: **Mathematical models for thermally coupled low speed flows.** *Adv Theor Appl Mech* 2009, **2**:93-112.
8. Rehm RG, Baum HR: **The equations of motion for thermally driven, buoyant flows.** *J Res Natl Bur Stand* 1978, **83**(3):297-308.
9. Paolucci S: **Filtering of sound from the Navier-Stokes equations.** *NASA STI/Recon Technical Report N* 1982, 83:26036.
10. Majda A, Sethian J: **The derivation and numerical solution of the equations for zero Mach number combustion.** *Combust Sci Technol* 1985, **42**(3-4):185-205.
11. Fedorchenko AT: **A model of unsteady subsonic flow with acoustics excluded.** *J Fluid Mech* 1997, **334**(1):135-155.
12. Métivier G, Schochet S: **The incompressible limit of the non-isentropic Euler equations.** *Arch Ration Mech Anal* 2001, **158**:61-90.
13. Kuchling H: *Taschenbuch der Physik*. Leipzig: Fachbuchverl; 1994.
14. Majda A: *Compressible Fluid Flow and Systems of Conservation Laws in Several Space Variables*. Volume 53. Berlin: Springer; 1984.
15. [http://www.klb-klimaleichtblock.de]
16. [http://www.schiedel.de]
17. Heskestad G: **Fire plumes, flame height and air entrainment.** In *Sfpe Handbook of Fire Protection Engineering*. Quincy: National Fire Protection Association; 2002.
18. Hu LH, Li YZ, Wang HB, Huo R: **An empirical equation to predict the growth coefficient of burning rate of wood cribs in a linear growth model.** *J Fire Sci* 2006, **24**(2):153.
19. Mizuno T, Kawagoe K: **Burning behaviour of upholstered chair: Part1.** *Fire Sci Technol* 1984, **4**(1):37-45.
20. Styahar A: **Modelling and simulation of chimney using one-dimensional solar updraft tower model.** *Master thesis*. Department of Mathematics, University of Hamburg, 2010.
21. Lions PL: *Mathematical Topics in Fluid Mechanics, Volume 2, Compressible Models*; 1998. [*Oxford Lecture Series in Mathematics and Its Applications*, vol 10.] reprinted 2007.
22. Schochet S: **The compressible Euler equations in a bounded domain: existence of solutions and the incompressible limit.** *Commun Math Phys* 1986, **104**:61-90.

23. Birken P, Meister A: **Stability of preconditioned finite volume schemes at low Mach numbers.** *BIT Numer Math* 2005, **45**(3):463-480.
24. Klein R: **Semi-implicit extension of a Godunov-type scheme based on low Mach number asymptotics I: one-dimesional flow.** *J Comput Phys* 1995, **121**(2):213-237.
25. Meister A: **Asymptotic single and multiple scale expansions in the low Mach number limit.** *SIAM J Appl Math* 1999, **60**(1):256-271.
26. Müller B: **Low mach number asymptotics of the Navier Stokes equations.** *J Eng Math* 1998, **34**:97-109.
27. Gasser I: **Modelling and simulation of a solar updraft tower.** *Kinet Relat Models* 2009, **2**(1):191-204.
28. Bauer M, Gasser I: **Modeling, asymptotic analysis, and simulation of an energy tower.** *SIAM J Appl Math* 2012, **72**(1):362-381.
29. Gasser I, Rybicki M: **Modeling and simulation of gas dynamics in an exhaust pipe.** *Appl Math Model* 2013, **73**(5):2747-2764.
30. Gasser I, Struckmeier J: **An asymptotic-induced one-dimensional model to describe fires in tunnels.** *Math Methods Appl Sci* 2002, **25**(14):1231-1249.
31. Batchelor GK: **The conditions for dynamical similarity of motions of a frictionless perfect-gas atmosphere.** *Q J R Meteorol Soc* 1953, **79**(340):224-235.

8

Simulation of floating potentials in industrial applications by boundary element methods

Dominic Amann[1], Andreas Blaszczyk[2], Günther Of[1*] and Olaf Steinbach[1]

*Correspondence: of@tugraz.at
[1]Institut für Numerische Mathematik, Technische Universität Graz, Steyrergasse 30, Graz, 8010, Austria
Full list of author information is available at the end of the article

Abstract

We consider the electrostatic field computations with floating potentials in a multi-dielectric setting. A floating potential is an unknown equipotential value associated with an isolated perfect electric conductor, where the flux through the surface is zero. The floating potentials can be integrated into the formulations directly or can be approximated by a dielectric medium with high permittivity. We apply boundary integral equations for the solution of the electrostatic field problem. In particular, an indirect single layer potential ansatz and a direct formulation based on the Steklov-Poincaré interface equation are considered. All these approaches are discussed in detail and compared for several examples including some industrial applications. In particular, we will demonstrate that the formulations involving constraints are vastly superior to the penalized formulations with high permittivity, which are widely used in practice.

MSC: 65N38; 78A30

Keywords: electrostatic field computations; boundary element methods; industrial applications

1 Introduction

For the solution of 3D electrostatic field problems, boundary element methods are widely used and are, in particular, advantageous in the presence of an unbounded domain. In addition to Dirichlet and Neumann boundary conditions, so-called floating potentials may occur. Isolated perfect electric conductors result in equipotential surfaces. The equipotential value of the surface is unknown, but, in addition, the flux through the closed surface must be zero in the absence of sources. Such floating electrodes are found in, e.g., some lightning protection systems and can modify the breakdown probability of air gaps [1].

While there are numerous papers on the solution of electrostatic field problems with floating potentials by boundary element methods and related methods, like the charge simulation method [2, 3], in the engineering literature, a detailed view based on a mathematically profound basis seems to be missing. In this paper, we try to close this gap and, in addition, compare several formulations in practical examples. In particular, we consider boundary element methods, see, e.g., [4–6], for the solution of electrostatic field problems with floating potentials in a multi-dielectric setting. We apply an indirect approach based on the single layer potential and a domain decomposition method based on symmetric approximations of the local Dirichlet to Neumann maps, the so-called Steklov-Poincaré operators, see e.g. [7–10]. These two methods have been compared for magnetostatic problems in [11, 12]. Here, we apply these formulations for electrodes at floating potentials.

As electrodes at floating potentials can be considered equivalent to dielectric bodies with infinite permittivity, a common strategy is to substitute electrodes at floating potential by a dielectric media with high permittivity, see e.g. [13]. This approximation can be interpreted as a penalty approach. We compare this strategy to the direct incorporation of the constant but unknown potential and of the zero flux constraint into the formulations. While the penalty approach of a dielectric media with high permittivity needs no additional implementation work in a code which can cope with jumping permittivities, the approach with constraints is highly preferable from a mathematical point of view. Our numerical examples will demonstrate that the results of the approach with constraints are superior to the ones of the penalty approach. In addition, we will show that the formulations based on the Steklov-Poincaré interface equation have advantages over the single layer potential ansatz in case of corners and edges.

The paper is organized as follows: A model problem of the electrostatic field computation with floating potential is introduced in Section 2. In Section 3, the formulations based on the Steklov-Poincaré interface equation and the single layer potential ansatz are presented, and the unique solvability of the variational formulations is proven. The boundary element discretization of both formulations is described in Section 4, and first academic examples in Section 5 show the advantages and disadvantages of the considered approaches. Finally we discuss several extensions of the model problem and apply the methods to examples of industrial applications like an arrester, a bushing, and an insulator with partial wetting in Section 6.

2 Floating potentials in electrostatic field problems

We apply the scalar potential ansatz for the computation of an electrostatic field $\mathbf{E} = -\nabla\varphi$. We consider the union $\Omega_0 = \Omega_E \cup \Omega_F \cup \Omega_D$ of several disjoint Lipschitz domains, a domain Ω_E of an electrode, a domain Ω_F with a floating potential, and a dielectric domain Ω_D. In addition we define the exterior domain $\Omega_0^c = \mathbb{R}^3 \setminus \overline{\Omega}_0$. For the ease of presentation we assume that the intersection of the closures of any two domains Ω_E, Ω_F, and Ω_D is empty. We will comment on more general situations in Section 6. The model problem reads: Find a scalar potential φ such that $\varphi_D = \varphi_{|\Omega_D}$, $\varphi_0 = \varphi_{|\Omega_0^c}$, and a constant $\alpha = \varphi_{|\Omega_F}$ are the solution of

$$-\Delta\varphi_D(x) = 0 \quad \text{for } x \in \Omega_D, \tag{1}$$

$$-\Delta\varphi_0(x) = 0 \quad \text{for } x \in \Omega_0^c, \tag{2}$$

$$\varphi_0(x) = g \quad \text{for } x \in \Gamma_E := \partial\Omega_E, \tag{3}$$

$$\varphi_D(x) = \varphi_0(x) \quad \text{for } x \in \Gamma_D, \tag{4}$$

$$\varepsilon_D \frac{\partial}{\partial n_D}\varphi_D(x) = \varepsilon_0 \frac{\partial}{\partial n_D}\varphi_0(x) \quad \text{for } x \in \Gamma_D, \tag{5}$$

$$\varphi_0(x) = \mathcal{O}\left(|x|^{-1}\right) \quad \text{as } |x| \to \infty, \tag{6}$$

$$\varphi_0(x) = \alpha \quad \text{for } x \in \Gamma_F := \partial\Omega_F, \tag{7}$$

$$\int_{\Gamma_F} \frac{\partial}{\partial n_F}\varphi_0(x)\, ds_x = 0. \tag{8}$$

Here n_i denotes the exterior unit normal vector on $\Gamma_i := \partial\Omega_i$, $i \in \{0, D, E, F\}$, and is defined almost everywhere. On the surface Γ_E of the electrode a constant potential g is given in (3), while we enforce continuity of the potential as well as of the flux for the dielectrics by (4) and (5). In addition, we introduce Γ as the union of all boundaries. For our simple model problem $\Gamma = \Gamma_0$. Note that we distinguish Γ_0 and Γ from the beginning, such that the developed boundary integral equations can be applied to more general settings which are discussed in Section 6. The dielectric domain is characterized by its relative permittivity ε_D and the exterior domain Ω_0^c by ε_0. For the floating potential, we assume a constant but unknown potential α on the boundary Γ_F in (7), but the total flux through this surface is zero, see (8).

We will consider two approaches to solve such boundary value problems with a floating potential numerically. The first approach is to solve the boundary value problem in the form (1)-(8), taking into account the constant but unknown potential α and the constraint (8) directly. The second approach, which is widely used in practice due to its simple implementation, is to approximate the floating potential by considering Ω_F to be a dielectric medium with high relative permittivity ε_F, i.e., to determine a potential φ_F instead of α. In this case we end up with a system consisting of (1)-(6) with additional transmission conditions on Γ_F:

$$\varphi_F(x) = \varphi_0(x) \quad \text{for } x \in \Gamma_F, \tag{9}$$

$$\varepsilon_F \frac{\partial}{\partial n_F} \varphi_F(x) = \varepsilon_0 \frac{\partial}{\partial n_F} \varphi_0(x) \quad \text{for } x \in \Gamma_F. \tag{10}$$

We will demonstrate by some numerical examples in Section 5 that the penalty approach of an additional dielectric medium with a high relative permittivity gives bad approximations of the floating potential in general.

3 Boundary integral equations

If we use $\Gamma_C = \Gamma \setminus \Gamma_E = \Gamma_D \cup \Gamma_F$ instead of Γ_D in (4) and (5) and set the permittivities ε correctly, the model with a high relative permittivity ε_F is the special case (1)-(6) of the full model (1)-(8) with a floating potential. Thus we will derive the boundary integral equations for the full model only. For the model with a high relative permittivity we just need to drop the boundary integral equations related to Γ_F and take into account the ones of Γ_D for Γ_F in addition.

We consider an approach which is based on the Steklov-Poincaré interface equation known from domain decomposition methods, see e.g. [14, 15], and an indirect ansatz leading to a single layer boundary integral equation. While the latter approach is popular due to the ease of implementation, the domain decomposition approach will result in better approximations for the examples in Sections 5 and 6.

3.1 Steklov-Poincaré interface equation

The solutions of the Laplace equations (1) and (2) are given by the representation formulae

$$\varphi_D(x) = \int_{\Gamma_D} U^*(x,y) t_D(y)\, ds_y - \int_{\Gamma_D} \frac{\partial}{\partial n_{D,y}} U^*(x,y) \varphi_D(y)\, ds_y \quad \text{for } x \in \Omega_D,$$

$$\varphi_0(x) = -\int_{\Gamma_0} U^*(x,y) t_0(y)\, ds_y + \int_{\Gamma_0} \frac{\partial}{\partial n_{0,y}} U^*(x,y) \varphi_0(y)\, ds_y \quad \text{for } x \in \Omega_0^c,$$

with $t_D := \frac{\partial}{\partial n_D}\varphi_D$, $t_0 := \frac{\partial}{\partial n_0}\varphi_0$, and the fundamental solution

$$U^*(x,y) = \frac{1}{4\pi}\frac{1}{|x-y|}.$$

Thus we need to determine the unknown parts of the Cauchy data $[t_i,\varphi_i]$, $i \in \{0,D\}$.

The interior Steklov-Poincaré operator $S^D : H^{1/2}(\Gamma_D) \to H^{-1/2}(\Gamma_D)$ maps some given Dirichlet datum φ_D onto the related Neumann datum $t_D = S^D\varphi_D$ of the corresponding solution of the Laplace equation (1). Analogously the exterior Steklov-Poincaré operator $S^0 : H^{1/2}(\Gamma_0) \to H^{-1/2}(\Gamma_0)$ gives $t_0 = -S^0\varphi_0$. These two operators can be defined in their so-called symmetric representation, see e.g. [6, Section 6.6.3, p. 148], by

$$S^D = D_D + \left(\frac{1}{2}I + K_D'\right)V_D^{-1}\left(\frac{1}{2}I + K_D\right),$$

$$S^0 = D_0 + \left(\frac{1}{2}I - K_0'\right)V_0^{-1}\left(\frac{1}{2}I - K_0\right).$$

The single layer boundary integral operator V_i, the double layer boundary integral operator K_i, its adjoint K_i', and the hypersingular operator D_i are defined with respect to Γ_i, $i \in \{0,D,E,F\}$, by

$$(V_i t_i)(x) = \int_{\Gamma_i} U^*(x,y)t_i(y)\,ds_y,$$

$$(K_i\varphi_i)(x) = \int_{\Gamma_i} \frac{\partial}{\partial n_{i,y}}U^*(x,y)\varphi_i(y)\,ds_y,$$

$$(K_i' t_i)(x) = \int_{\Gamma_i} \frac{\partial}{\partial n_{i,x}}U^*(x,y)\varphi_i(y)\,ds_y,$$

$$(D_i\varphi_i)(x) = -\frac{\partial}{\partial n_{i,x}}\int_{\Gamma_i} \frac{\partial}{\partial n_{i,y}}U^*(x,y)\varphi_i(y)\,ds_y.$$

As the two Steklov-Poincaré operators S^D and S^0 correspond to the solution of local Dirichlet boundary value problems, it remains to satisfy the boundary and transmission conditions. We need to find a global function $\varphi \in H^{1/2}(\Gamma)$ such that

$$\varphi(x) = g \quad \text{for } x \in \Gamma_E, \qquad \varphi(x) = \alpha \quad \text{for } x \in \Gamma_F,$$

i.e., the boundary conditions (3) and (7) as well as the transmission condition (4) are satisfied. Using $t_0 = -S^0\varphi$, $t_D = S^D\varphi_{|\Gamma_D}$, and the splitting $\varphi = \varphi_D + g1_E + \alpha 1_F$, where $1_i(x) = 1$ for $x \in \Gamma_i$ and 0 else, the remaining transmission condition (5) and the constraint (8) result in the final system: Find $\varphi_D \in H^{1/2}(\Gamma_D)$ and $\alpha \in \mathbb{R}$ such that

$$\left(\varepsilon_D S^D + \varepsilon_0 S^0\right)\varphi_D(x) + \alpha\varepsilon_0\left(S^0 1_F\right)(x) = -g\varepsilon_0\left(S^0 1_E\right)(x) \quad \text{for } x \in \Gamma_D, \tag{11}$$

$$\int_{\Gamma_F}\left(\left(S^0\varphi_D\right)(x) + \alpha\left(S^0 1_F\right)(x)\right)ds_x = -g\int_{\Gamma_F}\left(S^0 1_E\right)(x)\,ds_x. \tag{12}$$

3.2 Single layer boundary integral operator formulation

Due to its popularity in practice, see [16] and references given therein, we consider a global single layer potential ansatz by

$$\varphi(x) = \int_\Gamma U^*(x,y)w(y)\,ds_y \quad \text{for } x \in \mathbb{R}^3 \setminus \Gamma$$

for any single layer charge density $w \in H^{-1/2}(\Gamma)$ for the global solution φ of the boundary value problem (1)-(8). With this choice the local partial differential equations (1) and (2), the continuity condition (4) as well as the radiation condition (6) are satisfied. The remaining Dirichlet boundary condition (3), the floating potential condition (7), the flux transmission condition (5), and the scaling condition (8) provide the equations to determine the unknown density $w \in H^{-1/2}(\Gamma)$:

$$(Vw)(x) = g \quad \text{for } x \in \Gamma_E, \tag{13}$$

$$(Vw)(x) - \alpha = 0 \quad \text{for } x \in \Gamma_F, \tag{14}$$

$$\frac{1}{2}\frac{\varepsilon_D + \varepsilon_0}{\varepsilon_D - \varepsilon_0} w(x) + (K'w)(x) = 0 \quad \text{for almost all } x \in \Gamma_D, \tag{15}$$

$$\int_{\Gamma_F} \left(-\frac{1}{2} w(x) + (K'w)(x) \right) ds_x = 0, \tag{16}$$

where V denotes the global single layer boundary integral operator and K' is the global adjoint double layer boundary integral operator for $x \in \Gamma$:

$$(Vw)(x) = \int_\Gamma U^*(x,y)w(y)\,ds_y, \qquad (K'w)(x) = \int_\Gamma \frac{\partial}{\partial n_x} U^*(x,y)w(y)\,ds_y.$$

3.3 Unique solvability

Lemma 3.1 *There exists a unique solution* $(\varphi_D, \alpha) \in H^{1/2}(\Gamma_D) \times \mathbb{R}$ *satisfying* (11)-(12).

Proof Using the splitting $\varphi = \varphi_D + \alpha 1_F + g 1_E$, we can reformulate (11)-(12) as: Find $\varphi \in X := \{\psi \in H^{1/2}(\Gamma) : \psi_{|\Gamma_F} = \alpha, \alpha \in \mathbb{R}, \psi_{|\Gamma_E} = 0\}$:

$$\langle (\varepsilon_D S^D + \varepsilon_0 S^0)\varphi, \psi \rangle_\Gamma = -\langle g\varepsilon_0 S^0 1_E, \psi \rangle_\Gamma \quad \text{for all } \psi \in X.$$

This variational formulation admits a unique solution, as $X \subset H^{1/2}(\Gamma)$, the exterior Steklov-Poincaré operator S^0 is $H^{1/2}(\Gamma)$-elliptic, and the interior Steklov-Poincaré operator S^D is $H^{1/2}(\Gamma_D)$-semi-elliptic, see, e.g., [14, Lemma 1.83, p. 49f] and [6, Section 6.6.3, p.149]. □

Lemma 3.2 *Let* $(\varphi_D, \alpha) \in H^{1/2}(\Gamma_D) \times \mathbb{R}$ *be a solution of the Steklov-Poincaré interface equations* (11)-(12), *and let* $w \in H^{-1/2}(\Gamma)$ *be a solution of the indirect single layer approach* (13)-(16). *Then there holds the relation*

$$\varphi(x) = \varphi_D(x) + \alpha 1_F(x) + g 1_E(x) = (Vw)(x) \quad \textit{for } x \in \Gamma.$$

Proof Obviously, the statement holds true for all $x \in \Gamma_E$, as the condition (13) for the single layer potential approach coincide with the choice of φ for the Steklov-Poincaré interface equation. On Γ_F the assertion holds true due to (14). On Γ_D we start from the continuity (15) of the flux for the single layer potential approach, use $w = V^{-1}Vw$ and the symmetry relation $K'V^{-1} = V^{-1}K$, see e.g. [6, Corollary 6.19, p.138],

$$\begin{aligned} 0 &= \varepsilon_D\left(\frac{1}{2}I + K'\right)w(x) + \varepsilon_0\left(\frac{1}{2}I - K'\right)w(x) \\ &= \varepsilon_D V^{-1}\left(\frac{1}{2}I + K\right)Vw(x) + \varepsilon_0 V^{-1}\left(\frac{1}{2}I - K\right)Vw(x) \quad \text{for } x \in \Gamma_D. \end{aligned}$$

For the first term $u = V^{-1}(\frac{1}{2}I + K)z$ we can apply some simplifications using the splitting of the functions and operators

$$\begin{aligned} &(V_D u_D)(x) + (V_E u_E)(x) + (V_F u_F)(x) \\ &\quad = \frac{1}{2}z_D(x) + (K_D z_D)(x) + (K_E z_E)(x) + (K_F z_F)(x) \end{aligned}$$

for almost all $x \in \Gamma_D$, which reduces to

$$(V_D u_D)(x) = \left(\frac{1}{2}I + K_D\right)z_D(x) \quad \text{for almost all } x \in \Gamma_D$$

because there holds

$$(V_i t_i)(x) - (K_i z_i)(x) = 0 \quad \text{for } x \in \Omega_i^c, i \in \{E, F\},$$

for any solution z_i and $t_i = \frac{\partial z_i}{\partial n_i}$ of the local Laplace equations. With the so-called non-symmetric representations $S^D = V_D^{-1}(\frac{1}{2}I + K_D)$ and $S^0 = V_0^{-1}(\frac{1}{2}I - K_0)$ of the Steklov-Poincaré operators we end up with

$$\varepsilon_D S^D(Vw)(x) + \varepsilon_0 S^0(Vw)(x) = 0 \quad \text{for } x \in \Gamma_D.$$

Applying the same technique for the constraint (16) of the single layer potential ansatz, we end up with

$$\int_{\Gamma_F} S^0(Vw)(x)\, ds_x = 0.$$

Taking into account the floating potential (14), these two equations coincide with the formulation (11)-(12) of the Steklov-Poincaré interface equation, and hence we conclude $\varphi = Vw$ on Γ. □

Due to equivalence of the two formulations we conclude the unique solvability of the indirect approach (13)-(16) from Lemma 3.1.

4 Boundary element methods

For the discretization of the considered boundary integral formulations, we assume a quasi-uniform mesh of the surface Γ with N plane triangles and M nodes. The considered trial and ansatz spaces are the space $S_h^0(\Gamma) = \text{span}\{\psi_\ell^0\}_{\ell=1}^N$ of piecewise constant functions and the space $S_h^1(\Gamma) = \text{span}\{\psi_\ell^1\}_{\ell=1}^M$ of piecewise linear and continuous functions. We use Galerkin variational formulations for the discretization of the domain decomposition method (11)-(12) and of the single layer boundary integral equations (13)-(16).

4.1 Steklov-Poincaré interface equation

We transfer the splitting $\varphi = \varphi_D + \alpha 1_F + g 1_E$ of the solution of (11)-(12) to the Steklov-Poincaré operators such that S^0_{ij} indicates that the operator S^0 is applied to a function defined on Γ_j only and evaluated on Γ_i for $i, j \in \{D, E, F\}$.

Thus the discrete Galerkin variational formulation of (11)-(12) is to find $(\varphi_{D,h}, \alpha) \in S^1_h(\Gamma_D) \times \mathbb{R}$ such that

$$\langle (\varepsilon_D S^D_{DD} + \varepsilon_0 S^0_{DD}) \varphi_{D,h}, v_h \rangle_{\Gamma_D} + \alpha \varepsilon_0 \langle S^0_{DF} 1_F, v_h \rangle_{\Gamma_D} = -\varepsilon_0 g \langle S^0_{DE} 1_E, v_h \rangle_{\Gamma_D},$$

$$\varepsilon_0 \langle S^0_{FD} \varphi_{D,h}, 1_F \rangle_{\Gamma} + \alpha \varepsilon_0 \langle S^0_{FF} 1_F, 1_F \rangle_{\Gamma} = -\varepsilon_0 g \langle S^0_{FE} 1_E, 1_F \rangle_{\Gamma}$$

for all $v_h \in S^1_h(\Gamma_D)$. Due to the inverse of the single layer potential, a direct computation of S^0 and S^D is not possible in general. But we can use the approximations

$$S^D_h := D_{D,h} + \left(\frac{1}{2} M^\top_{D,h} + K^\top_{D,h} \right) V^{-1}_{D,h} \left(\frac{1}{2} M_{D,h} + K_{D,h} \right),$$

$$S^0_h := D_{0,h} + \left(\frac{1}{2} M^\top_{0,h} - K^\top_{0,h} \right) V^{-1}_{0,h} \left(\frac{1}{2} M_{0,h} - K_{0,h} \right).$$

These approximations are symmetric, positive semi-definite, and positive definite, respectively. The additional error caused by these approximations features the same quasi-optimal order of convergence as the Galerkin approximations of the exact operators, see e.g. [6, Lemma 12.11]. Thus, there is no substantial loss of accuracy. The Galerkin matrices are given by

$$D_{i,h}[k,\ell] = \langle D_i \psi^1_\ell, \psi^1_k \rangle_{\Gamma_i}, \qquad V_{i,h}[m,n] = \langle V_i \psi^0_n, \psi^0_m \rangle_{\Gamma_i},$$

$$K_{i,h}[m,\ell] = \langle K_i \psi^1_\ell, \psi^0_m \rangle_{\Gamma_i}, \qquad M_{i,h}[m,\ell] = \langle \psi^1_\ell, \psi^0_m \rangle_{\Gamma_i}$$

for $k, \ell = 1, \ldots, M_i$; $m, n = 1, \ldots, N_i$, and $i \in \{D, E, F\}$. Finally, we have to solve the following system of linear equations

$$\begin{pmatrix} \varepsilon_D S^D_{DD,h} + \varepsilon_0 S^0_{DD,h} & \underline{a} \\ \underline{a}^\top & \lambda \end{pmatrix} \begin{pmatrix} \underline{\varphi}_D \\ \alpha \end{pmatrix} = \begin{pmatrix} \underline{f}_D \\ f_F \end{pmatrix}, \tag{17}$$

where

$$\underline{a}[\ell] := \varepsilon_0 \langle S^0_{DF,h} 1_F, \psi^1_\ell \rangle_{\Gamma_D}, \qquad \lambda := \varepsilon_0 \langle S^0_{FF,h} 1_F, 1_F \rangle_{\Gamma_F}.$$

Due to the positive semi-definiteness of S^D_h and the positive definiteness of S^0_h the linear system (17) is uniquely solvable, see Lemma 3.1.

For the approach which approximates the floating potential by considering Ω_F to be a dielectric with a high relative permittivity ε_F, we have to solve the following system of linear equations

$$\begin{pmatrix} \varepsilon_D S^D_{DD,h} + \varepsilon_0 S^0_{DD,h} & \varepsilon_0 S^0_{DF,h} \\ \varepsilon_0 S^0_{FD,h} & \varepsilon_F S^F_{FF,h} + \varepsilon_0 S^0_{FF,h} \end{pmatrix} \begin{pmatrix} \underline{\varphi}_D \\ \underline{\varphi}_F \end{pmatrix} = \begin{pmatrix} \underline{f}_D \\ \underline{f}_F \end{pmatrix}. \tag{18}$$

The unique solvability of both discrete formulations (17) and (18) is a consequence of the positive definiteness of the approximation S^0_h of the exterior Steklov-Poincaré operator S^0 and the positive semi-definiteness of the approximation S^i_h of the other Steklov-Poincaré operator S^i ($i \in \{D, F\}$), see, e.g., [14, Lemma 1.93, p.58f] and [6, Lemma 12.11, p.285].

4.2 Single layer boundary integral operator formulation

We use piecewise constant functions from $S^0_h(\Gamma)$ as test and ansatz functions in the system (13)-(16). As before we apply the splitting of the unknown density $w_h \in S^0_h(\Gamma)$ into $(w_{F,h}, w_{E,h}, w_{D,h}) \in S^0_h(\Gamma_F) \times S^0_h(\Gamma_E) \times S^0_h(\Gamma_D)$. So the discrete variational formulation of the single layer boundary integral operator formulation (13)-(16) is to find $(w_{F,h}, w_{E,h}, w_{D,h}, \alpha) \in S^0_h(\Gamma_F) \times S^0_h(\Gamma_E) \times S^0_h(\Gamma_D) \times \mathbb{R}$ such that

$$\begin{aligned} \langle V_{EF} w_{F,h} + V_{EE} w_{E,h} + V_{ED} w_{D,h}, \psi_E \rangle_{\Gamma_E} &= g \langle 1_E, \psi_E \rangle_{\Gamma_E}, \\ \langle V_{FF} w_{F,h} + V_{FE} w_{E,h} + V_{FD} w_{D,h}, \psi_F \rangle_{\Gamma_F} - \alpha \langle 1_F, \psi_F \rangle_{\Gamma_F} &= 0, \\ \left\langle K'_{DF} w_{F,h} + K'_{DE} w_{E,h} + K'_{DD} w_{D,h}, \psi_D \right\rangle_{\Gamma_D} + \tfrac{1}{2} \tfrac{\varepsilon_D + \varepsilon_0}{\varepsilon_D - \varepsilon_0} \langle w_{D,h}, \psi_D \rangle_{\Gamma_D} &= 0, \\ \left\langle K'_{FF} w_{F,h} + K'_{FE} w_{E,h} + K'_{FD} w_{D,h}, 1_F \right\rangle_{\Gamma_F} - \tfrac{1}{2} \langle w_{F,h}, 1_F \rangle_{\Gamma_F} &= 0, \end{aligned} \tag{19}$$

for all test functions $\psi_i \in S^0_h(\Gamma_i)$ for $i \in \{F, E, D\}$. In the considered geometric setting, the equation (19) allows for some simplification utilizing the adjointness and the kernel properties of the double layer potential operator:

$$\begin{aligned} 0 &= \left\langle K'_{FF} w_{F,h} + K'_{FE} w_{E,h} + K'_{FD} w_{D,h}, 1_F \right\rangle_{\Gamma_F} - \frac{1}{2} \langle w_{F,h}, 1_F \rangle_{\Gamma_F} \\ &= \langle w_{F,h}, K_{FF} 1_F \rangle_{\Gamma_F} - \frac{1}{2} \langle w_{F,h}, 1_F \rangle_{\Gamma_F} = -\langle w_{F,h}, 1_F \rangle_{\Gamma_F}. \end{aligned}$$

This formulation is equivalent to the following system of linear equations

$$\begin{pmatrix} V_{EE,h} & V_{EF,h} & V_{ED,h} & \\ V_{FE,h} & V_{FF,h} & V_{FD,h} & -\underline{b} \\ \widetilde{K}^\top_{DE,h} & \widetilde{K}^\top_{DF,h} & \frac{1}{2} \frac{\varepsilon_D + \varepsilon_0}{\varepsilon_D - \varepsilon_0} \widetilde{M}_h + \widetilde{K}^\top_{DD,h} & \\ & \underline{b}^\top & & \end{pmatrix} \begin{pmatrix} \underline{w}_E \\ \underline{w}_F \\ \underline{w}_D \\ \alpha \end{pmatrix} = \begin{pmatrix} \underline{f}_E \\ \underline{0} \\ \underline{0} \\ 0 \end{pmatrix}, \tag{20}$$

where

$$\begin{aligned} V_{ij,h}[m,n] &= \left\langle V_j \psi^0_n, \psi^0_m \right\rangle_{\Gamma_i}, & \widetilde{K}_{ij,h}[m,n] &= \left\langle K_j \psi^0_n, \psi^0_m \right\rangle_{\Gamma_i}, \\ \widetilde{M}_{ij,h}[m,n] &= \left\langle \psi^0_n, \psi^0_m \right\rangle_{\Gamma_i}, & \underline{b}[m] &= \left\langle \psi^0_m, 1_F \right\rangle_{\Gamma_F} \end{aligned}$$

for $i, j \in \{D, E, F\}$, $m, n = 1, \ldots, N_i$ or N_j.

For the approach which approximates the floating potential by considering Ω_F as a dielectric with a high relative permittivity ε_F, the corresponding system reads

$$\begin{pmatrix} V_{EE,h} & V_{EF,h} & V_{ED,h} \\ \widetilde{K}_{FE,h}^{\top} & \frac{1}{2}\frac{\varepsilon_F+\varepsilon_0}{\varepsilon_F-\varepsilon_0}\widetilde{M}_h + \widetilde{K}_{FF,h}^{\top} & \widetilde{K}_{FD,h}^{\top} \\ \widetilde{K}_{DE,h}^{\top} & \widetilde{K}_{DF,h}^{\top} & \frac{1}{2}\frac{\varepsilon_D+\varepsilon_0}{\varepsilon_D-\varepsilon_0}\widetilde{M}_h + \widetilde{K}_{DD,h}^{\top} \end{pmatrix} \begin{pmatrix} \underline{w}_E \\ \underline{w}_F \\ \underline{w}_D \end{pmatrix} = \begin{pmatrix} \underline{f}_E \\ \underline{0} \\ \underline{0} \end{pmatrix}. \tag{21}$$

To our best knowledge, the stability of these indirect boundary element formulations is still an open problem for general Lipschitz surfaces due to the inconsistent, but widely used discretization of the adjoint double layer potential in $L_2(\Gamma)$.

5 Numerical examples

In this section, we consider a few rather academic examples to compare the introduced approaches to solve the electrostatic potential problem (1)-(8). We compare four formulations in total. We apply the Steklov-Poincaré (SP) operator formulation (18) and the indirect single layer potential (SL) ansatz (21) for the full dielectric approach (full dielectric) with a high relative permittivity $\varepsilon_F = 10{,}000$ to approximate the floating potential. For the direct incorporation (floating) of the floating potential we solve the Steklov-Poincaré (SP) system (17) and the indirect single layer potential (SL) ansatz (20), respectively.

For the computations, we used an implementation [17] of the proposed boundary element methods which is based on the Fast Multipole Method [18] for fast and data-sparse realizations of the involved boundary integral operators. The Steklov-Poincaré operator formulation is implemented by means of MPI and we used one process per active subdomain, i.e. two processes for (17) and three processes for (18). The implementation of the Fast Multipole Method utilizes OpenMP and we used two threads for each instance of the program. The computations were done on a Workstation with 2 Intel Xeon E5620 processors and 24 GB RAM.

We use the concept of operators of opposite order [19] for the preconditioning of the Steklov-Poincaré operator formulations (17) and (18). We apply the artificial multilevel preconditioning [20, 21] for the inner inversion of the Galerkin matrix of the single layer boundary integral operator in the Steklov-Poincaré operator formulations. For the systems (20) and (21) of the indirect single layer potential ansatz, we use the artificial multilevel preconditioning for the block of the single layer boundary integral operator and a diagonal scaling for the block of the adjoint double layer potential operator.

5.1 Two spheres

The two spheres of our first example [22] have the same diameter which is twice the distance of the two spheres. The first sphere Ω_E is an electrode with a given potential of $\varphi = g = 100$ on its surface. The second sphere Ω_F is either a floating potential or a dielectric with relative permittivity $\varepsilon_F = 10{,}000$, depending on the considered approach. The surrounding air has the relative permittivity $\varepsilon_0 = 1$.

In Table 1, we provide the approximations of the floating potential α and the computational times of the four formulations for several refinement levels. For this setting an approximate solution of an axial symmetric charge simulation solver (ELFI, [23]) is used for comparison. One purpose of the new solvers is to overcome the restrictions of ELFI to axial symmetric geometries. For the full dielectric approach, we do not determine α

Table 1 Approximate values of the floating potential α on Γ_F and computational times for the example of two spheres

Number of elements	256		1,040		4,160	
SP floating	32.41	1 s	33.63	8 s	33.86	35 s
SL floating	32.41	1 s	33.63	4 s	33.86	20 s
SP full dielectric	32.40	2 s	33.62	12 s	33.85	50 s
SL full dielectric	32.39	1 s	33.62	5 s	33.86	28 s
2D ELFI			33.9			

Table 2 Approximate values of the floating potential α on Γ_F and computational times for the example of a sphere and a bicone

Number of elements	384		1,536		6,144		24,576	
SP floating	44.512	2 s	45.339	15 s	45.572	79 s	45.637	329 s
SL floating	44.512	2 s	45.341	7 s	45.573	28 s	45.637	124 s
SP full dielectric	44.512	3 s	45.340	27 s	45.573	106 s	45.636	569 s
SL full dielectric	44.433	2 s	45.355	11 s	45.553	34 s	45.632	168 s
2D ELFI					45.7			

Table 3 Range of the floating potential φ_F for the sphere and the bicone

Number of elements	384	1,536	6,144	24,576
SP full dielectric	44.51-44.53	45.33-45.36	45.55-45.59	45.62-45.65
SL full dielectric	40.38-61.91	42.34-58.26	43.34-54.93	44.02-52.41

directly but provide the mean value of the potential on Γ_F. Even on the finest refinement level the potential φ_F is not constant, it has a range of 0.1983 for the indirect approach and 0.0209 for the Steklov-Poincaré operator formulation.

We notice that all four formulations result in good and similar approximate solutions. Only the indirect single layer formulation (21) for the full dielectric model gives a potential which is not quite constant although we consider approximations of smooth objects. We will encounter this behavior to a greater extend in the next example.

5.2 Sphere and bicone

Now we consider an example consisting of a sphere and a bicone [22]. Both have the same diameter and are arranged at a distance of one eighth of their diameter. One spike of the bicone points towards the sphere. The sphere Ω_E is an electrode with a given potential $\varphi = g = 100$. The bicone Ω_F has a floating potential and the exterior domain has a relative permittivity of $\varepsilon_0 = 1$. In Table 2, we present the approximations of the floating potential α and the computational times. Again an approximate solution of the potential on the surface Γ_F of the cone by an axial symmetric FEM solver is used for comparison.

Note that the mean value of the potential on the surface Γ_F is given for the full dielectric approaches. Therefore we analyze the range of φ_F for these approaches in Table 3 in details. The Steklov-Poincaré operator formulation gives an almost constant potential φ_F on the whole surface Γ_F. But we observe that the potential varies strongly for the single layer potential ansatz, even more than in the last example. The extremal values are taken at the spikes of the bicone. Such a behavior like artificial singularities can be observed for geometries with corners and edges and results in significant loss of accuracy, see, e.g., [11, 12]. The computational times of the formulations related to the Steklov-Poincaré interface equation are higher than those of the single layer potential ansatz for the same mesh, but we neglect the accuracy of the approximations (see results for 'SL full dielectric' in Table 3) in this comparison.

6 Extensions and applications

We observed poor approximations by the single layer potential ansatz for large jumps in the permittivities ε in the example of the sphere and the bicone due to artificial singularities in the discrete solution, see the formulation 'SL full dielectric' in Table 3. For such simple examples the single layer potential ansatz with direct realization of the floating potential ('SL floating') gives still good results. But for examples with large jumps of the coefficients ε_D and ε_0 this is not the case anymore. As the third row in (20) is similar to the second and third row in (21), we are facing the same problem as for the formulation 'SL full dielectric' in Table 3. We observed these problems with artificial singularities in the discrete solution in the presence of dielectric media already for relative permittivity ε_D of 800 and higher, see e.g. [11, 12].

But for more general examples we have to cope with such jumps in the relative permittivities. In such cases, the approximation error of the single layer approach is more than one order of magnitude larger than the one of the Steklov-Poincaré operator formulations for the same mesh, see [11, 12]. Thus one needs significantly finer meshes for the single layer potential ansatz to come up with the same accuracy for this class of problems. This results in larger computational times than for the Steklov-Poincaré operator formulations. Due to these significant drawbacks of the single layer potential ansatz, we will consider the Steklov-Poincaré operator formulations only.

For real world examples, we need to consider more general settings. For the ease of presentation we have restricted the description of the formulations to one representative of each kind of subdomains and to well separated subdomains. We will now comment on some extensions.

The extension to several electrodes and dielectric subdomains is straightforward. For each boundary Γ_{F_i} and Γ_{D_i} the corresponding boundary integral equations (11), (12), and (14)-(16), respectively, have to be considered separately. For each floating subdomain Γ_{F_i} a separate degree of freedom α_i and the corresponding constraint

$$\int_{\Gamma_{F_i}} \frac{\partial}{\partial n_{F_i}} \varphi(x)\, ds_x = 0$$

have to be incorporated.

If two subdomains are in contact, we have to make some additional modifications. If a dielectric subdomain is in contact with an electrode, we use a discrete extension $\widetilde{g}$ of the given potential g to the surface Γ_D of the dielectric and determine the unknown remainder $\varphi_D - \widetilde{g}$ of φ_D. If the floating potential is surrounded by a dielectric medium instead of the exterior air domain the vector $\underline{a}$ and the coefficient λ in (17) involve $\varepsilon_D S^D$ instead of $\varepsilon_0 S^0$. In (18), $\varepsilon_D S^D$ and $\varepsilon_0 S^0$ are interchanged.

If Ω_F has interfaces to more than one subdomain, the constraint of the floating potential has to be taken with care. In the case of an interface to Ω_D and to the exterior domain Ω_0^c, the constraint reads as

$$\varepsilon_D \int_{\Gamma_F \cap \Gamma_D} \frac{\partial}{\partial n_F} \varphi_D(x)\, ds_x + \varepsilon_0 \int_{\Gamma_F \cap \Gamma_0} \frac{\partial}{\partial n_F} \varphi_0(x)\, ds_x = 0.$$

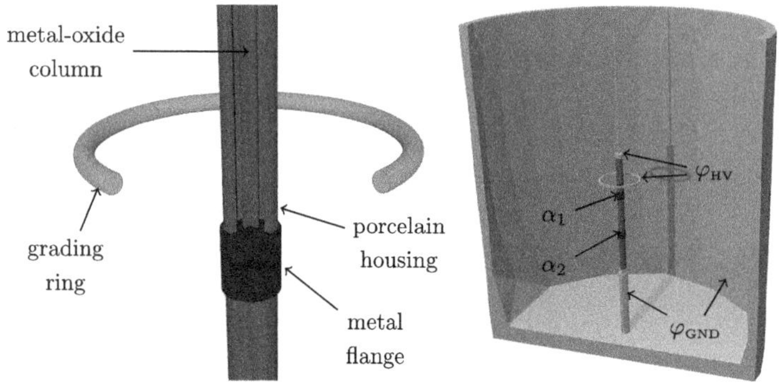

Figure 1 IEC arrester: the gray shades indicate the subdomains.

Table 4 Approximate values of the floating potentials and computational times for the IEC arrester

	α_1	α_2	Time
SP floating	57.44	24.28	4,559 s
SP full dielectric	57.41	24.23	9,659 s
2D ELFI	57.62	24.42	

This extended constraint can be transferred straightforward to the approach of the Steklov-Poincaré interface equation by the means of the related Steklov-Poincaré operators. For the indirect single layer potential ansatz, the simplification of the related constraint (19) seems not to be possible in general.

6.1 IEC arrester

The two remaining Steklov-Poincaré operator formulations are compared for the computation of the electric potential of the IEC surge arrester [24, Annex L] shown in Figure 1. Each of the three sections of the arrester (gray) consists of a metal-oxide cylindrical column with the equivalent relative permittivity $\varepsilon_D = 800$ surrounded by a porcelain housing with the relative permittivity $\varepsilon_D = 5$. In between the two dielectric domains there is a layer of air. The three sections are separated by two metal flanges (dark gray) at floating potentials. At the light gray parts the potential is given. The pedestal and the large surrounding cylinder are grounded electrodes with potential $\varphi_{\text{GND}} = 0$. The top high voltage lead and the toroidal grading ring are electrodes with potential $\varphi_{\text{HV}} = 100$. The exterior domain and the air inside the porcelain housing are modeled as dielectrics with $\varepsilon = 1$.

The difficulties in numerical 3D computations of the IEC arrester are related to the high permittivity of the metal-oxide as well as to the large differences between the radial and axial dimensions. Therefore, the accuracy of the potential computations using different approaches (FEM versus BEM or 2D versus 3D) is typically in the range of 2% as indicated in [24, Annex L]. The results in Table 4 show that the accuracy can be improved with the new Steklov-Poincaré operator formulation to the level of 0.5% (in spite of a relatively rough mesh). The values of the solution on the two floating potentials and the computational times are given in Table 4 for a surface mesh of 32,939 global nodes. As in the previous examples we observe a significant decrease of the computational time for the direct realization of the floating potential, while the solutions of the two approaches do not differ much. Therefore we dismiss the full dielectric model for the next examples.

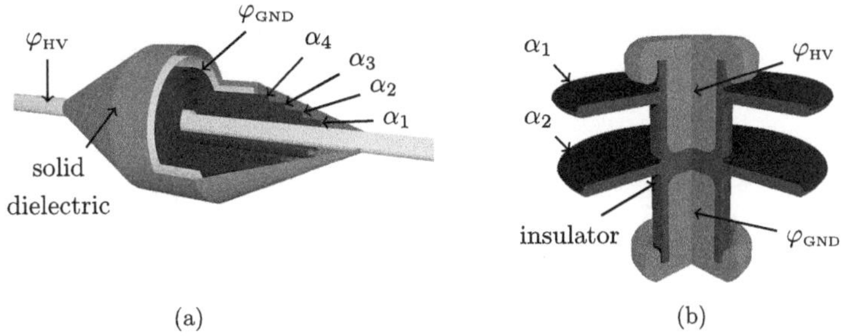

Figure 2 Geometric settings of (a) the bushing and (b) the partially wet insulator.

Table 5 Approximate values of the floating potentials for the bushing

	α_1	α_2	α_3	α_4
SP floating	70.7	51.4	35.1	19.0
2D ELFI	70.8	51.4	35.0	18.9

6.2 Bushing

The next example models a high voltage bushing [23] shown in Figure 2(a). It consists of a cylindrical conductor (light gray) with potential $\varphi_{HV} = 100$ surrounded by five thin metallic foils embedded in a solid dielectric material (gray) with the relative permittivity $\varepsilon_D = 5$. The most outer foil (light gray) is grounded $\varphi_{GND} = 0$ while the other four foils (dark gray) are at floating potentials. The role of the floating foils is enforcing a uniform potential distribution along the conical surface of the bushing. The difficult aspect of modeling bushings is the small thickness of the foils: for the bushing in Figure 2(a) the ratio between the foil thickness and its axial length is in the range of 10^{-3}. The computational mesh consists of 7,936 global nodes. Consequently the distance between elements created on the parallel foil surfaces is approximately 100 times smaller than the size of the elements. In spite of these extreme geometrical relations the floating potentials calculated for all foils with the Steklov-Poincaré operator approach show a good agreement with the 2D solution as presented in Table 5. The system (17) of linear equations is solved in 31 steps of a preconditioned CG method to a relative accuracy of 10^{-6}.

6.3 Insulator with partial wetting

The last example as depicted in Figure 2(b) is an insulator with embedded electrodes (light gray) at potentials $\varphi_{HV} = 100$ and $\varphi_{GND} = 0$. The relative permittivity of the insulator (gray) is $\varepsilon_D = 5$. The upper surface of both insulator sheds is covered by a water layer (dark gray). For the operational frequency of 50-60 Hz water behaves like a conducting material and can be approximated for a capacitive electrostatic field computation as an electrode. Consequently, the two very thin domains of water (dark gray) on the insulator sheds are modeled as electrodes at floating potentials. The geometrical dimensions of this arrangement are given in Table 6.

The solution of the floating version of the Steklov-Poincaré operator approach and the 2D solution are presented in Table 7. The computational mesh consists of 7,936 global nodes, and the system (17) of linear equations is solved in 49 steps of a preconditioned CG method to a relative accuracy of 10^{-6}.

Table 6 Geometrical dimensions of the insulator with partial wetting

Quantity	In mm
Electrodes distance	14
Insulation thickness between electrode and air	5
Shed diameter	160
Shed thickness	6
Water layer thickness	1

Table 7 Approximate values of the floating potentials for the insulator with partial wetting

	α_1	α_2
SP floating	80.93	51.71
2D ELFI	81.49	52.74

7 Conclusions

We insistently recommend the formulations which integrate the floating potential directly and the zero flux condition by a constraint. These formulations give better results and are faster than the approximation obtained by a dielectric media with high permittivity because of a smaller number of degrees of freedom and a smaller number of steps of the iterative solver. Thus the additional effort for the implementation of the modified system pays off.

In case of no or small jumps of the permittivity, the indirect approach by the single layer potential and the direct formulation based on the Steklov-Poincaré interface equation give good results. Here the computational times of the single layer approach are smaller.

In case of larger jumps of the permittivity, the direct formulation based on the Steklov-Poincaré interface equation turned out to be superior to the indirect single layer potential ansatz, as the results are of much better quality. In particular, the indirect approach shows unphysical singularities close to edges and corners in the case of large jumps of the permittivity. Comparing the computational times for a desired accuracy and not for a fixed mesh the Steklov-Poincaré formulations turns out to be faster.

Competing interests
The authors declare that they have no competing interests.

Authors' contributions
All authors contributed equally to the writing of this paper. All authors read and approved the final manuscript.

Author details
[1]Institut für Numerische Mathematik, Technische Universität Graz, Steyrergasse 30, Graz, 8010, Austria. [2]Corporate Research, ABB Switzerland Ltd., Baden-Dättwil, 5405, Switzerland.

Acknowledgements
This work was supported by the FP7 Marie Curie IAPP Project CASOPT (Controlled Component and Assembly Level Optimisation of Industrial Devices, www.casopt.com).

References

1. Roman F, Cooray V, Scuka V: **Comparison of the breakdown of rod-plane gaps with floating electrode.** *IEEE Trans. Dielectr. Electr. Insul.* 1998, **5**(4):622-624.
2. Blaszczyk A, Steinbigler H: **Region-oriented charge simulation.** *IEEE Trans. Magn.* 1994, **30**(5):2924-2927.
3. Singer H, Steinbigler H, Weiss P: **A charge simulation method for the calculation of high voltage fields.** *IEEE Trans. Power Appar. Syst.* 1974, **93**(5):1660-1668.
4. Hsiao GC, Wendland WL: *Boundary Integral Equations.* Berlin: Springer; 2008. [*Applied Mathematical Sciences*, vol. 164.]
5. Sauter SA, Schwab C: *Boundary Element Methods.* Berlin: Springer; 2011. [*Springer Series in Computational Mathematics*, vol. 39.]

6. Steinbach O: *Numerical Approximation Methods for Elliptic Boundary Value Problems. Finite and Boundary Elements*. New York: Springer; 2008.
7. Carstensen C, Kuhn M, Langer U: **Fast parallel solvers for symmetric boundary element domain decomposition equations**. *Numer. Math.* 1998, **79**(3):321-347.
8. Hsiao GC, Steinbach O, Wendland WL: **Domain decomposition methods via boundary integral equations**. *J. Comput. Appl. Math.* 2000, **125**(1-2):521-537.
9. Hsiao GC, Wendland WL: **Domain decomposition in boundary element methods**. In *Fourth International Symposium on Domain Decomposition Methods for Partial Differential Equations, Proc. Symp., Moscow/Russ. 1990*. Philadelphia: SIAM; 1991:41-49.
10. Hsiao GC, Wendland WL: **Domain decomposition via boundary element methods**. In *Numerical Methods in Engineering and Applied Sciences, Part I*. Edited by Alder H, Heinrich JC, Lavanchy S, Onate E, Suarez B. Barcelona: CIMNE; 1992:198-207.
11. Andjelic Z, Of G, Steinbach O, Urthaler P: **Boundary element methods for magnetostatic field problems: a critical view**. *Comput. Vis. Sci.* 2011, **14**:117-130.
12. Andjelic Z, Of G, Steinbach O, Urthaler P: **Fast boundary element methods for industrial applications in magnetostatics**. In *Fast Boundary Element Methods in Engineering and Industrial Applications*. Edited by Langer U, Schanz M, Steinbach O, Wendland WL. Berlin: Springer; 2012:111-143. [*Lecture Notes in Applied and Computational Mechanics*, vol. 63.]
13. Konrad A, Graovac M: **The finite element modeling of conductors and floating potentials**. *IEEE Trans. Magn.* 1996, **32**(5):4329-4331.
14. Pechstein C: *Finite and Boundary Element Tearing and Interconnecting Solvers for Multiscale Problems*. Berlin: Springer; 2013. [*Lecture Notes in Computational Science and Engineering*, vol. 90.]
15. Toselli A, Widlund O: *Domain Decomposition Methods - Algorithms and Theory*. Berlin: Springer; 2005. [*Lecture Notes in Computational Mathematics*, vol. 34.]
16. Andjelić Z, Smajić J, Conry M: **BEM-based simulations in engineering design**. In *Boundary Element Analysis*. Berlin: Springer; 2007:281-352. [*Lect. Notes Appl. Comput. Mech.*, vol. 29.]
17. Of G, Steinbach O, Wendland WL: **The fast multipole method for the symmetric boundary integral formulation**. *IMA J. Numer. Anal.* 2006, **26**(2):272-296.
18. Greengard L, Rokhlin V: **A fast algorithm for particle simulations**. *J. Comput. Phys.* 1987, **73**:325-348.
19. Steinbach O, Wendland WL: **The construction of some efficient preconditioners in the boundary element method**. *Adv. Comput. Math.* 1998, **9**(1-2):191-216.
20. Steinbach O: **Artificial multilevel boundary element preconditioners**. *Proc. Appl. Math. Mech.* 2003, **3**:539-542.
21. Of G: **An efficient algebraic multigrid preconditioner for a fast multipole boundary element method**. *Computing* 2008, **82**(2-3):139-155.
22. Blaszczyk A: **Region-oriented BEM formulation for numerical computation of electric fields**. In *Scientific Computing in Electrical Engineering SCEE2008*. Edited by Roos J, Costa LRJ. Berlin: Springer; 2010:69-76. [*Mathematics in Industry*, vol. 14.]
23. Andjelic Z, Kristajic B, Milojkovic S, Blaszczyk A, Steinbigler H, Wohlmuth M: **Integral methods for the calculation of electric fields, for application in high voltage engineering**. *Technical report* 10, Scientific Series if the International Bureau, Forschungszentrum Jülich; 1992.
24. International Electrotechnical Commission: IEC Technical Standard 60099-4: *Surge Arresters - Part 4: Metal-Oxide Surge Arresters Without Gaps for A.C. Systems*. 2.2 edition; 2009.

9

Adjoint methods for car aerodynamics

Carsten Othmer*

*Correspondence:
carsten.othmer@volkswagen.de
Vehicle Technology, Group
Research, Volkswagen AG, 38436
Wolfsburg, Germany

Abstract

The adjoint method has long been considered as the tool of choice for gradient-based optimisation in computational fluid dynamics (CFD). It is the independence of the computational cost from the number of design variables that makes it particularly attractive for problems with large design spaces. Originally developed by Lions and Pironneau in the 70's, the adjoint method has evolved towards a standard tool within the development processes of the aeronautical industries. Its uptake in the automotive industry, however, lags behind. The first systematic applications of adjoint methods in automotive CFD have interestingly not taken place in the classical shape design arena, but in a relatively young discipline of sensitivity-based optimisation: fluid dynamic topology optimisation. While being an established concept in structure mechanics for decades already, its transfer to fluid dynamics took place just ten years ago. We demonstrate that specifically for ducted flow applications, like airducts for cabin ventilation or engine intake ports, it constitutes a very powerful tool and has matured over the last years to a level that allows its systematic usage for various automotive applications. To drive adjoint-based shape optimisation to the same degree of maturity and robustness for car applications is the subject of ongoing research collaborations between academia and the car industry. Achievements and challenges encountered during these efforts are presented.

1 Background

Computational Fluid Dynamics (CFD) is a central element of the automotive development process. Besides the classical external aerodynamics for the prediction of drag and lift coefficients, there is a whole plethora of applications for ducted flows: airducts for cabin ventilation, engine intake ports, exhaust systems including catalytic converters, air intakes, water jackets of cylinder heads, cooling plates for electric vehicle battery packs and many more (Figure 1). The standard procedure of developing these parts still consists of manual iteration loops between designer and computational engineer, and automatic optimisation methods are being used systematically only for some selected CFD applications (e.g. for airducts, Figure 2). Via an automatic process chain consisting of a CAD (Computer-Aided Design) system, meshing software, CFD solver and post-processor a black-box optimisation algorithm, typically of the evolutionary strategy type, is employed to drive a parameterised CAD-model into an optimal state.

Due to the high computational effort associated with black-box optimisation, where the number of CFD evaluations scales roughly linearly with the number of design variables, the explorable design space is very limited: In our current practice, these methods cannot afford more than ten design parameters. An additional obstacle that inhibits the further

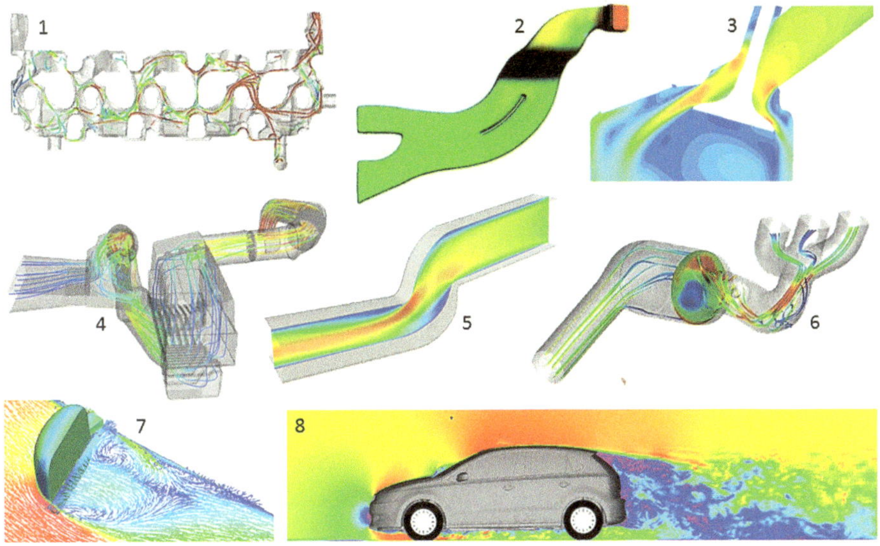

Figure 1 The ubiquity of CFD in automotive development. External aerodynamics (**7** and **8**) is only one aspect of automotive CFD. The many ducted flow applications include, among others, water jackets for cylinder heads (**1**), airducts for cabin ventilation (**2** and **5**), engine intake ports (**3**), raw air intakes (**4**) and exhaust systems (**6**).

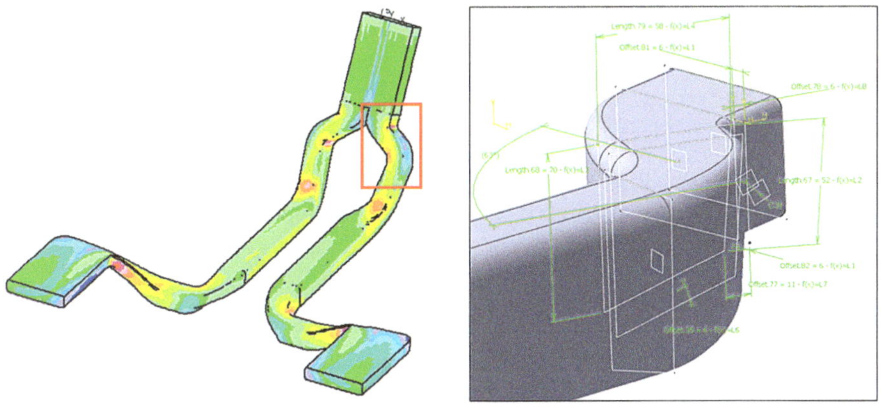

Figure 2 Automotive CFD optimisation today. Example of a CAD-parameterisation of an airduct segment (provided by S Brück and C Ehlers, Volkswagen AG) as input to a black-box optimisation process chain.

adoption of black-box optimisation within the regular automotive development process is the necessity of a versatile but stable CAD parameterisation. For e.g. an engine intake, the setup of the parameterised model might take an experienced CAD engineer several weeks before it comes up to the stability requirements of an automatic optimisation process chain. It is these limitations that call for a radically different approach to automotive CFD optimisation - especially in view of applications like entire vehicle aerodynamics that entail shape optimisation of highly complex free-form geometries with a practically unlimited design space.

Adjoint-based optimisation [1–4] has long been recognized to be efficient for tasks with large numbers of design variables. Its underlying principle, the adjoint method, allows for a very elegant computation of sensitivities, i.e. of the derivative of the objective function

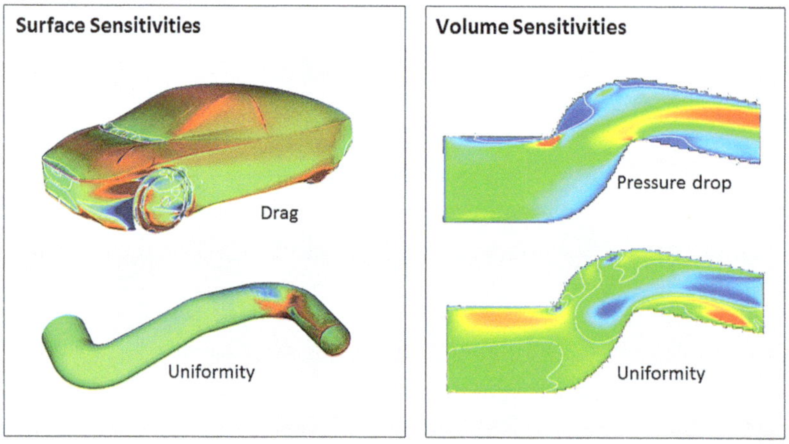

Figure 3 Sensitivity maps. Surface sensitivity maps (left) display the gradient of the cost function w.r.t. normal displacements of the surface. In red areas, a movement away from the fluid (i.e. inwards for the car, outwards for the pipe) would result in an improvement of the cost function. Contrarily, blue areas indicate the regions where a surface perturbation towards the fluid improves performance, while modifications in greenish surface sections have little effect on the cost function. The white lines on the car body are the isolines of zero sensitivity, i.e. the borderline between favourable inward and outward movement. Topological or volume sensitivities (shown on the right-hand side for a cut through the S-bend of Figure 2) represent the cost function gradient w.r.t. changes of the individual cell permeability. Blue volume sections are those where decreasing the cell permeability (by adding a porosity-based friction term) would improve the cost function. Those areas are thus counterproductive to the component's fluid dynamic performance and should be removed - the basis of CFD topology optimisation.

w.r.t. the design variables. The computational effort is thereby independent of the number of design variables and just involves one solve of the adjoint counterparts of the governing equation system, in our case the Navier-Stokes equations. When applied to a surface mesh representation or a volume mesh of the part to be optimised, information like the one depicted in Figure 3, for which we coined the term 'sensitivity maps', can be generated: A surface sensitivity map tells us for each and every surface node how the objective function changes w.r.t. an infinitesimally small normal displacement of this surface node. On the other hand, a volume sensitivity map represents for each volume cell the sensitivity of the objective function w.r.t. a change of permeability of this volume cell. Both sensitivity maps give precise indications on where and how to change the geometry - perturb the surface inwards or outwards, and remove counterproductive cells from the flow domain, respectively - in order to improve the objective function. Obviously, this wealth of information would be inaccessible without the adjoint method, and, as will be shown in the following, allows to develop very efficient optimisation methods for automotive part design.

Despite being already a long-established tool in the aerospace sector, the adjoint method has only recently started to also enter the automotive development processes. This backlog is mainly due to two obstacles specific to the automotive industry.

Obstacle #1 *Rather than on in-house codes, the car industry relies almost exclusively on commercial CFD software.*

The implementation of the adjoint method requires access to the source code - either to implement the analytically derived adjoint equations into the code (so-called continuous adjoint) or to perform a manual or automatic differentiation of the primal code (discrete adjoint, see [5] for a comparison). The sources of commercial CFD codes are, however, inaccessible for the user, and by the time we started our project to industrialise the adjoint method, the software vendors of the CFD solvers that are operational at Volkswagen declined to implement it into their codes. As a remedy, we selected the well-validated and versatile CFD toolbox OpenFOAM® [6] as our platform to develop adjoint-based optimisation methods for car applications. In addition to its open source character, OpenFOAM comes with a high-level symbolic programming style, which makes it ideally suited to implement additional partial differential equations - like the adjoint Navier-Stokes equations.

The first continuous adjoint implementation in OpenFOAM [7] is based on the derivation outlined in [8]. It focused on steady-state, incompressible ducted flows as described by the RANS (Reynolds-averaged Navier-Stokes) equations:

$$(\mathbf{v} \cdot \nabla)\mathbf{v} = -\nabla p + \nabla \cdot \big(2\nu \mathsf{D}(\mathbf{v})\big) - \alpha \mathbf{v}, \tag{1}$$

$$\nabla \cdot \mathbf{v} = 0, \tag{2}$$

where $\mathbf{v}$ and p stand for velocity and pressure, respectively, and α for the porosity distribution. The effective kinematic viscosity ν is the sum of molecular and turbulent viscosity, and the rate of strain tensor $\mathsf{D}(\mathbf{v})$ is defined as $\mathsf{D}(\mathbf{v}) = \frac{1}{2}(\nabla\mathbf{v} + (\nabla\mathbf{v})^T)$. For a given cost function J composed of contributions from the design domain boundary $\Gamma = \partial\Omega$ and from the interior of Ω,

$$J = \int_\Gamma J_\Gamma \, d\Gamma + \int_\Omega J_\Omega \, d\Omega, \tag{3}$$

the implemented adjoint field equations are

$$-(\mathbf{v} \cdot \nabla)\mathbf{u} - \nabla\mathbf{u} \cdot \mathbf{v} = -\nabla q + \nabla \cdot \big(2\nu \mathsf{D}(\mathbf{u})\big) - \alpha \mathbf{u} - \frac{\partial J_\Omega}{\partial \mathbf{v}}, \tag{4}$$

$$\nabla \cdot \mathbf{u} = \frac{\partial J_\Omega}{\partial p}, \tag{5}$$

with $\mathbf{u}$ and q denoting adjoint velocity and adjoint pressure, respectively. While the cost function contribution from the domain interior J_Ω gives rise to source terms in the adjoint field equations, the surface contribution J_Γ enters the adjoint boundary conditions:

$$\mathbf{n} \cdot \nabla q = 0, \qquad \mathbf{u}_t = 0, \qquad u_n = -\frac{\partial J_\Gamma}{\partial p} \tag{6}$$

for boundaries with Neumann conditions for the primal velocity and a Dirichlet condition for the primal pressure (usually inlet and wall boundaries), and

$$q = \mathbf{u} \cdot \mathbf{v} + u_n v_n + \nu(\mathbf{n} \cdot \nabla)u_n + \frac{\partial J_\Gamma}{\partial v_n}, \tag{7}$$

$$0 = v_n \mathbf{u}_t + \nu(\mathbf{n} \cdot \nabla)\mathbf{u}_t + \frac{\partial J_\Gamma}{\partial \mathbf{v}_t} \tag{8}$$

for boundaries with Dirichlet conditions for the primal velocity and a Neumann condition for the primal pressure (e.g. outlet boundaries, for details we refer again to [8]). By solving both primal and adjoint equation system, the desired (volume-normalised) topological and (surface-normalised) shape sensitivities can be computed as

$$\frac{\partial J}{\partial \alpha} = \mathbf{u} \cdot \mathbf{v} \quad \text{and} \quad \frac{\partial J}{\partial \beta} = -\nu \partial_n \mathbf{u} \cdot \partial_n \mathbf{v}, \tag{9}$$

where β denotes the normal node displacement.

Within the still ongoing collaboration with Prof. Giannakoglou's group at the National Technical University of Athens (NTUA) and Dr. de Villiers at Engys®, this basic implementation was subsequently enhanced towards an inclusion of an adjoint turbulence model [9], optimisation of external flows [10], sensitivities for flow control [11], adjoint wall functions [12], the inclusion of heat conduction and constraints [13] and mildly compressible flows [14]. In addition, the uptake and further development of the basic adjoint code by other researchers lead to remarkable advances [15–18], especially in the area of topology optimisation for exhaust systems [19, 20]. As a result of these development efforts, a versatile continuous adjoint code in OpenFOAM for steady-state RANS is now available.

Obstacle #2 *The geometries of entire vehicles and even of single automotive parts are highly complex.*

The complexity of typical automotive geometries, especially of entire vehicles, which consist of an assembly of several hundred parts, necessitates the use of an automatic process for the generation of the computational mesh. Even though the sophistication of automatic meshing software is continuously increasing, the achievable mesh quality is still inferior to a hand-made mesh, especially concerning the creation of layers along the surface of the flow domain. While this quality suffices for obtaining reasonable results for primal computations, it has been found that the adjoint equation system reacts much more delicately to imperfections of the computational mesh. This is particularly the case in the presence of flow separation, which is commonly encountered in flows in and around automotive geometries.

This lack of robustness of the adjoint when running on typical industrial meshes is a problem that has been accompanying the adjoint development in OpenFOAM from the very beginning - and for certain applications it is still inhibiting the systematic use of the adjoint method in the regular computational processes. The source of this instability is the so-called 'adjoint transpose convection' $\nabla \mathbf{u} \cdot \mathbf{v}$ - the second convection term on the left-hand side of Eqn. (4). By developing several formulations for this term, as well as explicit and implicit treatments, major steps towards increased stability were made. As it introduces a high degree of cross-coupling between the three components of the velocity, recent efforts concentrate on solving the adjoint equation system as a single block rather than in a segregated manner [21].

While those developments are very promising, the current practice of running the adjoint - at least on meshes of the typically encountered quality in the car industry environment - still requires to trade robustness against accuracy: by using limiters, relaxed adjoint boundary conditions or by damping the adjoint transpose convection in an intelligent way that does not harm the *qualitative* correctness of the sensitivities. To report the achievements as well as the challenges of this pragmatic approach to solving adjoints for car applications is the intention of this article. It is organized as follows: The concepts of topology optimisation and shape optimisation for ducted flows are introduced in the next section, along with an application of these two complementary methodologies to engine port flows. Section 3 focuses on external aerodynamics, comprising a full-vehicle validation study, an example of a component optimisation and a comparison of steady-state RANS sensitivities with those based on a time-averaged transient primal flow. The article finally closes with a comparatively young adjoint application area - car drag reduction by flow control.

2 Ducted flows

Geometric optimisation of ducted flows in cars, like airducts for cabin climatisation, engine air intakes or exhaust systems, are commonly subject to severe packaging constraints. This gave rise to the development and adoption of topology optimisation methods for automotive applications. After a concise retrospective of these development efforts, it will be shown that especially for ducted flows, topology optimisation is a perfect complement to the classical shape optimisation.

2.1 Topology optimisation

Topology optimisation is a well-established tool in computational structure mechanics [22] with widespread industrial use. In its simplest realisation, a topological optimisation starts from the available design domain filled up entirely with solid material of a certain density. In an iterative fashion, the given loads are applied, the stresses are computed all over the domain and the areas with low stresses are weakened by assigning a lower density to them. After several iterations, this method retains high-density material only in regions that are critical to fulfill the structural task, and in this way generates optimal lightweight structural designs. It delivers an un-biased design from scratch that automatically fulfills the installation space constraints.

It almost comes as a surprise that it took until 2003 until this elegant concept was transferred to fluid dynamics: Independently of each other Borrvall and Petersson [23] and Klimetzek [24] presented the first topological optimisation methods for ducted flows. Analogously to its structural mechanic archetype, their fluid dynamic topology optimisation starts from a completely flooded computational domain (Figure 4), uses some local criterion to identify those areas that are counterproductive w.r.t. the chosen objective function and iteratively 'punishes' or removes them from the fluid domain. As a result, the remaining areas constitute the optimal duct between inlet(s) and outlet(s) of the given installation space.

Despite being based on the same concept of flow optimisation, these two approaches exhibit significant differences. Klimetzek's method is a rather pragmatic solution: The local criterion he employs to identify counterproductive cells is the deviation between the actual velocity direction and a desired direction computed as polygonal traces between inlet and outlet. If this deviation exceeds a certain angle, the corresponding cells are blocked for the fluid. His criterion allows to efficiently remove unwanted recirculations from the flow and - albeit being used for other cost function as well - is therefore ideally suited to reduce the pressure drop. From the date of its original publication, Klimetzek's method has already been applicable to three-dimensional turbulent Navier-Stokes flows and is seeing a continuously increasing number of successful applications in industrial flow optimisation.

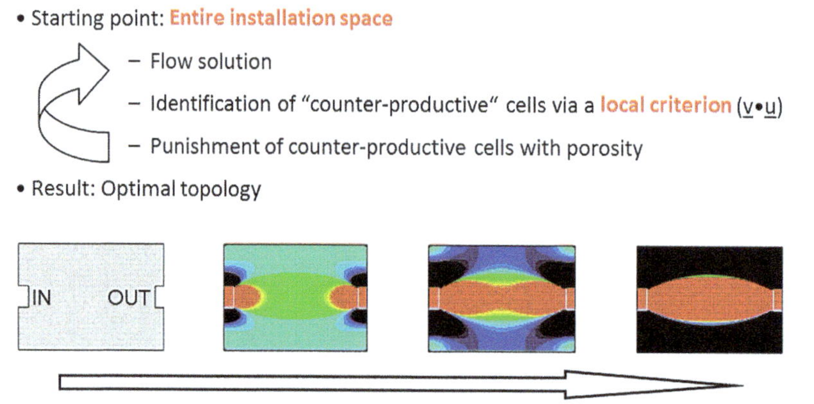

Figure 4 The principle of CFD topology optimisation. Identified counter-productive cells are iteratively punished (black cells).

In contrast, the method proposed by Borrvall and Petersson was restricted to two-dimensional Stokes flows between parallel plates, and the punishment of counterproductive areas is performed by locally decreasing the distance between the plates - until zero upon convergence. What inspired a series of subsequent research works is, however, the versatility of their approach that lies in the way how they identify counterproductive fluid cells: by computing actual sensitivities via the adjoint method.

In order to extend the Borrvall and Petersson method towards three-dimensional flows, the punishment via plate distance was replaced by introducing a Darcy porosity term $-\alpha \mathbf{v}$ into the governing Eqn. (1) (see [25] and [26] for a finite-element and a finite-volume implementation, respectively). The individual cell porosities allow a continuous transition between fluid ($\alpha = 0$) and solid ($\alpha = \alpha_{max}$) and act as the actual design variables of the optimisation problem. With this setting, topological sensitivity maps like those shown in Figure 3 are nothing but the derivative of the cost function w.r.t. an increase in the cell porosity α.

Another requirement for the industrial applicability of adjoint-based topology optimisation was its generalisation from pure Stokes flows to turbulent Navier-Stokes flows. After the principal feasibility of laminar Navier-Stokes topology optimisation was first demonstrated via Automatic Differentiation of an academic CFD code [27], the concept was implemented via a continuous adjoint also for turbulent flows [8] - albeit under the assumption of 'frozen turbulence', i.e. fully neglecting the variation of the turbulent flow quantities. Finally, the development of a continuous adjoint turbulence model [9] now allows to run topology optimisation for turbulent Navier-Stokes flows under full consideration of variations in the turbulence fields [13].

As topology optimisation is not only very fast but also mimics the classical way of part design when driven by installation space constraints, the uptake of this method in the regular automotive design process was quite straightforward. A number of ducted flow applications are today routinely optimised via adjoint-based topology optimisation (see Figure 5 for examples from the Volkswagen Group and [19, 20] for others). Exemplarily, one of them - the optimisation of an engine intake port - will be presented in more detail in the following.

The main components of the cylinder intake port static test configuration geometry that was the optimisation target for this study can be seen in Figure 6. Only the volume

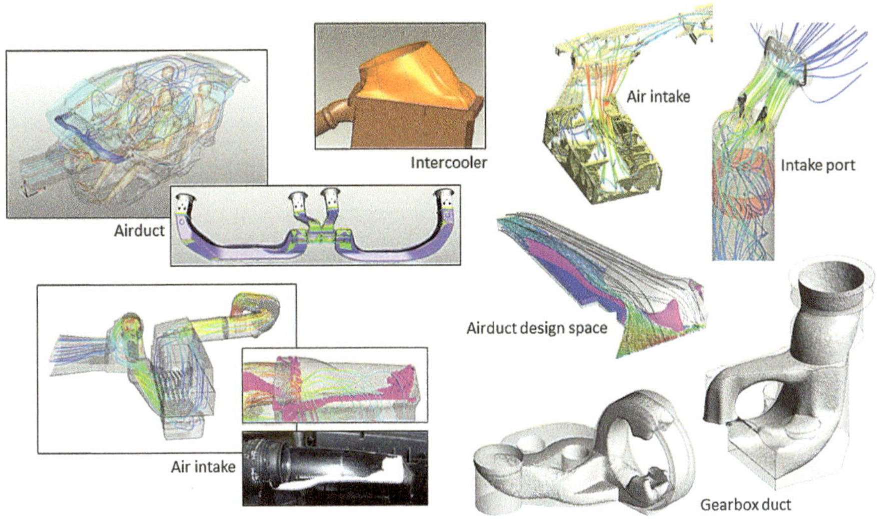

Figure 5 Sample applications of CFD topology optimisation. Kindly provided by U Giffhorn, W Py, M Tomecki, C Ehlers, M Towara from Volkswagen AG and N. Peller from Audi AG.

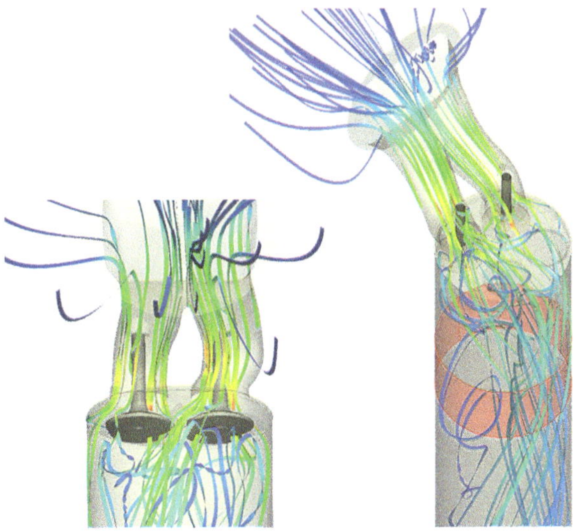

Figure 6 Intake port geometry and streamlines of the original flow. Only the two arms connecting the inlet plenum and the barrel were subject to geometrical modifications upon optimising for (1) pressure drop between inlet and outlet and (2) swirl within the cylindrical slice indicated by the red mantle area.

enclosed by the two intake port arms was subjected to modification. The barrel, valves, valve seats and upstream inlet geometry (not shown) had to be kept clear of blockage via geometric and proximity based constraints. The working fluid for all the simulations was air modeled as a compressible perfect gas. Pressure at the outlet of the extended barrel was fixed, and at the inlet a fixed mass flow at ambient temperature was specified. For the adjoint system, density and turbulence was assumed 'frozen', thus no governing equations were solved for the adjoint systems beyond the momentum and pressure analogues. For the port flow the objectives were (1) to minimise the pressure loss between the inlet and

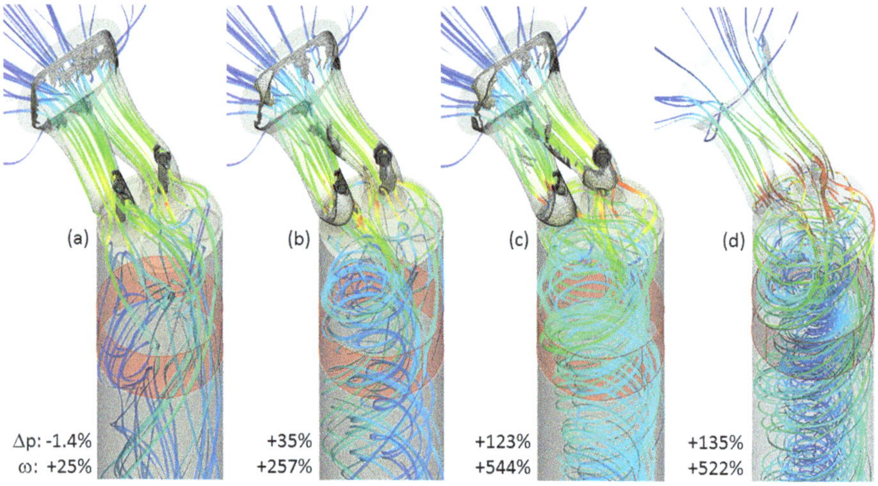

Figure 7 Intake port topology optimisation results. The three figures (**a**)-(**c**) show the Pareto-optimal states for low, mid- and high swirl weighting, respectively. The grey areas within the intake port arms are the cells that were blocked during the topology optimisation. Note the obvious improvement of swirl (ω, bottom numbers) from (a) to (c) at the cost of increasing pressure drop (Δp, top numbers), as compared to the original flow of Figure 6. The topology of the high swirl case (c) was recasted into a new CAD description and recomputed (**d**). Flow field, swirl and pressure drop confirm the results of the topology optimisation (details see [14]).

the outlet and (2) to maximise the swirl, i.e. the angular momentum $\mathbf{r} \times \rho\mathbf{v}$ around the barrel axis in the cylindrical sub-zone indicated in Figure 6.

One of the main issues with multi-objective adjoint optimisation is the choice of a sensible relative weighting when combining the sensitivity fields from the two objectives for input into the topology update. As a first step, we therefore performed an exploration by varying the weighting factors dynamically within a wide range. After choosing three design points with different weightings (low, mid- and high swirl) that represented a likely range for desirable fluid dynamic performances of the intake port, we actually ran the topology optimisation for these three trade-offs between swirl and pressure loss until convergence and thus obtained three Pareto-optimal states (Figure 7a-c, details see [14]).

For the low swirl case, it can be assumed that the geometry is only driven by the swirl sensitivities where the modifications have a very small detrimental effect on the pressure loss. As a result, we see a small reduction in pressure loss along with a modest increase in swirl. Predictably, the pressure loss starts to go up as the swirl weighting is increased and the fluid looses additional energy due to the highly swirling flow and the blockages required to produce it. What is encouraging is the extent to which the swirl can be reliably increased - by more than 500% - compared to the relatively modest relative increase in pressure loss (high swirl case of Figure 7c).

For topology optimisation, it is critical that the design extracted from the optimisation process be verified with a high accuracy simulation. To this end, the blocked element description from the high swirl case was extracted and incorporated into a new CAD description of the port flow geometry (Figure 7d). It is immediately apparent that the modified design is rather complex in its detail and not something a designer would likely create through intuition or experience. Neither could these geometries easily be arrived at via a process of parameter-based stochastic optimisation. The most important point to note is that the accurately rebuilt geometry very nearly reproduces the results from the topology optimisation run. While we do not contend that this will consistently be the case, it certainly proves the efficacy of the process in this instance.

2.2 Shape optimisation

The topology optimisation method as introduced above, acts on the fluid domain like a sculptor on a piece of rock: It removes the unwanted portions and reduces the original geometry to its optimal shape. Since the original volume mesh is not changed during the optimisation process, the final geometry consequently consists of a subset of those cells that made up the original shape. Thus, the inherent geometrical resolution of topology optimisation corresponds to the cell size, and the final shape will inevitably have a ragged surface. Topology optimisation is therefore to be regarded as a *drafting* method. For *fine-tuning* geometries that are already close to their optimal shape, it is ideally complemented by so-called shape optimisation. Rather than on topological or volume sensitivities, shape optimisation is based on the detailed information contained in surface sensitivity maps in order to morph the shape towards a further improvement of the cost function.

An example of a surface sensitivity map of an exhaust port is shown in Figure 8. Red areas are those that have to be moved outwards (away from the air) in order to reduce the pressure drop while blue regions need to be pushed inwards. It seems straightforward to translate this information into a shape modification that improves the performance of the exhaust port: Simply move every surface node in a steepest descent manner inwards or outwards by a distance corresponding to its sensitivity magnitude. Due to the inherent 'roughness' of the sensitivity distribution this would, however, quickly result in a collapse of the surface mesh. Therefore, a clever smoothing process has to be applied to the surface sensitivities before they are actually translated into a shape update. This smoothing process does not only have to generate a good compromise between ironing out the sharpest discontinuities of the sensitivity map and at the same time conserving the essen-

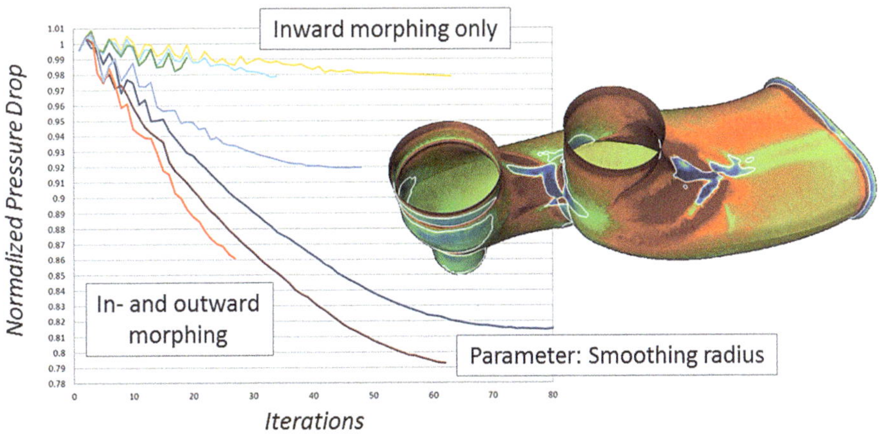

Figure 8 Exhaust port shape optimisation. Based on a pressure drop surface sensitivity map (colours like in Figure 3 left), a one-shot shape optimisation was conducted with the morphing routines from TUM [28, 29]. Expectedly, the achievable improvement is much bigger when shape deformations in both inward and outward direction are allowed (four bottom curves) as compared to enforcing only inward morphing (top three curves). Due to the small spatial structures of the sensitivity map, also the amount of smoothing applied to the shape morphing had a significant effect on the optimisation. The meshed geometry and the morphing results were kindly provided by F Kunze from SEMCON® and M Hojjat and E Stavropoulou from TUM, respectively.

tial sensitivity information, but also has to be capable of keeping certain feature lines of the geometry.

A surface regularisation method that comes up to these requirements has been developed by Prof. Bletzinger and his team at the Technical University of Munich (TUM) [28, 29]. The results of applying their method to the exhaust port above are shown in Figure 8. Depending on the size of the smoothing radius and on whether morphing is allowed only in the inward direction or both inwards and outwards, different pressure drop reductions between 2% and 20% are achieved. Since this shape optimisation process is set up such that primal and adjoint calculation are closely coupled to the shape update itself ('one-shot optimisation'), the overall cost of the optimisation amounts to not more than four primal computations.

As demonstrated above, topology optimisation and shape optimisation are complementary methods: the former for drafting, the latter for fine-tuning. Especially for the optimisation of car components that are subject to design space constraints - and in the increasingly tighter getting installation spaces within the engine compartment and the passenger cabin this is the case for the majority of components - the combination of both methods suits the requirements of automotive component design very well. Starting from the feasible installation space, they are capable of delivering a fine-tuned optimal geometry, thereby each of them making use of the full potential of the available design space.

3 External aerodynamics

The adjoint-based topology and shape optimisation methods described above are obviously not restricted to ducted flows but can equally well be applied to external flows. The external aerodynamics of entire vehicles is, however, a peculiar application for optimisation methods: Except for dedicated low emission cars, vehicle shapes are to a high degree driven by aesthetic considerations rather than aerodynamic performance. This has two implications: (1) Given the rather limited design freedom, external aerodynamic optimisation is more about fine-tuning. Topology optimisation is therefore not an adequate tool. (2) Since aesthetic requirements can impossibly be casted into mathematical constraints, the *automatic* optimisation of vehicle shapes forbids itself - unless restricted to small portions like spoilers, mirrors or, of course, to the underbody.

Under these circumstances, the mere computation of surface sensitivity maps - with neither a topological nor a shape update - turned out to be a tool that fits very nicely into the specific requirements of the development of external vehicle shapes: The information contained in these maps provides the designer with concrete suggestions for aerodynamic improvements, and it is in his/her hand to translate this information into an aerodynamically improved shape without compromising the aesthetics.

3.1 Validation of RANS shape sensitivities for external aerodynamics

In order to validate sensitivity maps and to explore their potential for external aerodynamics, the adjoint method was first applied to the Volkswagen XL1 - a dedicated low emission vehicle developed by Volkswagen in 2011 (Figure 9). Based on a steady-state RANS solution on a low-Reynolds mesh of a half-model of the XL1, our adjoint OpenFOAM code with adjoint turbulence model according to Zymaris et al. [9] was employed to compute a sensitivity map for drag (Figure 10).

Various areas are identified as being very sensitive (i.e. dark blue or dark red). In relation to the experience of a well-versed aerodynamicist, these areas are of different qualities:

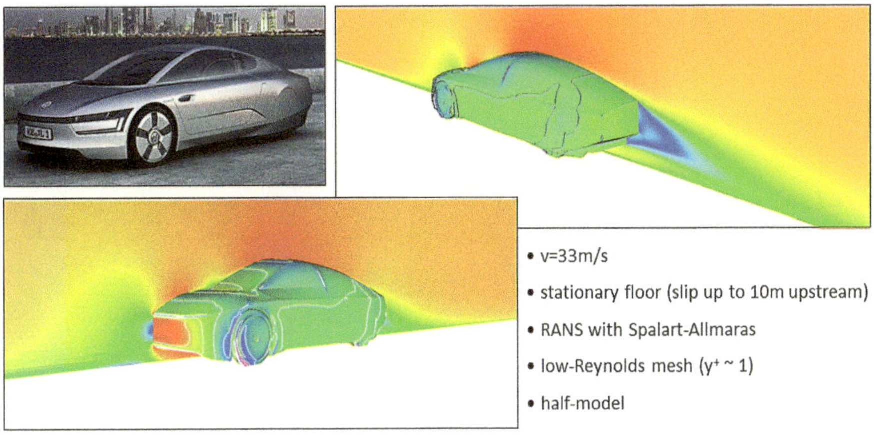

Figure 9 The primal simulation for the Volkswagen XL1. The symmetry plane is coloured according to the velocity magnitude, while the car surface shows the pressure distribution.

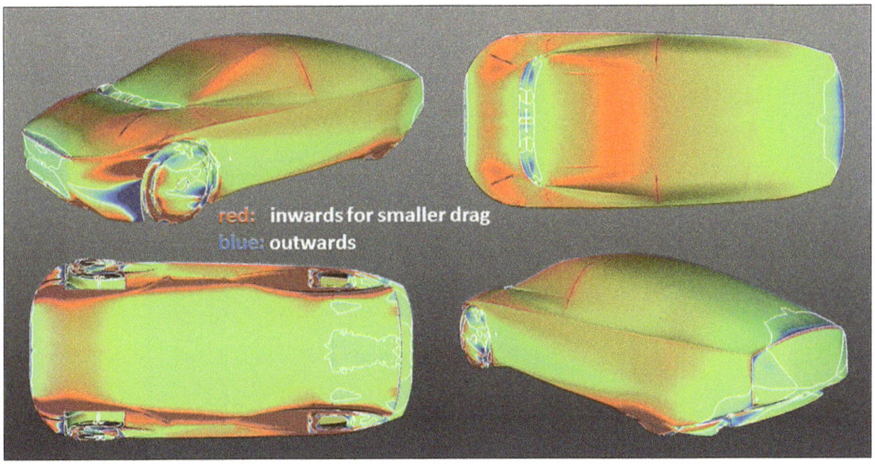

Figure 10 Drag sensitivity maps for the XL1. Isometric front and back view, bottom view (bottom left) and top view (top right).

First of all, the 'usual suspects' of external car aerodynamics (the roof, the high-curvature regions of the bonnet, the areas around the wheels and the rear sillboard) are confirmed. For other sensitive areas like the spanwise blue stripe at the transition between the front-end and the bonnet (region 2 in Figure 11), it can be seen in the primal solution already that the flow detaches and requires the curvature to be reduced *locally* - an adjoint is not needed to get this piece of information. The added-value of the adjoint, however, becomes obvious for regions 1, 3 and 4: the rear end of the car where the adjoint 'suggests' a spoiler, the blue area of the front wing which should be further exposed, and the wheel spoiler area in front of the front tyres. This information cannot be deduced by looking only at the primal. The effect of changing the car shape here is *non-local* and happens further downstream of the sensitive regions themselves. It can therefore only be identified by the adjoint.

For the validation study, the four regions already mentioned and shown in Figure 11 were parameterised with free-form deformation boxes (Figure 12), and the car shape was

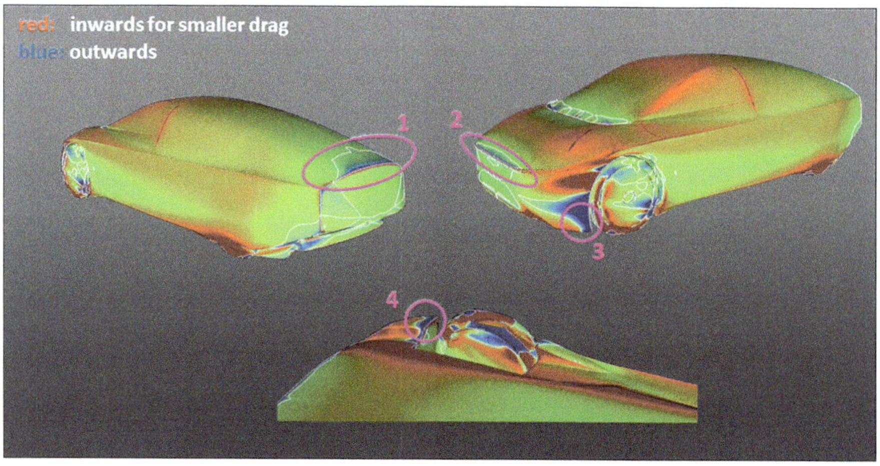

Figure 11 Selected areas for the validation of RANS surface sensitivity maps. Rear spoiler (**1**), front end of bonnet (**2**), front wing (**3**) and wheel spoiler area (**4**).

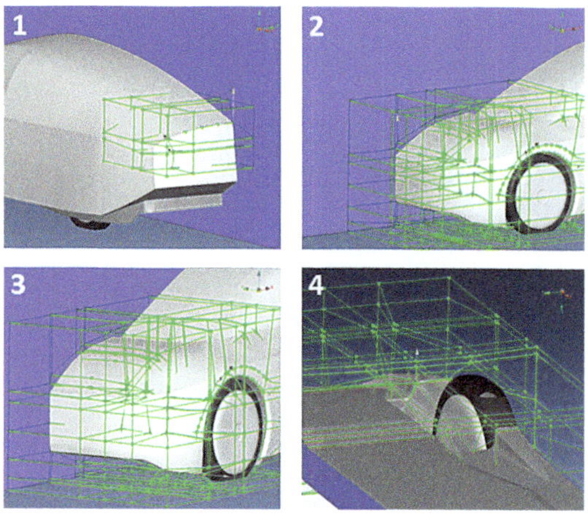

Figure 12 Definition of morphing boxes. The numbers refer to the region numbering of Figure 11 (kindly set up by BETA CAE Systems within their pre-processor ANSA®).

morphed by 'macroscopic' distances (5 or 10 mm, respectively) both according to and opposite to the computed sensitivity map. Recomputations of the morphed car shapes confirmed the correctness of the sensitivity signs in all cases: Morphing in the direction of the sensitivities resulted in a decrease of drag, while drag increased upon morphing opposite to the local sensitivities (Figure 13). Additionally, the drag reduction resulting from a rear spoiler (region 1) and from a wheel spoiler (region 4) were confirmed in wind tunnel experiments.

The automation of vehicle shape morphing - at least for limited portions of the entire vehicle shape - can be accomplished by coupling the adjoint OpenFOAM solver with the free-form deformation tool of ANSA® [30]: Via the chain rule, the node sensitivities as computed by the adjoint solver are mapped onto the morphing control points of ANSA (see [10]). Their positions can then be optimised by any gradient-based algorithm. Exemplarily, this method was applied to the rear spoiler area of the XL1 (region 1). The rear edge of the half-model was parameterised by 5 morphing control points whose move-

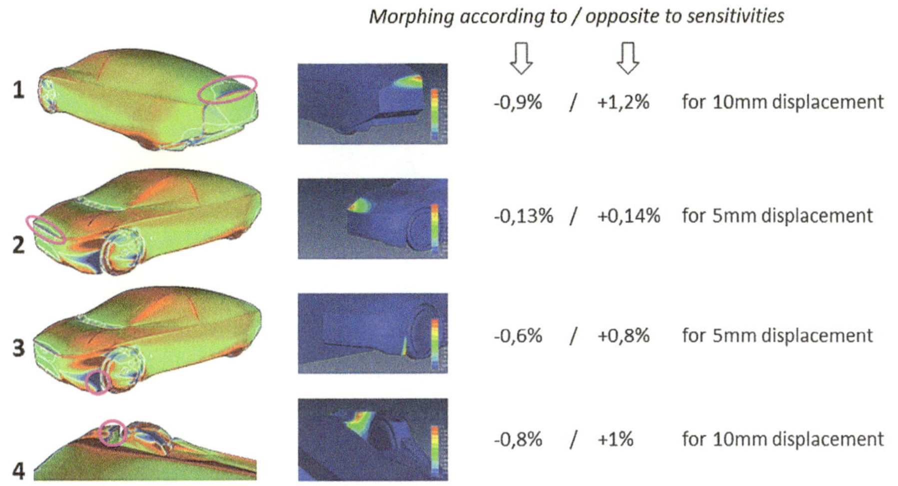

Figure 13 Recomputation of morphed geometries. Given are the drag variations upon morphing in the direction suggested by the sensitivity map (left column) and in the opposite direction (right column) for a deformation by 5 or 10 mm (assumed as sensible macroscopic distances for the respective region of the car).

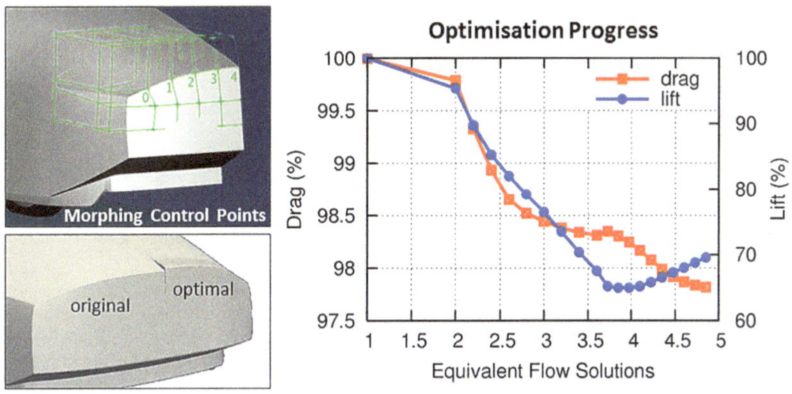

Figure 14 One-shot optimisation of the XL1 spoiler shape. The rear edge of the XL1 half-model is parameterised with five morphing control points (top left). After twenty steepest descent driven shape updates in one-shot fashion, drag reduced by 2% and lift by 30% at a total cost of five equivalent flow solutions (right). The bottom left figure compares the original and the optimal shape of the rear car edge.

ments were restricted to the z-direction (i.e. up and down, Figure 14). Within a one-shot optimisation driven by a simple steepest descent algorithm the z-coordinates of the morphing control points were moved towards their optimal position after roughly 20 iterations (Figure 14). The overall cost of the optimisation corresponded to less than five primal flow solutions and resulted in a drag decrease of 2%. Given the fact that the XL1 with a drag coefficient of less than 0.2 was aerodynamically nearly perfect already, this was a significant reduction. As a beneficial side-effect, lift decreased by 30%.

Rather than with free-form deformation boxes, which allow to modify shapes in a very controlled manner but are quite tedious to set up, another option of translating surface sensitivities into an improved shape are of course the node-based morphing algorithms from TUM as introduced in the previous section. Since they are capable of moving each

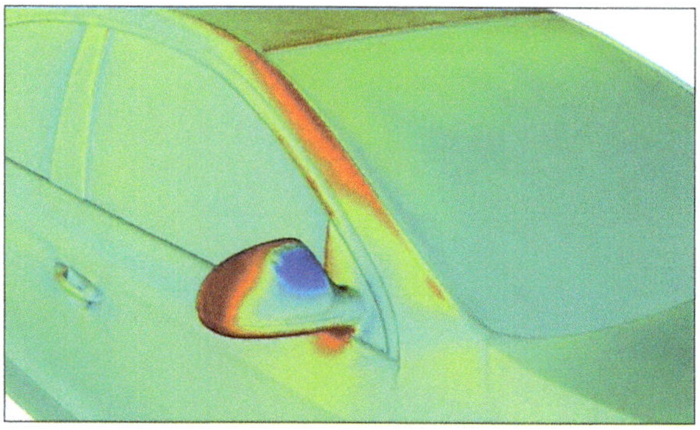

Figure 15 Shape optimisation of an external mirror. Displayed are RANS-based drag sensitivites in the mirror region of the previous Volkswagen Passat. The colouring is as usual - red regions have to be pushed inwards and blue to be pulled out in order to reduce drag.

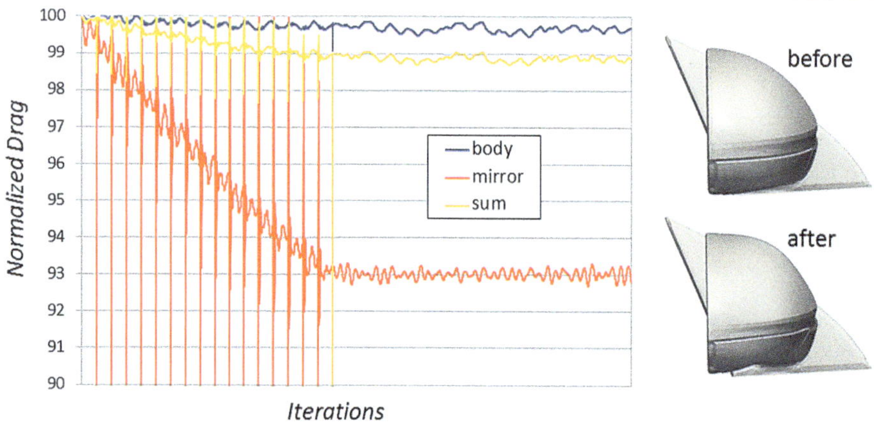

Figure 16 Shape optimisation of an external mirror. Based on the drag sensitivities of Figure 15, the TUM morphing routines were employed to conduct a one-shot shape optimisation. Only the mirror and the triangle mounted to the car were subject to deformation. The evolution of drag on the mirrors themselves (orange curve), the remaining car body (blue) and their sum effect on the entire vehicle (yellow) are shown. Due to a missing constraint on the mirror glass area, the optimisation would ultimately generate inapplicably small mirror glasses and was therefore aborted before that happened. A constrained version of this optimisation will be published in [29]. Note how the morphing affected the overall shape of the mirror (e.g. the bulgier upstream shape and the more horizontal upper egde), but was capable of keeping the essential feature lines of the original design. An illustrative movie of the shape evolution during optimisation is provided as *Additional file* 1.

and every node into its optimal position, they are ideally suited for the optimisation of vehicle details - like external mirrors.

Figure 15 depicts an extract of the drag sensitivity map of the previous Volkswagen Passat. It was used in combination with the TUM morphing routines in order to optimise the shape of the external mirrors. The evolution of drag against the iterative progress of the one-shot optimisation is shown in Figure 16. A remarkable 1% drag reduction of the entire car was achieved by only optimising the mirror shapes and leaving the rest of the car unchanged. The bulgier upstream part and the more horizontal top edge reduce the size of the mirror wake and thus the drag (Figure 16).

Figure 17 The current Audi A7. As simplifications for the primal flow model, the engine bay was completely sealed (i.e. no underhood flow) and the floor was assumed stationary.

3.2 RANS sensitivities vs. approximate time-averaged DES sensitivities

The preceding section summarised the achievements in the development of adjoint-based shape optimisation methods for external aerodynamics of vehicles. These methods rest upon steady-state RANS simulations. Meanwhile, however, the common practice of external aerodynamics computations has moved - at least in the largest part of the Volkswagen Group - from steady-state RANS to unsteady Detached-Eddy Simulations (DES, [31]). To create an adjoint to a time-varying primal like DES entails particular difficulties: Since the adjoint runs backwards in time, all primal states for the time period of interest have either to be stored or recomputed when running the adjoint. The huge amount of storage space and computational time can, in principle, be reduced to a manageable amount by clever checkpointing algorithms [32, 33], but for the typical model sizes of car external aerodynamics an adjoint to a DES run is still not feasible yet.

Our provisional solution to compute sensitivity maps for primal DES computations therefore looks as follows: We plug the time-averaged velocity $\mathbf{v}$ and pressure p from the primal DES computation into a RANS turbulence model in order to solve for a turbulent viscosity ν_t. With these quantities - time-averaged $\mathbf{v}$ and p as well as RANS-ν_t - we run the existing RANS adjoint code to obtain the sensitivities. A validation study similar to the one above for the XL1 revealed that we can only expect qualitative accuracy from this procedure. Quantitatively, the sensitivities are not correct. Still, the following comparison will demonstrate that these approximate DES sensitivities are useful and actually have an added value against pure RANS sensitivities.

Drag sensitivities for the current Audi A7 (Figure 17) as computed by two different procedures are compared in Figures 18 to 20: RANS adjoint based on a RANS primal on the one hand, and the same RANS adjoint but run on a DES primal as described above on the other hand. For both RANS and DES primal the Spalart-Allmaras turbulence model was employed, while the adjoint made use of the frozen turbulence assumption. On the largest part of the car body, the sensitivities can be seen to be very similar. This does not come as a surprise, because here the flow is attached to the surface and RANS and DES flows do not differ dramatically for attached flows. Significant differences occur in the rear part of the car where the flow is close to detachment: While the RANS-based adjoint fails to predict the experimentally well-confirmed beneficial effect of a rear spoiler, it is clearly indicated by the DES-based adjoint. Moreover, the productive effect of a further 'boat-tailing', i.e. tapering of the car back, is almost missed by the RANS-based adjoint while

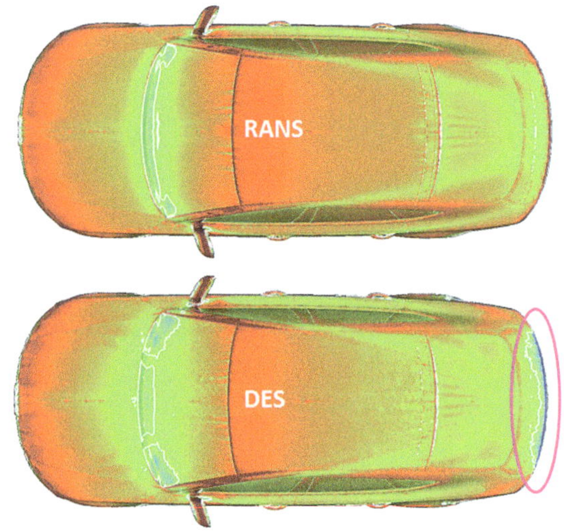

Figure 18 RANS vs. approximate DES sensitivities. Note the similarity between RANS- and DES-based drag sensitivities for the largest part of the car hat - and the dramatic difference at the rear part. While the RANS computation misses the favourable effect of a rear spoiler, it is clearly present in the DES results.

Figure 19 RANS vs. approximate DES sensitivities. A rear view of the sensitivity differences in the spoiler area, and the actual shape of the active spoiler in the current Audi A7.

clearly present in the DES counterpart. It is this kind of results that - despite the lack of their quantitative accuracy - gave a lot of credibility to our provisional solution of computing sensitivities based on DES primals and supported their integration into the regular computational processes for external aerodynamics (see Figure 21).

3.3 Adjoint methods for flow control

Geometrical modifications, be it via shape or via topology optimisation, are not the only option of improving the fluid dynamic performance of car components or the entire vehicle. Recently there is a growing interest in qualifying active flow control devices for vehicle aerodynamics: Brunn et al. [34] demonstrated how steady blowing and oscillating jets can be successfully applied to reduce drag of the well-known Ahmed body [35]. Bideaux et al. [36] used the same bluff body shape to investigate the effect of a line of pulsed jet actuators

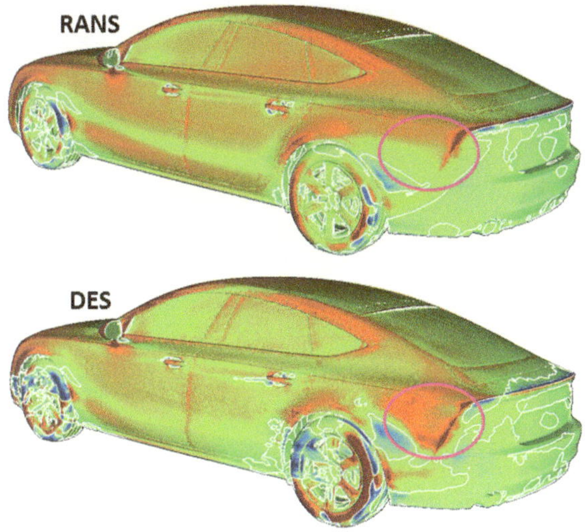

Figure 20 RANS vs. approximate DES sensitivities. The RANS-based sensitivities also fail to predict the beneficial effect of further lateral tapering of the car's back ('boat-tailing').

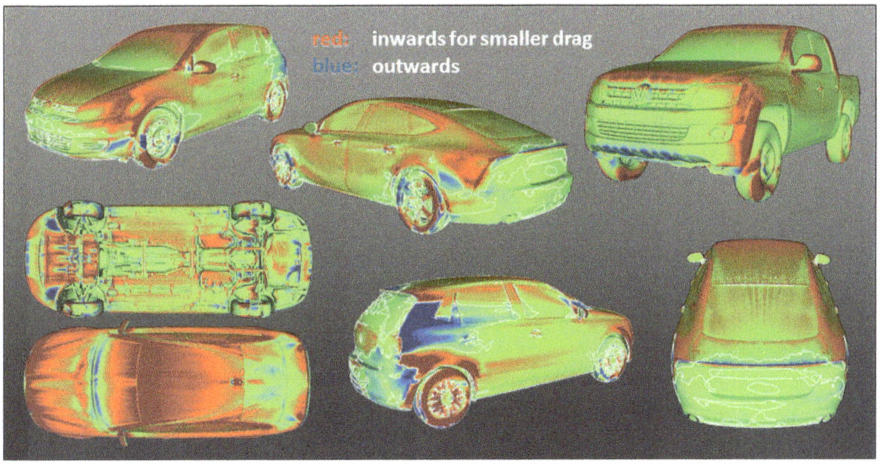

Figure 21 Drag sensitivity maps based on DES. Further examples of drag sensitivity maps for various car models of Volkswagen and Audi as computed by the outlined DES-based method.

on drag and lift, while Heinemann et al. [37] experimentally investigated continuous jets on a 1:4 model of a passenger production car.

Since flow control of entire vehicles is a rather young subject, experience for the optimal layout of jet configurations is still very limited. The questions to be answered when designing a flow control concept are: (1) What kind of jets should be applied - blowing or suction? (2) Where should the jets be positioned to have a maximum effect? (3) What is the achievable aerodynamic improvement?

To answer these questions, the adjoint method is again the tool of choice, as it allows to compute a sensitivity map of the whole car for the positioning of jets. While for shape sensitivity maps, the design variables were the normal displacements of all surface nodes, it is now the normal flow velocity v_n on each surface element that constitutes the design space. In other words, the car is regarded as having a perforated surface with initially zero normal

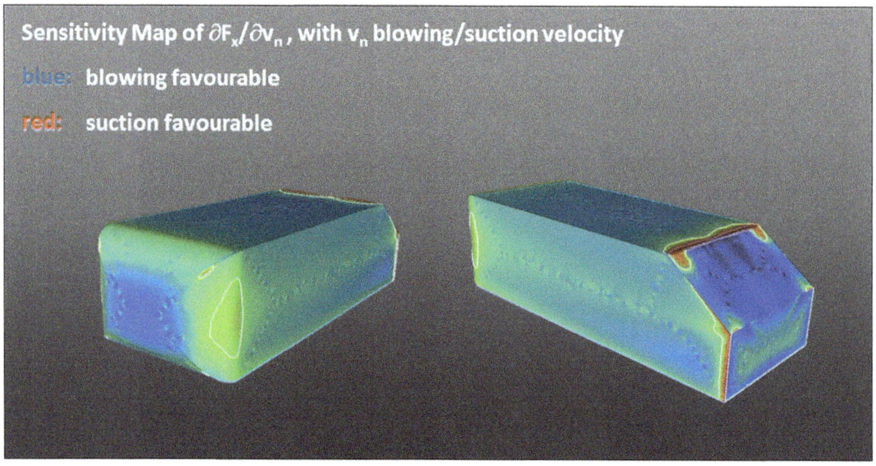

Figure 22 Flow control sensitivities for the Ahmed body. Based on a RANS computation, a flow control sensitivity map of drag F_x w.r.t. normal jet velocity v_n was computed for the 35° case of the Ahmed body. The highest sensitivities are expectedly encountered at the rear part. Suction is favourable at the rear separation edge and at the origin of the longitudinal vortices, whereas on the slant surface blowing would reduce the drag by filling the wake.

mass flux through it, and the adjoint method is then applied to compute the sensitivity of drag F_x or lift F_z w.r.t. changing the mass flux through each of the 'holes' [11].

Figure 22 shows such a sensitivity map for the Ahmed body, i.e. the distribution of $\partial F_x/\partial v_n$. Blue areas are those where blowing ($v_n > 0$) would result in drag reduction, while in red areas a suction jet would help to reduce drag. The colour intensity indicates the magnitude of the effect on the drag F_x. The computation was carried out on a low-Reynolds mesh ($y^+ \approx 1$) using a RANS primal with the Spalart-Allmaras turbulence model and its adjoint counterpart.

The validation of these sensitivities is shown in Figure 23. At two different locations, one in the blue and another in the red region, various normal velocities ranging from −40 m/s (suction) to +40 m/s (blowing) were applied in order to compute their effect on drag. As expected, for small jet velocities of up to 5 m/s, the sensitivity captures the gradient $\Delta F_x/\Delta v_n$ very well. But surprisingly, still for higher jet velocities the sensitivity constitutes a good approximation to the actual effect of the jet. In conclusion, flow control sensitivity maps do not only indicate whether a suction jet or a blowing jet is favourable and where to place it, but also give a useful estimate of the achievable effect.

For the Volkswagen XL1, apart from pure shape optimisation as shown above, also the possibility of further drag reduction by blowing jets was investigated. A flow control sensitivity map revealed several promising areas for the application of blowing jets (blue areas in Figure 24). On the car hat, they correspond to a large extent to the blue areas of Figure 10, i.e. where an outward perturbation of the surface was found beneficial. The actual application of flow control, however, concentrated on the underbody - more precisely in a region behind the rear axis (pink circle in Figure 24), where we installed a number of blowing jets. Force measurements, oil film visualisations as well as Particle-Image-Velocimetry (PIV) measurements on a 1:4 model (Figure 25, details see [38]) revealed the effect of these jets on the flow around the XL1: The recirculation area behind the car became smaller, and the flow in the symmetry plane was directed in the upward direction (Figure 26). This was found to be due to a weakening of the longitudinal vortices that emanate from the rear

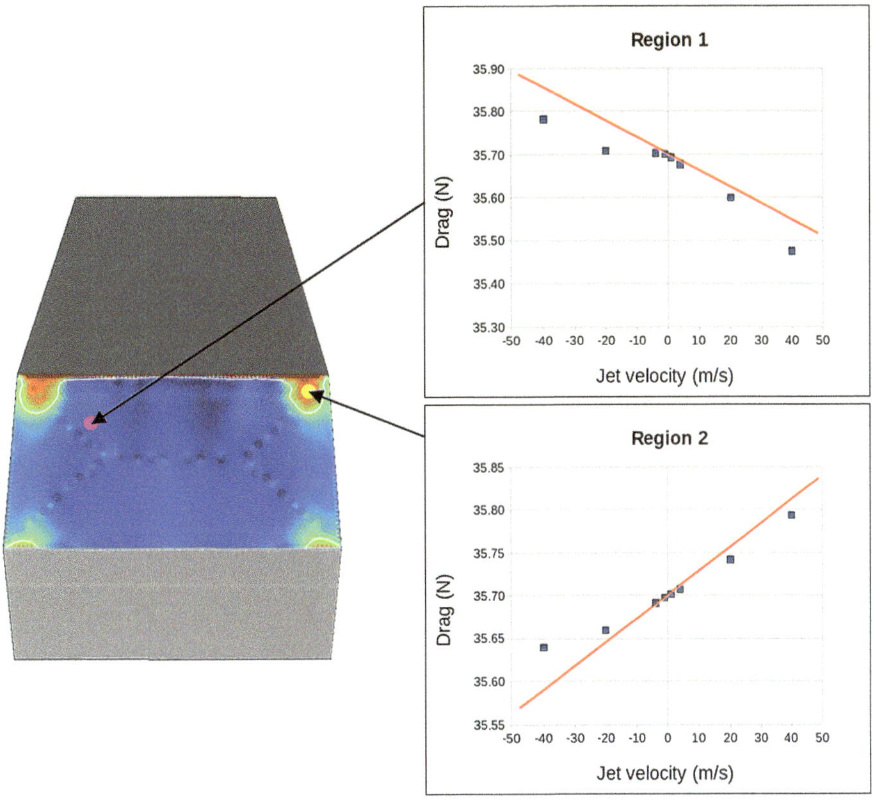

Figure 23 Flow control sensitivities for the Ahmed body. For validation purposes, normal jet velocities between –40 m/s (suction) and +40 m/s (blowing) were applied on surface elements at two different locations of opposite sensitivity signs (pink and yellow spot, respectively). The blue squares indicate the computed drag values, and the red lines correspond to the sensitivity of the respective locations. Note the qualitative and quantitative agreement even for high jet velocities.

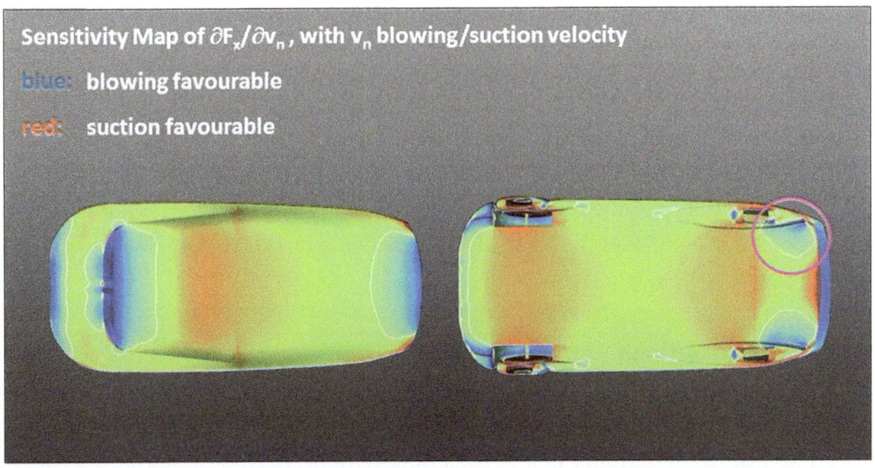

Figure 24 Flow control sensitivities for the Volkswagen XL1. Note the correspondence between 'blue' flow control sensitivities and shape sensitivities (Figure 10) on large parts of the car hat. In the hardware 1:4 model, blowing jets were applied symmetrically in the region within the pink circle.

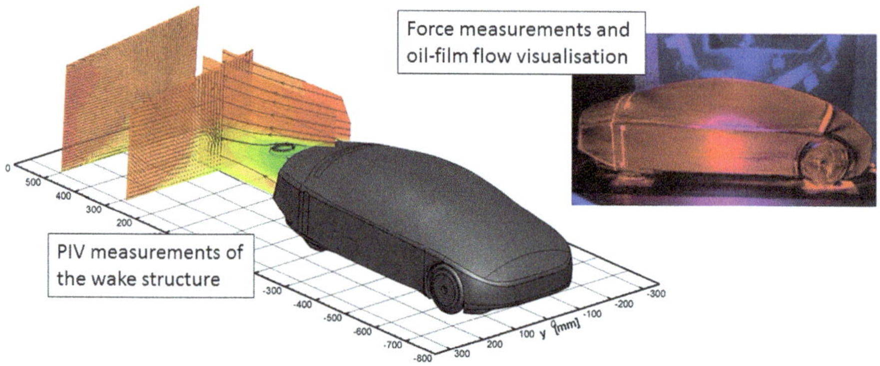

Figure 25 Wind tunnel measurements of the XL1. In addition to force measurements and oil film visualisation, PIV measurements were conducted in four different planes in the car's wake, including the symmetry plane (data kindly provided by R Petzold and P Scholz from the Technical University of Braunschweig [38]).

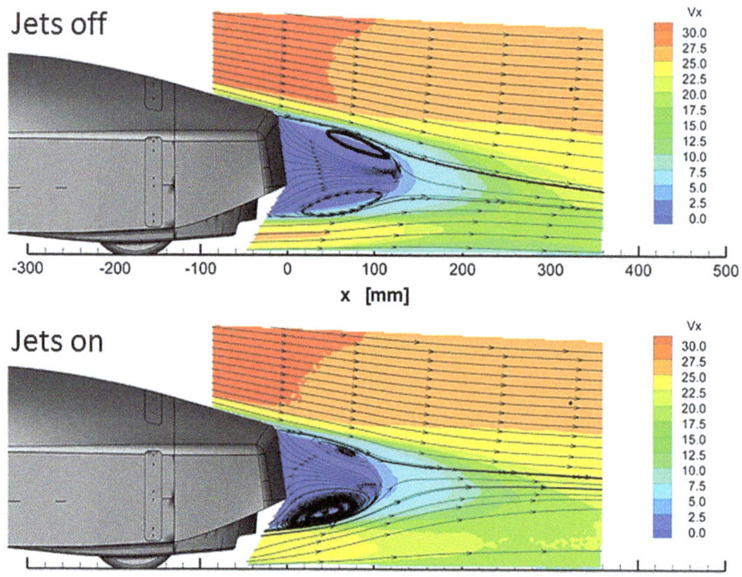

Figure 26 PIV measurements in the symmetry plane of the XL1 wake. Note the slightly smaller recirculation area and the upwardly directed flow when the jets are switched on (see [38] for details).

corners of the car (Figure 27). As a result, lift dropped by 20%, and drag - in accordance with the computed sensitivities - decreased by about 2%. Even though this additional drag reduction of 2% is significant for a car like the XL1, it is unfortunately not enough to incorporate a jet system into a car and run it productively.

4 Summary and outlook

Starting from the obstacles of introducing the adjoint method into automotive CFD, we have reported on the efforts undertaken with various partners to develop an industrially viable adjoint solver. Based on OpenFOAM, powerful optimisation methods - topology optimisation for ducted flows, and shape optimisation for ducted as well as external flows - have been developed. While topology optimisation is being used systematically already, the implementation of shape optimisation into the regular automotive development pro-

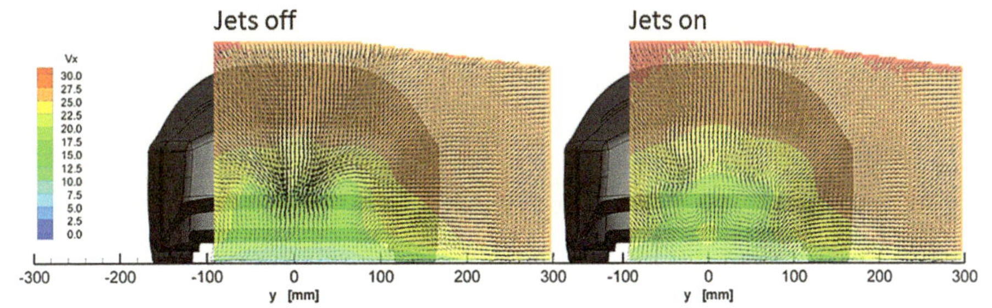

Figure 27 PIV measurements in a perpendicular plane behind the XL1. Upon switching on the jets, the two counter-rotating longitudinal vortices are effectively weakened. This goes along with less downwash and hence decreased lift (see [38] for details).

cess has just started with some initial promising results. Both methods will most surely have an impact on overall vehicle performance and consumption in the future.

The adjoint development efforts have so far concentrated on steady-state phenomena. For an extension towards inherently transient flow applications, like aeroacoustics, flow control with pulsating jets and transient aerodynamics with DES, the adjoint has to become transient as well. This is a subject that is under investigation by several research groups already (e.g. [39–41]) and is currently addressed in the Marie-Curie International Training Network 'aboutFlow' [42]. Its industrialisation can be regarded as the next challenge in developing adjoint methods for the automotive industry.

Additional material

Additional file 1: External mirror shape optimisation. How the shape of the external mirror of Figures 15 and 16 evolves from the original to the optimal shape and back again is illustrated by this movie. The colour represents the magnitude of node displacement as compared to the original (details see [28, 29]).

Competing interests

The author declares that they have no competing interests.

Acknowledgements

The author is truly grateful to E de Villiers and his colleagues at Engys, KC Giannakoglou and his team at NTUA (above all EM Papoutsis-Kiachagias and AS Zymaris), E Stavropoulou, M Hojjat and KU Bletzinger from TUM, JD Müller (Queen Mary University London) and the FlowHead team, as well as E Skaperdas and K Haliskos from BETA CAE Systems. The support from many CFD colleagues of the Volkswagen Group in the form of continuous interest in the adjoint method, provision of resources and of valuable feedback is also gratefully acknowledged.

References

1. Lions JL: *Optimal Control of Systems Governed by Partial Differential Equations*. New York: Springer; 1971.
2. Pironneau O: **On optimum design in fluid mechanics.** *J Fluid Mech* 1974, **64**:97-110.
3. Jameson A: **Aerodynamic design via control theory.** *J Sci Comput* 1988, **3**:233-260.
4. Löhner R, Soto O, Yang C: **An adjoint-based design methodology for CFD optimization problems.** *AIAA-03-0299* 2003.
5. Nadarajah S, Jameson A: **A comparison of the continuous and discrete adjoint approach to automatic aerodynamic optimization.** *AIAA-00-0667* 2000.
6. **OpenFOAM® - The Open Source CFD Toolbox** [http://www.openfoam.com]
7. Othmer C, de Villiers E, Weller HG: **Implementation of a continuous adjoint for topology optimization of ducted flows.** *AIAA-2007-3947* 2007.
8. Othmer C: **A continuous adjoint formulation for the computation of topological and surface sensitivities of ducted flows.** *Int J Numer Methods Fluids* 2008, **58**:861-877.
9. Zymaris AS, Papadimitriou DI, Giannakoglou KC, Othmer C: **Continuous adjoint approach to the Spalart-Allmaras turbulence model for incompressible flows.** *Comput Fluids* 2009, **38**(8):1528-1538.
10. Othmer C, Papoutsis-Kiachagias EM, Haliskos K: **CFD optimization via sensitivity-based shape morphing.** In *Proceedings of the 4th ANSA & μETA Internat. Conf., Thessaloniki, Greece, BETA CAE Systems S.A.* 2011.

11. Zymaris AS, Papadimitriou DI, Giannakoglou KC, Othmer C: **Optimal location of suction or blowing jets using the continuous adjoint approach.** In *Proceedings of the 5th European Conf. on Computational Fluid Dynamics, Lisbon, Portugal, ECCOMAS CFD* 2010.
12. Zymaris AS, Papadimitriou DI, Giannakoglou KC, Othmer C: **Adjoint wall functions: a new concept for use in aerodynamic shape optimization.** *J Comput Phys* 2010, **229**(13):5228-5245.
13. Kontoleontos EA, Papoutsis-Kiachagias EM, Zymaris AS, Papadimitriou DI, Giannakoglou KC: **Adjoint-based constrained topology optimization for viscous flows, including heat transfer.** *Eng Optim* 2013, **45**(8):941-961.
14. de Villiers E, Othmer C: **Multi-objective adjoint optimization of intake port geometry.** *SAE Technical Paper 2012-01-0905* 2012.
15. Helgason E, Krajnovic S: **Aerodynamic shape optimization of a pipe using the adjoint method.** In *Proceedings of the ASME 2012 International Mechanical Engineering Congress & Exposition, Houston, Texas, USA, IMECE* 2012.
16. Lincke A, Rung T: **Adjoint-based sensitivity analysis for Buoyancy-driven incompressible Navier-Stokes equations with heat transfer.** In *Proceedings of the Eighth Internat. Conf. on Engineering Computational Technology, Dubrovnik, Croatia* 2012.
17. Jakubek D, Wagner C: **Shape optimization of train head cars using adjoint-based computational fluid dynamics.** In *Proceedings of the First International Conference on Railway Technology, Las Palmas de Gran Canaria, Spain* 2012.
18. Towara M: **Numerical optimization of an oil intake duct with adjoint topological methods.** *Diploma thesis.* RWTH Aachen University, Dept. of Software and Tools for Computational Engineering; 2011.
19. Hinterberger C, Olesen M: **Automatic geometry optimization of exhaust systems based on sensitivities computed by a continuous adjoint CFD method in OpenFOAM.** *SAE Technical Paper 2010-01-1278* 2010.
20. Hinterberger C, Olesen M: **Industrial application of continuous adjoint flow solvers for the optimization of automotive exhaust systems.** In *Proceedings of the ECCOMAS Thematic Conf. - CFD & Optimization, Antalya, Turkey, ECCOMAS* 2011.
21. Giannakoglou KC: **Continuous adjoint methods in shape, topology, flow-control and robust optimization.** In *Open Source CFD International Conference, London, UK, ICON-CFD* 2012.
22. Bendsoe MP, Sigmund O: *Topology Optimization: Theory, Methods and Applications.* Berlin: Springer; 2004.
23. Borrvall T, Petersson J: **Topology optimization of fluids in Stokes flow.** *Int J Numer Methods Fluids* 2003, **41**:77-107.
24. Moos O, Klimetzek F, Rossmann R: **Bionic optimization of air-guiding systems.** *SAE Technical Paper 2004-01-1377* 2004.
25. Gersborg-Hansen A, Sigmund O, Haber R: **Topology optimization of channel flow problems.** *Struct Multidiscip Optim* 2005, **30**:181-192.
26. Othmer C, Grahs T: **Approaches to fluid dynamic optimization in the car development process.** In *Proceedings of the EUROGEN Conference, Munich, Germany* 2005.
27. Othmer C, Kaminski T, Giering R: **Computation of topological sensitivities in fluid dynamics: cost function versatility.** In *Proceedings of the ECCOMAS CFD Conference, Delft, The Netherlands* 2006.
28. Stavropoulou E, Hojjat M, Bletzinger KU: **In-plane mesh regularization for node-based shape optimization problems.** *Comput Methods Appl Mech Eng*, in press [http://www.sciencedirect.com/science/article/pii/S004578251400070X]
29. Hojjat M, Stavropoulou E, Bletzinger KU: **The vertex morphing method for node-based shape optimization.** *Comput Methods Appl Mech Eng* 2014, **268**:494-513.
30. **ANSA® pre-processor – BETA CAE Systems S.A.** [http://www.beta-cae.gr/ansa.htm]
31. Islam M, Decker F, de Villiers E, Jackson A, Gines J, Grahs T, Gitt-Gehrke A, Comas i Font J: **Application of detached-Eddy simulation for automotive aerodynamics development.** *SAE Technical Paper 2009-01-0333* 2009.
32. Griewank A: **Achieving logarithmic growth of temporal and spatial complexity in reverse automatic differentiation.** *Optim Methods Softw* 1992, **1**:33-54.
33. Wang Q, Moin P, Iaccarino G: **Minimal repetition dynamic checkpointing algorithm for unsteady adjoint calculation.** *SIAM J Sci Comput* 2009, **31**(4):2549-2567.
34. Brunn A, Wassen E, Sperber D, Nitsche W, Thiele F: **Active drag control for a generic car model.** *Notes Numer Fluid Mech Multidiscipl Des* 2007, **95**:247-259.
35. Ahmed SR, Ramm R, Falting G: **Some salient features of the time-averaged ground vehicle wake.** *SAE Technical Paper 84-0300* 1984.
36. Bideaux E, Bobillier P, Fournier E, Gillieron P, El Hajem M, Champagne JY, Gilotte P, Kourta A: **Drag reduction by pulsed jets on strongly unstructured wake: towards the square back control.** *Int J Aerodyn* 2011, **1**(3/4):282-298.
37. Heinemann T, Springer M, Lienhart H, Kniesburges S, Becker S: **Active flow control on a 1:4 car model.** In *Proceedings of the 16th Int. Symp. on Applications of Laser Techniques to Fluid Mechanics, Lisbon, Portugal* 2012.
38. Petzold R: **Experimentelle Untersuchungen zur Strömungsbeeinflussung an einem 1:4-Fahrzeugmodell.** *Diploma thesis.* TU Braunschweig, Inst. für Strömungsmechanik; 2011.
39. Torbert S, Löhner R: **Checkpointing schemes for adjoint methods and strongly unsteady flow.** *AIAA-2012-0572* 2012.
40. Economon TD, Palacios F, Alonso JJ: **A coupled-adjoint method for aerodynamic and aeroacoustic optimization.** *AIAA-2012-5598* 2012.
41. Carnarius A, Thiele F, Özkaya E, Nemili A, Gauger N: **Optimal control of unsteady flows using a discrete and a continuous adjoint approach.** In *Proceedings of the 25th Conference on System Modeling and Optimization, Berlin, Germany*; 2011:318-327.
42. **aboutFlow - Adjoint-based optimisation of industrial and unsteady flows** [http://aboutflow.sems.qmul.ac.uk]

10

Simultaneous estimation of mass and aerodynamic rotor imbalances for wind turbines

Jenny Niebsch[1*] and Ronny Ramlau[1,2]

*Correspondence:
jenny.niebsch@oeaw.ac.at
[1] Austrian Academy of Sciences, RICAM, Altenbergerstr. 69, Linz, 4040, Austria
Full list of author information is available at the end of the article

Abstract

The safe operation of wind turbines requires a well-balanced rotor. The balancing of the rotor requires a method to determine its imbalances. We propose an algorithm for the reconstruction of two types of imbalances, i.e., mass and aerodynamic imbalances from pitch angle deviation. The algorithm is based on the inversion of the (nonlinear) operator equation that links the imbalance distribution of the rotor to its vibrations during operation of the wind turbine. The algorithm requires a simple finite element model of the wind turbine as well as the minimization of a Tikhonov functional with a nonlinear operator. We propose the use of a gradient-based minimization routine. The approach is validated for artificial vibration data from a model of a Nordwind NTK 500 wind turbine.

Keywords: rotor imbalance; inverse problem; nonlinear Tikhonov regularization

1 Introduction

In the growing field of wind energy extraction, the topic of rotor imbalances of wind turbines (WT) is of vital importance for the operation, safety, and durability of the turbines. Approximately 20%-50% of WT have significant rotor imbalances. This leads to damages of important components, high repair expenses, and reduced output [1]. Rotor imbalances can be categorized into mass and aerodynamic imbalances. Mass imbalances arise from inhomogeneous mass distributions of the rotor including blades and hub. They induce centrifugal forces that depend on the square of the rotational frequency Ω_{rot}. The consequences are harmonic vibrations with Ω_{rot} in the rotor plane (lateral). The vibration amplitude depends on the ratio between Ω_{rot} and the resonance frequency (first eigenfrequency) Ω_0 of the turbine. Mass imbalances can be eliminated or reduced using balancing counterweights. Aerodynamic imbalances evolve from deviations in the aerodynamic properties of the blades. These lead to different thrust and tangential forces and moments, and they depend not only on the rotational frequency but also on the wind speed. Aside from lateral vibrations, they induce axial and torsional vibrations of WT. The main reason for aerodynamic imbalance is a relative deviation of the pitch angles of the blades. This can only be eliminated by a correction of those angles. Additional masses are not helpful here [1]. In practice, both types of imbalances are frequently observed both separately and in combination.

Usually, the detection of imbalances is based on spectral analysis and order analysis methods [2]. Another technique is to monitor the power characteristic [3], whereby power mean values are observed, and deviations from faultless conditions are used for the calculation of alarm limits. Unfortunately, this requires a learning phase under faultless conditions. This also holds for other signal processing methods, cf. [2]. Another disadvantage is the fact that the amount of the imbalance, e.g. absolute value and location of a mass imbalance or the deviation of one or more pitch angles, is not computable.

The state of the art in rotor imbalance determination still requires an expert team on-site. Initially, aerodynamic imbalances must be detected using optical methods. After the elimination of the aerodynamic imbalances, the mass imbalance is determined using vibration measurements with and without reference weights.

In the last few years, the authors have developed methods which allow for imbalance determination using data that can be collected by a condition monitoring system (CMS). As a consequence, the amount and location of rotor imbalance can be computed without the expensive measurements using reference weights on-site. The methods are model-based, but they use a very simple model of the WT that can be obtained easily for each turbine type from elementary geometrical and physical parameters. The general basics of our approach are summarized in Section 2. Essentially, the method consists of the following steps: 1. modeling of the WT; 2. description of the loads (forces, moments) arising from rotor imbalance; 3. solution of the forward problem, i.e. the computation of the resulting vibrations in every node of the WT model for a given imbalance situation and 4. solution of the inverse problem where the unknown imbalance is determined from vibration data obtained in the nacelle. In our new method we began by determining the mass imbalance [4]. Later aerodynamic imbalance from pitch angle deviations was included [5].

In case of pure mass imbalances, only lateral vibrations need to be considered. Accordingly, in the model of the WT only degrees of freedom (DOF) in the lateral direction needed to be taken into account. Let us call this restricted approach the mass imbalance approach (MIA). The presence of aerodynamic imbalances in addition to mass imbalances would lead to additional lateral forces and therefore to incorrect reconstructions using MIA since the assumption in this model is that aerodynamic imbalances are not present. The model and the description of the forces were expanded in [5] in order to include the effect of pitch angle deviations as a main reason for aerodynamic imbalances. Nevertheless, MIA can serve to provide an initial guess for the solution of the problem for combined mass and aerodynamic imbalances. In the combined imbalance case, the inverse problem is nonlinear. As the problem is ill-posed, Tikhonov regularization is used. This includes the minimization of the Tikhonov functional which was done in [5] using a direct search method. Although we could obtain first test results, the routine was far to slow and not stable enough for extensive test computations. The acceleration of the minimization routines requires the computation of the Frechet derivative of the forward problem operator **A**. As most of the theory for Tikhonov regularization is formulated for real valued vector spaces, the forward operator **A** has to be formulated in a real setting.

This paper focuses on a stable reconstruction method of mass imbalance and pitch angle deviation based on the minimization of the Tikhonov functional. To this end, the nonlinear forward operator **A** is derived in detail and its Frechet derivative is computed which is new compared to [5]. Now the minimization of the Tikhonov functional is performed using the steepest decent method. The resulting reconstruction algorithm is stable and reasonable fast. Additionally, a suitable metric is defined to measure the reconstruction quality. We present test results obtained with the gradient-based algorithm for artificial data. The paper is organized as follows: In Section 3 we derive the forward problem operator **A**. We will be very brief if there are no changes to the preceding paper [5] and more detailed in the description of the new parts. In particular, the Frechet derivative of **A** is computed. Section 4 is concerned with the inverse problem. Regularization and minimization techniques using gradient-based methods are presented as well as the final algorithm that reconstructs the imbalances from measurement data. In Section 5 we present numerical test computations for the reconstruction of pitch angle deviations only and a combination of mass imbalances and pitch angle deviations as well as results for the reconstruction quality in the presented metric.

2 General approach

The procedure to determine imbalances based on a model of the WT can be summarized as follows: The load $p(\mathbf{x})$ arising from an imbalance $\mathbf{x}$ induces displacements or vibrations u of the WT in different directions. Here we deal with displacements in axial direction (along the y-axis), lateral displacements (along the z-axis), and torsion around the tower center axis (x-axis), cf. Figure 1. The displacements and the imbalance loads are coupled via a Partial Differential Equation (PDE) in time and space which is seldom explicitly solvable. Therefore, we use the Finite Element Method (FEM) to transform the PDE into a system of Ordinary Differential Equations (ODE) in time. To this end we consider the rotor and nacelle as point masses and the WT tower as a flexible shaft that is divided into n elements with $n + 1$ nodes (see Figure 1). The ODE system reads

$$\mathbf{M}\ddot{\mathbf{u}}(t) + \mathbf{S}\mathbf{u}(t) = \mathbf{p}(t), \tag{1}$$

where **M** and **S** represent the mass and stiffness matrices, **u** the vector of displacements or degrees of freedom (DOF) of each node, and **p** the vector of loads for each DOF. There could be an additional damping term $\mathbf{D}\dot{\mathbf{u}}(t)$ on the left hand side of (1), but so far we have achieved sufficiently good results with damping neglected.

The matrices **M** and **S** need to be derived for each different type of WT separately. This is the *first step* in our procedure. The *second step* pertains to the description of the load $\mathbf{p}(\mathbf{x})$ as a function of the imbalances $\mathbf{x}$. In the *third step*, the now fully described ODE system (1) is solved. The solution operator $\tilde{\mathbf{A}}$ maps the imbalances $\mathbf{x}$ to the displacement $\mathbf{u}$ at each node of the FE model: $\tilde{\mathbf{A}}\mathbf{x} = \mathbf{u}$. Now we restrict the displacement vector $\mathbf{u}$ that contains the displacement of each model DOF to the vector $\mathbf{y}$ of the DOF that are accessible for measurements. The operator $\tilde{\mathbf{A}}$ is accordingly restricted to **A**. The resulting operator equation is $\mathbf{A}\mathbf{x} = \mathbf{y}$ where the operator **A** maps the imbalance $\mathbf{x}$ of the rotor in terms of mass imbalance and pitch angle deviation to the displacement $\mathbf{y}$ in the nodes that can be observed by measurements. This operator equation represents the forward problem. Hence *step 1-3* are connected with the solution of the forward problem. Now in a *fourth step* we solve the inverse problem, i.e., we determine the imbalance $\mathbf{x}$ from measurements $\mathbf{y}$. Since the inverse problem is ill-posed, we use regularization techniques, [6]. This approach has been applied successfully in the more simple case of mass imbalances only [4]. First steps in the direction of reconstructing mass imbalances and pitch angle deviation simultaneously were done in [5].

Figure 1 Model of a wind turbine and real wind turbine. Top: Finite-Element model of a WT with coordinate system. Bottom: NTK500/41 in Denmark.

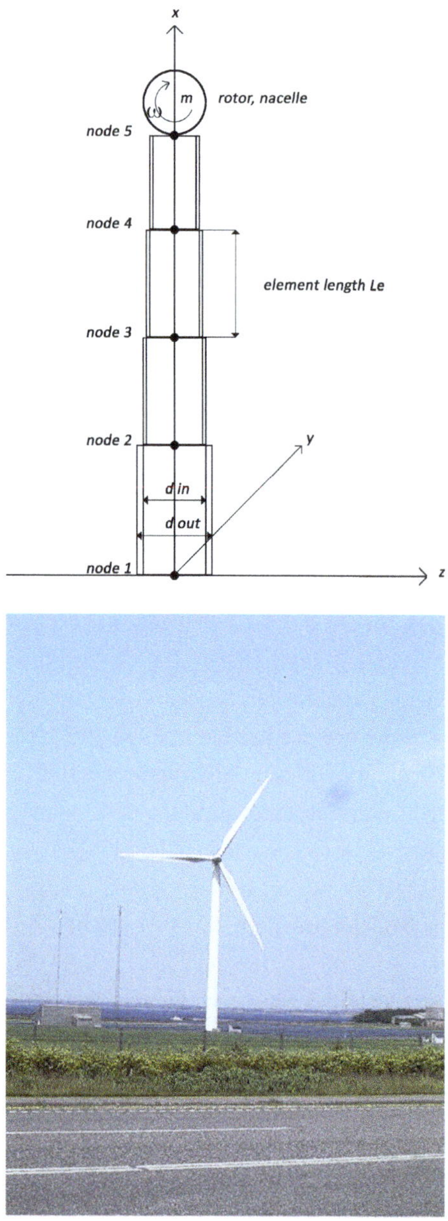

3 Forward problem

3.1 Mass and stiffness matrices

The derivation of **M** and **S** for the simple model of MIA was taken from [7] and presented in [4]. In the situation of mass and aerodynamic imbalances the model needs to be expanded to displacements in axial direction and torsion around the tower central axis. Still, the derivation of **M** and **S** follows exactly the Ritz method described in [7, p.155-160]. Once the model matrices are derived for a certain WT type they need to be optimized for each individual turbine of this type with respect to the first eigenfrequency Ω_0. The general optimization procedure is described in [4].

Thus far we have modeled six different types of WT in this way: Vestas V80, V82, V90, Südwind S77, NTK500/41 and GE1.5. For the test computations presented in this paper, we have used the model of a Nordtank turbine NTK500/41, see Figure 1 (bottom), with 35 m hub height because we are provided with aerodynamic data for the blades of this WT.

3.2 Forces and moments from imbalances – the load vector p

The imbalance of the rotor creates forces and moments that affect the top node of the FE model. The forces and moments are ordered in accordance to the DOF in the displacement vector **u**:

$$\mathbf{p}^T(t) = (0, \ldots, 0, F_y, F_z, M_x, M_y, M_z), \tag{2}$$

where F_y and F_z denote forces in y- and z-direction, M_x, M_y, and M_z are moments of forces or torques w.r.t. the x-, y- and z-axis, respectively. In particular, we have $M_y = 0$ as we assume that mechanical energy related to M_y is transformed into electrical energy. The forces and moments are the sum of the forces and moments from mass imbalance and pitch angle deviation:

$$\begin{aligned} F_y &= F_1 + F_2 + F_3, \\ F_z &= T_z + (F_c)_z, \\ M_x &= M_x^{(m)} + M_x^{(F)} + M_x^{(T)}, \\ M_z &= M_z^{(m)} + M_z^{(F)} + M_z^{(T)}. \end{aligned} \tag{3}$$

Here, the F_1, F_2, F_3 are the thrust forces at each blade. If there is no deviation in the pitch angle of the blades all three forces are equal. T_z is the projection of the sum of the tangential forces at each blade $T = T_1 + T_2 + T_3$ onto the z-axis. It is zero if there is no deviation in the pitch angles. F_c denotes the centrifugal force induced by the mass imbalance. Its projection onto the z-axis $(F_c)_z$ adds to F_z. The moments $M_x^{(m)}$, $M_z^{(m)}$ are induced by F_c of the mass imbalance, $M_x^{(F)}$, $M_z^{(F)}$, $M_x^{(T)}$, $M_z^{(T)}$ correspond to the thrust force and tangential force. Whereas the mass imbalance force can be directly computed, the aerodynamic imbalance forces need to be determined via, e.g., the Blade-Element-Momentum (BEM) method, a nonlinear procedure. In the BEM method, the blades are divided into a finite number of elements which are annulus segments with the center at the root of the blades. The cross section of each element is called an 'airfoil'. If the airfoil data, the angle of attack of the wind, and the relative wind velocity as well as a lift and drag coefficient table are given, the normal or thrust forces F and the tangential forces T can be calculated according to the BEM method, see, e.g., [8]. The BEM method is used in a lot of programs for the simulation of aerodynamic effects, cf. [9], which were tested on real WT.

For the presentation of the forces and moments, we will use the following notations and abbreviations (cf. Figures 2 and 3):

- mr, ϕ_m – absolute value and angle w.r.t. blade A of the mass imbalance,
- θ_i – deviation of the pitch angle of ith blade to an adjusted pitch angle θ,
- $\omega = 2\pi\Omega$ – angular frequency (note: WT rotates clockwise seen from the wind),
- $\varphi = 2/3\pi$ – angle between blades,
- ϕ – initial location of blade A w.r.t. a zero mark (equals 0 if the mark can be chosen at blade A),
- L – distance from hub to nacelle midpoint (x-axis),

Figure 2 Notations and force from mass imbalance.

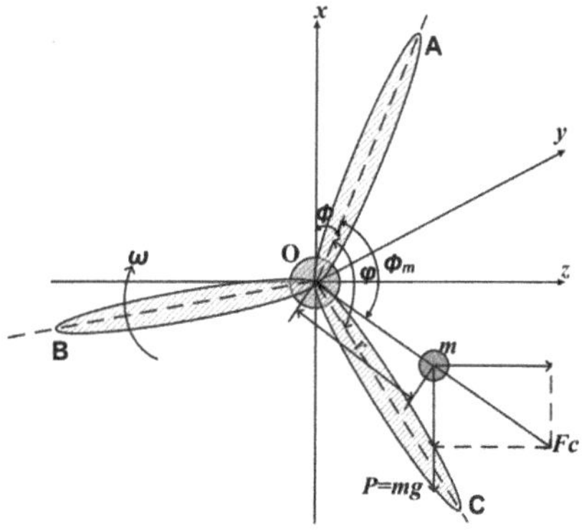

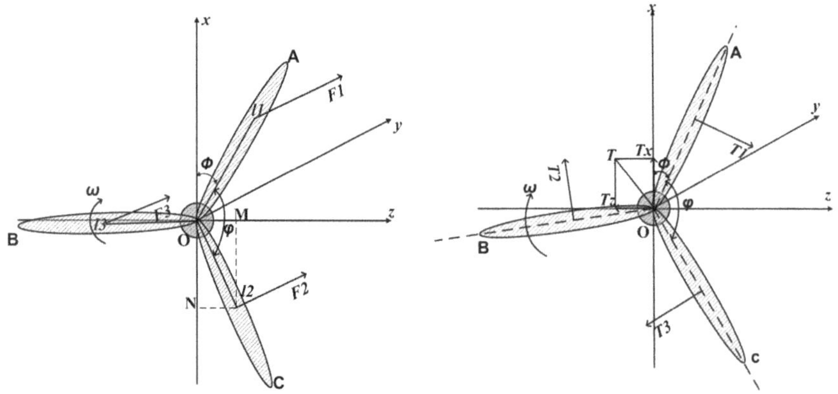

Figure 3 Thrust and tangential forces. Left: Thrust forces F_1, F_2, F_3 and corresponding lengths. Right: Tangential forces T_1, T_2, T_3.

- $F_i := F(x_i)$ – thrust force at blade i with pitch angle deviation θ_i, $i = 1, 2, 3$,
- $T_i := T(x_i)$ – tangential force at blade i with pitch angle deviation θ_i, $i = 1, 2, 3$,
- $l_i := l(x_i)$ – corresponding length (distance of resulting force vector from nacelle) for pitch angle θ_i, $i = 1, 2, 3$.

3.2.1 Forces and moments from mass imbalance

The centrifugal force of a mass m with the distance r from the hub rotating with the frequency ω has the absolute value

$$|F_c| = \omega^2 mr.$$

The projections of F_c onto the z- and x-axis are

$$\begin{aligned} (F_c)_z &= |F_c| \sin(\omega t + \phi + \phi_m), \\ (F_c)_x &= |F_c| \cos(\omega t + \phi + \phi_m), \end{aligned} \tag{4}$$

cf. Figure 2. The rotor has a small distance L to the tower center, the x-axis. Thus the forces induce moments given by

$$M_x^{(m)} = (F_c)_z \cdot L, \qquad M_z^{(m)} = (F_c)_x \cdot L. \tag{5}$$

3.2.2 Forces and moments from pitch angle deviation

The thrust force F and the tangential force T on a blade depend on the geometry and aerodynamics of the blade, the wind speed, the rotational speed, and the pitch angle. If the pitch angles are not the same for every blade the forces are also different. The derivation of the thrust forces F_1, F_2, F_3, the tangential forces T_1, T_2, T_3 and the corresponding lever arms l_1, l_2, l_3 (cf. Figure 3) uses the nonlinear Blade Element Momentum (BEM) method. This is described in detail in [5, 8]. The BEM method is used to compute the forces for each blade element. For our purpose the distributed forces are converted into one equivalent force at a corresponding lever arm l_i, cf. Figure 3. Also from Figure 3 we observe that the thrust forces F_i act in y direction. Thus

$$F_y = F_1 + F_2 + F_3. \tag{6}$$

The forces create moments around the x- and the z-axis:

$$\begin{aligned} M_x^{(F)} &= F_1 l_1 \sin(\omega t + \phi) + F_2 l_2 \sin(\omega t + \phi + \varphi) + F_3 l_3 \sin(\omega t + \phi + 2\varphi), \\ M_z^{(F)} &= F_1 l_1 \cos(\omega t + \phi) + F_2 l_2 \cos(\omega t + \phi + \varphi) + F_3 l_3 \cos(\omega t + \phi + 2\varphi). \end{aligned} \tag{7}$$

The total tangential force

$$T = T_1 + T_2 + T_3$$

has components in x- and z-direction:

$$\begin{aligned} T_z &= T_1 \sin(\omega t + \phi) + T_2 \sin(\omega t + \phi + \varphi) + T_3 \sin(\omega t + \phi + 2\varphi), \\ T_x &= T_1 \cos(\omega t + \phi) + T_2 \cos(\omega t + \phi + \varphi) + T_3 \cos(\omega t + \phi + 2\varphi). \end{aligned} \tag{8}$$

As mentioned before we have a small distance L between rotor plane and tower center, thus T_z and T_x also induce moments around the x- and the z-axes:

$$M_x^{(T)} = T_z \cdot L, \qquad M_z^{(T)} = T_x \cdot L. \tag{9}$$

3.2.3 Combination and rewriting of forces and moments – the final load vector

We observe that F_y is the only constant force, whereas all other forces and moments are harmonic with the rotating angular frequency ω. The system's answer to a constant force are vibrations with the natural eigenfrequencies $\omega_0, \omega_1, \dots$ of the WT centered around a constant displacement:

$$\mathbf{u}_1(t) = \left[\mathbf{I} - \cos\left(\sqrt{\mathbf{M}^{-1}\mathbf{S}}t\right)\right]\mathbf{S}^{-1}\mathbf{p}_1(\mathbf{x}),$$

where $\mathbf{p}_1^T := (0, \dots, 0, F_y, 0, 0, 0, 0)$, and the ω_i are the roots of the eigenvalues of $\mathbf{M}^{-1}\mathbf{S}$. For WT the only eigenfrequency close to the operating frequency ω is ω_0. For safety reasons, the operation of the WT with $\omega = \omega_0$ is avoided. As only vibrations with rotating frequency ω will be extracted from the measurement, the influence of F_y can be neglected. Therefore we can omit F_y by setting it to zero.

In detail we have

$$
\begin{aligned}
F_z &= T_1 \sin(\omega t + \phi) + T_2 \sin(\omega t + \phi + \varphi) + T_3 \sin(\omega t + \phi + 2\varphi) \\
&\quad + \omega^2 mr \sin(\omega t + \phi + \phi_m), \\
M_x &= F_1 l_1 \sin(\omega t + \phi) + F_2 l_2 \sin(\omega t + \phi + \varphi) + F_3 l_3 \sin(\omega t + \phi + 2\varphi) + LF_z, \\
M_z &= F_1 l_1 \cos(\omega t + \phi) + F_2 l_2 \cos(\omega t + \phi + \varphi) + F_3 l_3 \cos(\omega t + \phi + 2\varphi) \\
&\quad + L\big[T(x_1)\cos(\omega t + \phi) + T(x_2)\cos(\omega t + \phi + \varphi) + T(x_3)\cos(\omega t + \phi + 2\varphi) \\
&\quad + \omega^2 mr \cos(\omega t + \phi + \phi_m)\big].
\end{aligned}
$$

We use the trigonometric identities to decompose F_z, M_x, and M_z into cosine and sine parts of the argument ωt:

$$
\begin{aligned}
F_z &= F_z^c \cos(\omega t) + F_z^s \sin(\omega t), \\
M_x &= M_x^c \cos(\omega t) + M_x^s \sin(\omega t), \\
M_z &= M_z^c \cos(\omega t) + M_z^s \sin(\omega t).
\end{aligned}
$$

With the definitions

$$
\begin{aligned}
c_\phi &:= \cos(\phi), \qquad c_1 := \cos(\varphi) = \cos(2\pi/3) = \cos(4\pi/3) = \cos(2\varphi) = -0.5, \\
s_\phi &:= \sin(\phi), \qquad s_1 := \sin(\varphi) = \sin(2\pi/3) = -\sin(4\pi/3) = -\sin(2\varphi) \approx 0.866,
\end{aligned}
$$

we get

$$
\begin{aligned}
F_z^c &= T_1 s_\phi + T_2[s_\phi c_1 + c_\phi s_1] + T_3[s_\phi c_1 - c_\phi s_1] + \omega^2[s_\phi x_4 + c_\phi x_5], \\
F_z^s &= T_1 c_\phi + T_2[c_\phi c_1 - s_\phi s_1] + T_3[c_\phi c_1 + s_\phi s_1] + \omega^2[c_\phi x_4 - s_\phi x_5], \\
M_x^c &= F_1 l_1 s_\phi + F_2 l_2[s_\phi c_1 + c_\phi s_1] + F_3 l_3[s_\phi c_1 - c_\phi s_1] + LF_z^c, \\
M_x^s &= F_1 l_1 c_\phi + F_2 l_2[c_\phi c_1 - s_\phi s_1] + F_3 l_3[c_\phi c_1 + s_\phi s_1] + LF_z^s, \\
M_z^c &= (F_1 l_1 + LT_1)c_\phi + (F_2 l_2 + LT_2)[c_\phi c_1 - s_\phi s_1] + (F_3 l_3 + LT_3)[c_\phi c_1 + s_\phi s_1] \\
&\quad + L\omega^2[c_\phi x_4 - s_\phi x_5], \\
M_z^s &= -(F_1 l_1 + LT_1)s_\phi - (F_2 l_2 + LT_2)[s_\phi c_1 + c_\phi s_1] - (F_3 l_3 + LT_3)[s_\phi c_1 - c_\phi s_1] \\
&\quad - L\omega^2[s_\phi x_4 + c_\phi x_5].
\end{aligned}
\tag{10}
$$

According to our model and the corresponding nodes the load vector from aerodynamic and mass imbalances is given by

$$
\begin{aligned}
&\mathbf{p}(t) = \mathbf{p}_c \cos(\omega t) + \mathbf{p}_s \sin(\omega t) \quad \text{with} \\
&\mathbf{p}_c := \left(0, \ldots, 0, F_z^c, M_x^c, 0, M_z^c\right)^T, \\
&\mathbf{p}_s := \left(0, \ldots, 0, F_z^s, M_x^s, 0, M_z^s\right)^T.
\end{aligned}
\tag{11}
$$

3.3 Derivation of the forward operator A

3.3.1 The imbalance vector **x** *and the data vector* **y**

The imbalance vector **x** contains the pitch angle deviations as well as the information on mass imbalances. It is given by

$$\mathbf{x} = (x_1, x_2, x_3, x_4, x_5) = \left(\theta_1, \theta_2, \theta_3, mr\cos(\phi_m), mr\sin(\phi_m)\right), \tag{12}$$

where the θ_i are the pitch angle deviations of each blade, and the mass imbalance is split into real part $x_4 = \Re(mr\exp(i\phi_m))$ and imaginary part $x_5 = \Im(mr\exp(i\phi_m))$. The forward operator **A** maps the imbalance **x** onto the data **y**. We obtain the data by measuring the amplitude c and phase shift γ of the vibration with frequency ω at each sensor position. We have at most 3 sensors located at the last (upper) node of the model. The corresponding DOF at this node have the numbers $\{N-4, N-3, N-2\}$. We can measure

1. the vibration $c_y(\mathbf{x})\cos(\omega t - \gamma_y(\mathbf{x}))$ at DOF $(N-4)$ (displacement in y-direction),
2. the vibration $c_z(\mathbf{x})\cos(\omega t - \gamma_z(\mathbf{x}))$ at DOF $(N-3)$ (displacement in z-direction),
3. the vibration $c_{\beta_x}(\mathbf{x})\cos(\omega t - \gamma_{\beta_x}(\mathbf{x}))$ at DOF $(N-2)$ (torsion around the x-axis).

For all vibrations we record the cosine and sine part of the vibration, cf. (13). Using

$$c\cos(\omega t - \gamma) = c\cos(\gamma)\cos(\omega t) + c\sin(\gamma)\sin(\omega t), \tag{13}$$

we can arrange the data vector as

$$\mathbf{y} = \begin{pmatrix} c_y(\mathbf{x})\cos(\gamma_y(\mathbf{x})) \\ c_z(\mathbf{x})\cos(\gamma_z(\mathbf{x})) \\ c_{\beta_x}(\mathbf{x})\cos(\gamma_{\beta_x}(\mathbf{x})) \\ c_y(\mathbf{x})\sin(\gamma_y(\mathbf{x})) \\ c_z(\mathbf{x})\sin(\gamma_z(\mathbf{x})) \\ c_{\beta_x}(\mathbf{x})\sin(\gamma_{\beta_x}(\mathbf{x})) \end{pmatrix}$$

for all three sensors.

3.3.2 Solution of the ODE system

In order to describe the operator **A** that maps the vector of imbalance **x** onto the data **y** we solve (1) for the cosine and the sine part of **p**, cf. (11), separately. The solution **u** is the sum of the solutions $\mathbf{u}_c$ and $\mathbf{u}_s$ of the two inhomogeneous systems

$$\mathbf{M}\ddot{\mathbf{u}}_c(t) + \mathbf{S}\mathbf{u}_c(t) = \mathbf{p}_c(\mathbf{x})\cos(\omega t),$$

$$\mathbf{M}\ddot{\mathbf{u}}_s(t) + \mathbf{S}\mathbf{u}_s(t) = \mathbf{p}_s(\mathbf{x})\sin(\omega t).$$

Although we have neglected the damping in the model of the WT, there will always be some small damping in every system. Therefore the homogeneous solution is decaying. As the vibration data are obtained over a long time period, we can assume that the homogeneous solution has already died away. With the abbreviation

$$\mathbf{B} := \left(-\omega^2\mathbf{M} + \mathbf{S}\right)^{-1}$$

the solutions are given by

$$\mathbf{u}_c(t) = \mathbf{B}\mathbf{p}_c \cos(\omega t),$$

$$\mathbf{u}_s(t) = \mathbf{B}\mathbf{p}_s \sin(\omega t)$$

with $\mathbf{p}_c$ and $\mathbf{p}_s$ from (11). Since both vectors $\mathbf{p}_c$ and $\mathbf{p}_s$ have only entries at the positions $\{(N-3),(N-2),N\}$ and $\mathbf{u}(t)$ can only be measured at DOF $\{(N-4),(N-3),(N-2)\}$, we restrict $\mathbf{B}$, $\mathbf{p}_c$, $\mathbf{p}_s$, and $\mathbf{u}$ by withdrawing the relevant rows and columns (subscript r, see (15)) and get

$$\mathbf{u}_r(t) = (\mathbf{B}_r\mathbf{p}_{c,r})\cos(\omega t) + (\mathbf{B}_r\mathbf{p}_{s,r})\sin(\omega t).$$

It follows that $\mathbf{A}$ is given by

$$\mathbf{A}(\mathbf{x}) = \mathbf{y} = \begin{pmatrix} \mathbf{B}_r\mathbf{p}_{c,r} \\ \mathbf{B}_r\mathbf{p}_{c,s} \end{pmatrix}, \tag{14}$$

where

$$\mathbf{p}_{c,r}(\mathbf{x}) := \begin{pmatrix} F_z^c \\ M_x^c \\ M_z^c \end{pmatrix}, \qquad \mathbf{p}_{s,r}(\mathbf{x}) := \begin{pmatrix} F_z^s \\ M_x^s \\ M_z^s \end{pmatrix}, \quad \text{and} \quad \mathbf{B}_r := \mathbf{B}_{\{[N-4:N-2];[N-3,N-2,N]\}}. \tag{15}$$

3.4 Derivative

We recall that parts of $\mathbf{p}_{c,r}$ and $\mathbf{p}_{s,r}$ occurring in the operator $\mathbf{A}$ are derived with the nonlinear BEM method. Hence $\mathbf{A}$ is a nonlinear operator. The Frechet derivative $\mathbf{A}'$ of $\mathbf{A}$ is given by the Jacobi matrix

$$\mathbf{A}'(\mathbf{x}) = \begin{pmatrix} \mathbf{B}_r \\ \mathbf{B}_r \end{pmatrix} \begin{pmatrix} \frac{\partial \mathbf{p}_{c,r}}{\partial x_1} & \cdots & \frac{\partial \mathbf{p}_{c,r}}{\partial x_5} \\ \frac{\partial \mathbf{p}_{s,r}}{\partial x_1} & \cdots & \frac{\partial \mathbf{p}_{s,r}}{\partial x_5} \end{pmatrix}, \tag{16}$$

with

$$\frac{\partial \mathbf{p}_{c,r}}{\partial x_i} = \begin{pmatrix} \frac{\partial F_z^c}{\partial x_i} \\ \frac{\partial M_x^c}{\partial x_i} \\ \frac{\partial M_z^c}{\partial x_i} \end{pmatrix}, \qquad \frac{\partial \mathbf{p}_{s,r}}{\partial x_i} = \begin{pmatrix} \frac{\partial F_z^s}{\partial x_i} \\ \frac{\partial M_x^s}{\partial x_i} \\ \frac{\partial M_z^s}{\partial x_i} \end{pmatrix}.$$

We denote by ∂T_i the partial derivative $\partial T_i/\partial x_i$ and by $\partial(Fl)_i$ the partial derivative $\partial(F_i \cdot l_i)/\partial x_i$. Starting from (10) we have, e.g.

$$\frac{\partial F_z^c}{\partial x_1} = s_\phi \partial T_1, \qquad \frac{\partial F_z^c}{\partial x_2} = (s_\phi c_1 + c_\phi s_1)\partial T_2, \qquad \frac{\partial F_z^c}{\partial x_3} = (s_\phi c_1 - c_\phi s_1)\partial T_3,$$

$$\frac{\partial F_z^c}{\partial x_4} = \omega^2 s_\phi, \qquad \frac{\partial F_z^c}{\partial x_5} = \omega^2 c_\phi, \qquad \frac{\partial M_x^c}{\partial x_1} = s_\phi\big(\partial(Fl)_1 + L\partial T_1\big),$$

$$\frac{\partial M_x^c}{\partial x_2} = (s_\phi c_1 + c_\phi s_1)\big(\partial(Fl)_2 + L\partial T_2\big), \qquad \ldots.$$

It remains to compute the partial derivatives ∂T_i and $\partial(Fl)_i$, $i = 1,2,3$. Since T, F, and l are obtained via a nonlinear numerical routine we cannot compute the partial derivatives explicitly. Instead, we use the differential quotients as an approximation provided Δ is small, e.g.

$$\frac{\partial T_j}{\partial x_j}(x_j) \approx \frac{T_j(x_j + \Delta) - T_j(x_j)}{\Delta}, \qquad \ldots,$$
$$\frac{\partial (F_j l_j)}{\partial x_j}(x_j) \approx \frac{\partial F_j}{\partial x_j}(x_j) l_j + F_j \frac{\partial l_j}{\partial x_j}(x_j), \quad j = 1,2,3.$$

Numerical experiments show, that $\Delta \approx 0.1$ provides a sufficiently accurate approximation of the partial derivatives. Thus, the matrix

$$\partial \mathbf{p} = \begin{pmatrix} \frac{\partial \mathbf{p}_{c,r}}{\partial x_1} & \cdots & \frac{\partial \mathbf{p}_{c,r}}{\partial x_5} \\ \frac{\partial \mathbf{p}_{s,r}}{\partial x_1} & \cdots & \frac{\partial \mathbf{p}_{s,r}}{\partial x_5} \end{pmatrix} \tag{17}$$

can be computed.

4 Regularization

The inverse problem of (14) is to find the imbalance vector $\mathbf{x}$ for given data $\mathbf{y}$ that, in case of real measurements, are contaminated by noise. We assume that the noisy data $\mathbf{y}^\delta$ fulfill the inequality $\|\mathbf{y} - \mathbf{y}^\delta\| \leq \delta$, i.e., a bound on the noise is known. The operator $\mathbf{A}$ is nonlinear and ill-posed. To obtain a stable solution, one has to use regularization methods, [6, 10]. Theoretical results for the regularization of nonlinear problems were derived in the last decade, cf. [11–16]. We have used the most prominent Tikhonov regularization.

4.1 Tikhonov functional

In Tikhonov regularization we determine an approximation $\mathbf{x}_\alpha^\delta$ of the solution $\mathbf{x}$ as a minimizer of the Tikhonov functional $J_\alpha(\mathbf{x})$ with a rule to choose the regularization parameter α. We use Morozov's discrepancy principle, i.e., we define a sequence of α_k, e.g., $\alpha_{k+1} = q\alpha_k$, $q < 1$, and compute $\mathbf{x}_{\alpha_k}^\delta$ as minimizer of $J_{\alpha_k}(\mathbf{x})$ for each k and stop the iteration if for the first time

$$\|\mathbf{A}(\mathbf{x}_{\alpha_k}^\delta) - \mathbf{y}^\delta\| \leq \tau\delta, \quad \tau > 1 \text{ fixed},$$

holds. In order to minimize the Tikhonov functional for each α_k we will use a gradient-based method. The Tikhonov functional is given by

$$J_\alpha(\mathbf{x}) = \|\mathbf{A}(\mathbf{x}) - \mathbf{y}^\delta\|^2 + \alpha\|\mathbf{x} - \bar{\mathbf{x}}\|^2,$$

where $\|\cdot\|$ is the usual $l_2(n)$-norm with $n = 5$, and $\bar{\mathbf{x}}$ is an a priori guess of the solution. In our case, $\bar{\mathbf{x}}$ can be the vector $(0, 0, 0, \overline{mr}\cos(\bar{\varphi}_m), \overline{mr}\sin(\bar{\varphi}_m))$ where the mass imbalance is computed using MIA. In view of the different absolute values of the entries in the vector $\mathbf{x}$ it is appropriate to choose a weighted norm instead of $\|\mathbf{x} - \bar{\mathbf{x}}\|^2$. According to the common applications we chose $\mathbf{c} = (1, 1, 1, 0.01, 0.01)$ and define

$$\|\mathbf{x}\|_c^2 := \sum_{i=1}^{5} c_i x_i^2 = (\mathbf{c}.*\mathbf{x})^T\mathbf{x},$$

$$\langle \mathbf{x}, \mathbf{z}\rangle_c := \sum_{i=1}^{5} c_i x_i z_i = (\mathbf{c}.*\mathbf{x})^T\mathbf{z},$$

where $\mathbf{c}.*\mathbf{x} = (c_1x_1, \ldots, c_5x_5)^T$ denotes the component-by-component multiplication. We remark that $\langle \mathbf{x}, \mathbf{z}\rangle_c = \langle \mathbf{c}.*\mathbf{x}, \mathbf{z}\rangle$.

Hence the Tikhonov functional now reads

$$\begin{aligned} J_\alpha(\mathbf{x}) &= \left\|\mathbf{A}(\mathbf{x}) - \mathbf{y}^\delta\right\|^2 + \alpha\|\mathbf{x} - \bar{\mathbf{x}}\|_c^2 \\ &= \left\|\mathbf{A}(\mathbf{x}) - \mathbf{y}^\delta\right\|^2 + \alpha\left\|\mathbf{c}.*(\mathbf{x} - \bar{\mathbf{x}})\right\|^2. \end{aligned} \tag{18}$$

4.2 The gradient of Tikhonov functional

In order to minimize the functional, the gradient has to be computed. Since $\mathbf{A}$ is nonlinear we use the Taylor expansion of $\mathbf{A}(\mathbf{x} + \mathbf{h})$,

$$\mathbf{A}(\mathbf{x} + \mathbf{h}) = \mathbf{A}(\mathbf{x}) + \mathbf{A}'(\mathbf{x})\mathbf{h} + O(\mathbf{h}^2),$$

where $\mathbf{A}'$ is the Frechet derivative of $\mathbf{A}$ computed in (16) and (17). We have

$$\begin{aligned} J_\alpha(\mathbf{x} + \mathbf{h}) &= \left\|\mathbf{A}(\mathbf{x}) - \mathbf{y}^\delta\right\|^2 + \alpha\|\mathbf{x} - \bar{\mathbf{x}}\|_c^2 \\ &\quad + 2\left\langle 1./\mathbf{c}.*\left\{\left[\mathbf{A}'(\mathbf{x})\right]^T\left(\mathbf{A}(\mathbf{x}) - \mathbf{y}^\delta\right)\right\} + \alpha(\mathbf{x} - \bar{\mathbf{x}}), \mathbf{h}\right\rangle_c + O(\mathbf{h}^2) \\ &= J_\alpha(\mathbf{x}) + 2\left\langle\left[\mathbf{A}'(\mathbf{x})\right]^T\left(\mathbf{A}(\mathbf{x}) - \mathbf{y}^\delta\right) + \alpha\left(\mathbf{c}.*(\mathbf{x} - \bar{\mathbf{x}})\right), \mathbf{h}\right\rangle + O(\mathbf{h}^2). \end{aligned}$$

It follows that

$$\begin{aligned} \frac{1}{2}\nabla J_\alpha(\mathbf{x}) &= 1./\mathbf{c}.*\left\{\left[\mathbf{A}'(\mathbf{x})\right]^T\left(\mathbf{A}(\mathbf{x}) - \mathbf{y}^\delta\right)\right\} + \alpha(\mathbf{x} - \bar{\mathbf{x}}) \\ &= \left[\mathbf{A}'(\mathbf{x})\right]^T\left(\mathbf{A}(\mathbf{x}) - \mathbf{y}^\delta\right) + \alpha\mathbf{c}.*(\mathbf{x} - \bar{\mathbf{x}}). \end{aligned}$$

4.3 Minimization with steepest decent and algorithm

The functional (18) can be minimized for each fixed α by the steepest decent method. Starting from an initial value $\mathbf{x}_0$ the minimizer is computed iteratively by updating the current iterate in the direction of steepest descent, i.e., the direction of the negative gradient:

$$\begin{aligned} \mathbf{x}_{k+1} &= \mathbf{x}_k - \beta\nabla J_\alpha(\mathbf{x}) \\ &= \mathbf{x}_k + \beta 2\left\{\left[\mathbf{A}'(\mathbf{x}_k)\right]^*\left(\mathbf{y}^\delta - \mathbf{A}(\mathbf{x}_k)\right) - \alpha\mathbf{c}.*(\mathbf{x}_k - \bar{\mathbf{x}})\right\}. \end{aligned}$$

There are theoretical results on how to choose the step size β, [6, 13]. In the numerical realization, we have chosen $\beta = 1/(2\alpha^2)$. If α becomes very small, β is chosen fixed, e.g. $\beta = 10$. The final minimization algorithm is shown in Algorithm 4.

Algorithm 4

CHOOSE

$\alpha, q < 1, \tau > 1, x_0$, estimate δ

WHILE $\|\mathbf{A}(\mathbf{x}) - \mathbf{y}^\delta\| > \tau\delta$

  $\alpha = q\alpha,\ \beta = \frac{1}{2\alpha^2}$

  if $\beta > 10$ **then** $\beta = 10$ **end**

  WHILE $\frac{|J_\alpha(\mathbf{x}_0) - J_\alpha(\mathbf{x}_1)|}{|J_\alpha(\mathbf{x}_0)|} > \varepsilon$

    Compute $\mathbf{A}'(\mathbf{x}_0)$

    $\mathbf{x}_1 = \mathbf{x}_0 + \beta 2\{[\mathbf{A}'(\mathbf{x}_0)]^*(\mathbf{y}^\delta - \mathbf{A}(\mathbf{x}_0)) - \alpha\mathbf{c}. * (\mathbf{x}_0 - \bar{\mathbf{x}})\}$

    $\mathbf{x}_0 = \mathbf{x}_1$

  end

end

5 Numerical tests

5.1 Settings

5.1.1 Model data

We have tested the imbalance reconstruction algorithm for a wind turbine of the type NTK 500 with constant rotational frequency, since the aerodynamic data for this WT can be found in the literature. Namely, we use the following system parameters:

- Rotational speed $\Omega_{rot} = 0.45$ Hz = 27 rpm.
- Aerodynamic data: NACA63-421 (lift and drag coefficients: http://airfoiltools.com/airfoil/details?airfoil=naca634421-il), NKT500-41 (profile data: http://www.readbag.com/130-226-17-201-extra-web-docs-nordtank-wt-description), blade length = 20.5 m, wind speed = 8 m/s.
- Model data:
  - Masses: $m_r = 9{,}030$ kg (rotor), $m_n = 15{,}400$ kg (nacelle), $m_t = 22{,}500$ kg (tower).
  - Geometry: Tower height: $h = 33.8$ m, outer diameter at tower top: $d_{out}^{top} = 1.69$ m, outer diameter at tower bottom: $d_{out}^{bot} = 2.4$ m, distance from rotor to nacelle midpoint $L = 1.5$ m.
  - Eigenfrequency: $\Omega_0 = 0.925$ Hz.

We have set the zero mark at blade A. Thus blade B correspond to an angle of 240° and blade C corresponds to 120°, resp. For example, assuming a pitch angle deviation of blade B of 2° compared to blade A and C and a mass imbalance of 50 kgm at blade A, the corresponding imbalance vector is $\mathbf{x} = (0, 2, 0, 50, 0)^T$; if the imbalance is at blade C, we have $\mathbf{x} = (0, 2, 0, 50\cos(2/3\pi), 50\sin(2/3\pi))^T$. We obtain radial, axial and torsion vibration data $\mathbf{y}$ by applying the forward operator to $\mathbf{x}$. To simulate real measurements, the data $\mathbf{y}$ is contaminated with Gaussian noise. The disturbed data $\mathbf{y}^\delta$ as well as the noise $\delta = \|\mathbf{y} - \mathbf{y}^\delta\|$ are used as input data for the reconstruction process.

5.1.2 Reconstruction quality

Due to the different magnitudes of mass imbalance and pitch angle deviation we have to choose an appropriate metric in order to evaluate the reconstruction quality of our method. Additionally, we observe that classes of pitch angle vectors actually represent the same pitch angle deviation, namely

$$(x_1, x_2, x_3) \sim (x_1 + s, x_2 + s, x_3 + s), \quad s \in R.$$

In order to obtain accuracy results we chose from the class of the reconstructed pitch angle deviation the representative $(x^\delta_{\alpha,1} + s, x^\delta_{\alpha,2} + s, x^\delta_{\alpha,3} + s)$ with the smallest distance to the original pitch angle deviation (x_1, x_2, x_3), i.e. we chose s s.t.

$$s = \arg\min_{t \in R} \left\| (x_1, x_2, x_3) - \left(x^\delta_{\alpha,1} + t, x^\delta_{\alpha,2} + t, x^\delta_{\alpha,3} + t\right) \right\|_2.$$

As one can easily verify s is given by

$$s = \frac{1}{3}\left[\left(x_1 - x^\delta_{\alpha,1}\right) + \left(x_2 - x^\delta_{\alpha,2}\right) + \left(x_3 - x^\delta_{\alpha,3}\right)\right]. \tag{19}$$

In practice the original pitch angle deviation is not known. Here we chose the number s in a way that at least one of the x_i, $i = 1, 2, 3$, is zero. The residual vibrations are the same for all representatives of the class.

We split the norm of the deviation of the reconstruction $\mathbf{x}^\delta_\alpha$ and the original imbalance $\mathbf{x}$ into two parts. The first part is given by the pitch angle deviations,

$$\varepsilon_{\text{pitch}}\left(\mathbf{x}^\delta_\alpha, \mathbf{x}\right) := \left\| (x_1, x_2, x_3) - \left(x^\delta_{\alpha,1} + s, x^\delta_{\alpha,2} + s, x^\delta_{\alpha,3} + s\right) \right\|_2,$$

where s is given by (19). The second part is the difference for the mass imbalance in the usual l_2-norm:

$$\varepsilon_{\text{mass}}\left(\mathbf{x}^\delta_\alpha, \mathbf{x}\right) := \left\| \left(x^\delta_{\alpha,4}, x^\delta_{\alpha,5}\right) - (x_4, x_5) \right\|_2.$$

In this way we can give a sophisticated account of the reconstruction accuracy of both imbalances types separately. The relative errors are defined by

$$\varepsilon^{\text{rel}}_{\text{pitch}}\left(\mathbf{x}^\delta_\alpha, \mathbf{x}\right) := \frac{\varepsilon_{\text{pitch}}(\mathbf{x}^\delta_\alpha, \mathbf{x})}{\|(x^\delta_{\alpha,1} + s, x^\delta_{\alpha,2} + s, x^\delta_{\alpha,3} + s)\|_2},$$

$$\varepsilon^{\text{rel}}_{\text{mass}}\left(\mathbf{x}^\delta_\alpha, \mathbf{x}\right) := \frac{\varepsilon_{\text{mass}}(\mathbf{x}^\delta_\alpha, \mathbf{x})}{\|(x^\delta_{\alpha,4}, x^\delta_{\alpha,5})\|_2},$$

provided the denominator is not zero.

5.2 Aerodynamic imbalance caused by pitch angle deviation

The first test was concerned with the reconstruction of pitch angle deviation imbalances in the absence of mass imbalances. Here we present two examples.

Example 1 We have chosen a pitch angle deviation of 2° at blade B, i.e., the imbalance vector is given by $\mathbf{x} = (0, 2, 0, 0, 0)^T$. Figure 5, top picture, shows the original and the reconstructed pitch angle deviations for a relative noise of $\delta_{\text{rel}} = 10\%$. In Table 1 the reconstructed pitch angle deviation as well as the reconstruction error is given for several noise levels. Since the mass imbalance is zero, the absolute mass imbalance error is given. In Figure 6 the reconstruction error as a function of the noise level and the development of the residual $\|\mathbf{A}(\mathbf{x}_n) - \mathbf{y}^\delta\|$ during the iteration process are presented. The original vibrations in the imbalanced case and the residual vibrations after balancing with the reconstructed balancing setting are shown in Figure 7. The original vibrations are relatively small due to

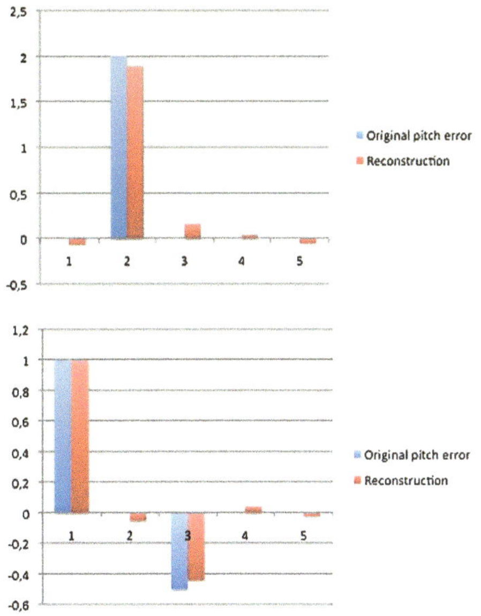

Figure 5 Comparison of reconstruction and original pitch angle deviation. Top picture: (0, 2, 0, 0, 0) with 10% noise level; bottom picture: (1, 0, −0.5, 0, 0) with 7% noise level.

Table 1 Reconstruction results for imbalance $\mathbf{x} = (0, 2, 0, 0, 0)^T$.

δ_{rel}	s	$\mathbf{x}_{\alpha^*+s}^{\delta}$	$\varepsilon_{pitch}^{rel}(\mathbf{x}_{\alpha}^{\delta}, \mathbf{x})$	$\varepsilon_{mass}(\mathbf{x}_{\alpha}^{\delta}, \mathbf{x})$
3%	0.67	$(-0.03, 1.93, -0.05, 0.04, -0.06)^T$	5.9%	0.07
5%	0.67	$(0.06, 1.97, -0.04, 0.03, -0.05)^T$	4.6%	0.07
8%	0.67	$(0.10, 1.94, -0.04, 0.04, -0.05)^T$	8.5%	0.06
10%	0.67	$(-0.06, 1.89, 0.16, 0.04, -0.05)^T$	13.2%	0.06
12%	0.67	$(-0.06, 1.88, 0.18, 0.02, -0.06)^T$	15%	0.07
15%	0.67	$(0.25, 1.82, -0.07, 0.02, -0.06)^T$	21.5%	0.07
17%	0.67	$(-0.09, 1.83, 0.25, 0.03, -0.06)^T$	21.7%	0.07

the fact that the turbine rotates with a fixed rotor frequency which is only about 70% of the tower eigenfrequency. The remaining vibrations after balancing are almost invisible.

Example 2 The reconstruction result for an imbalance $\mathbf{x} = (1, 0, -0.5, 0, 0)^T$ for a reconstruction with noisy data and noise level of $\delta_{rel} = 7\%$ is shown in Figure 5, bottom picture. Here, the parameter s is computed as $s = 0.167$ and the relative pitch angle reconstruction error is $\varepsilon_{pitch}^{rel}(\mathbf{x}_{\alpha}^{\delta}, \mathbf{x}) = 7\%$.

The initial approximation $\mathbf{x}_0$ for both examples was chosen as $\mathbf{x}_0 = (0, 0, 0, 0, 0)^T$.

5.3 Combined mass and aerodynamic imbalance

As a second test we have considered a combined imbalance setting, a pitch angle deviation of 2° of blade B and a mass imbalance of 50 kgm at blade A. Blade A corresponds to an initial angle of 0°, hence $\mathbf{x} = (0, 2, 0, 50, 0)^T$. The data were disturbed with a noise level of 7%. To test the performance regarding the initial value for the iteration process we have chosen two different initial approximations: firstly the zero vector $\mathbf{x}_0 = (0, 0, 0, 0, 0)^T$ and secondly $\mathbf{x}_0 = (0, 0, 0, 51.5, -4.8)^T$ that represents zero pitch angle deviation and a mass imbalance of 51.8 kgm at an angle of −5°. This mass imbalance value was reconstructed from the noisy data with MIA. The reconstruction results for both start vectors are presented in Figure 8 and Table 2.

Figure 6 Reconstruction performance.
Reconstruction error versus noise level (top picture), residual during iteration (bottom picture).

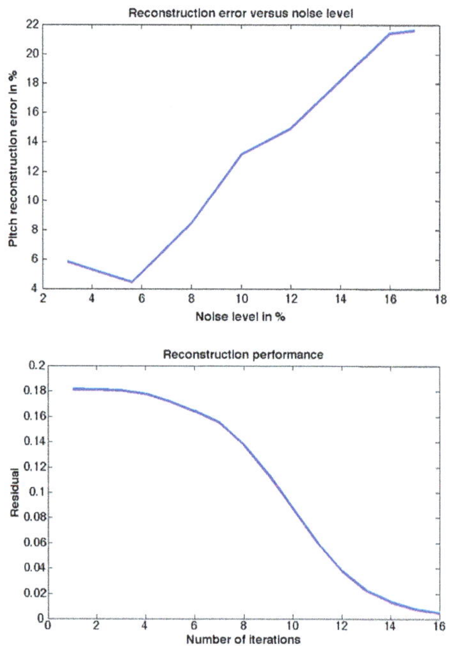

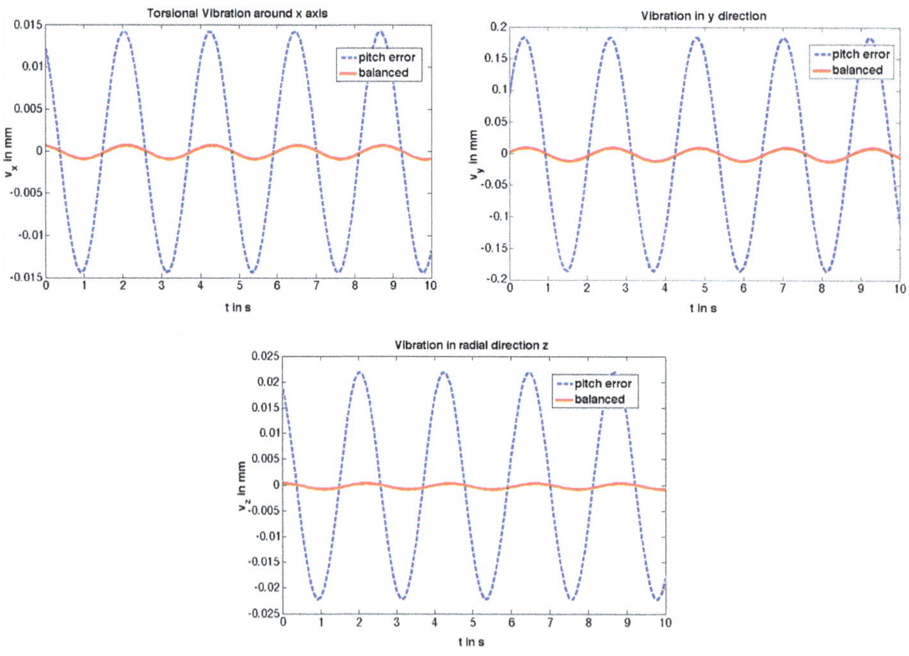

Figure 7 Imbalance vibrations and residual vibrations after balancing.

We observe that the reconstruction process with zero vector as initial imbalance vector converges more slowly but results in a better pitch angle reconstruction. The reconstruction with the precomputed mass imbalance is faster and gives good results for both imbalance types. The original imbalance vibrations and residual vibrations after balancing are plotted in Figure 9 for all vibration directions. The amplitude of the vibrations is reduced

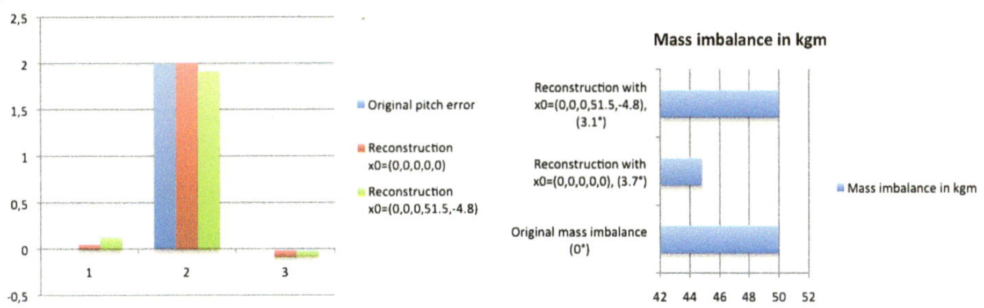

Figure 8 Pitch angle (left) and mass imbalance reconstruction for different initial values.

Table 2 Reconstruction results for imbalance $x = (0, 2, 0, 50, 0)^T$.

Start vector	t	$x^\delta_{\alpha^*} + s$	$\varepsilon^{rel}_{pitch}(x^\delta_\alpha, x)$	$\varepsilon^{rel}_{mass}(x^\delta_\alpha, x)$
$(0,0,0,0,0)^T$	403 s	$(0.06, 2.02, -0.08, 44.7, 2.88)^T$	6%	13%
$(0,0,0,51.5,-4.8)^T$	120 s	$(0.14, 1.93, -0.07, 50.1, 0.67)^T$	11%	1.4%

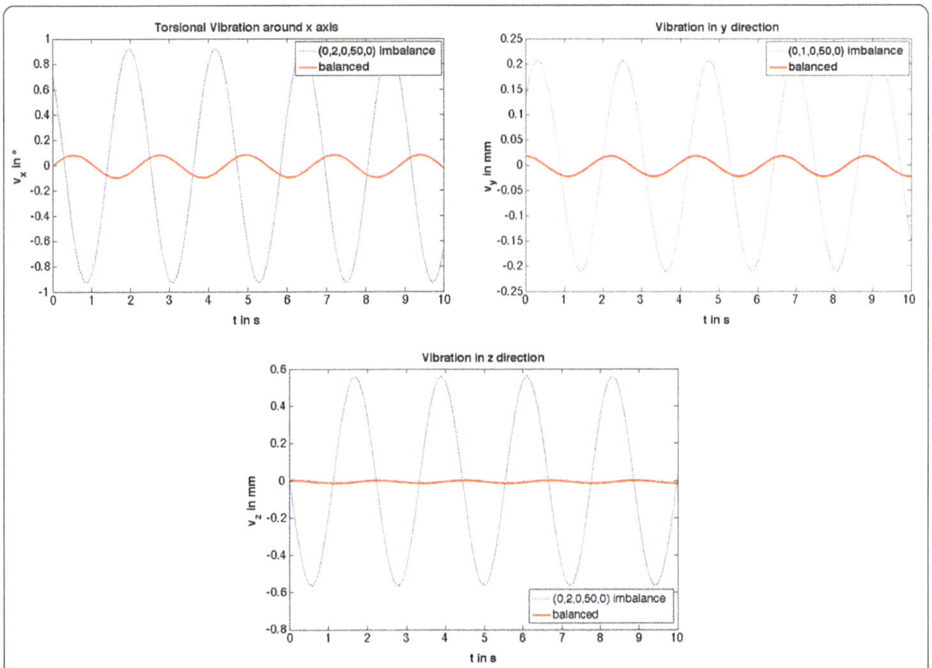

Figure 9 Imbalance vibrations and residual vibrations after balancing. Imbalance $(0, 2, 0, 50, 0)^T$, start vector $(0, 0, 0, 51.5, -4.8)^T$.

to less than 10% in case of torsional and axial vibrations and even 2% fin case of radial vibration.

As mentioned in the Introduction, the problem of simultaneous reconstruction of mass imbalances and pitch angle deviation was already treated in [5]. In this first attempt, a direct search method for minimizing the nonlinear Tikhonov functional was used, which was far too slow for extensive test computations. Also convergence was not always be achieved. With our gradient-based reconstruction method the computation time was reduced significantly to a more reasonable time of about 2 minutes on 2.8 GHz Intel Core2 Duo processor. Convergence could also be observed in all tests.

6 Summary and outlook

The paper deals with the mathematical determination of two types of imbalances in a WT: aerodynamic imbalances from pitch angle deviation of the blades and mass imbalances of the rotor. Presently, both types have to be determined on-site by an expert team performing time consuming measurement procedures. The mathematical approach provides a reconstruction method that can be implemented into a condition monitoring system. Instead of the expert team, the method requires the knowledge of geometrical and physical parameters of the WT and aerodynamic airfoil data of the blades. Whereas the former usually can be obtained from inspection reports, the latter represent very sensible data and are hardly available for newer WT. For that reason, so far the method could only be tested for an older NTK500 turbine. Unfortunately, our industrial partners could not provide the necessary vibration data for this type. Thus we had to restrict all tests to artificial data. For future applicability, the WT owners and manufacturers need to be persuaded that tit in beneficial for the safe operation of the turbine to provide the sensible airfoil data in a confined way.

The problem of imbalance determination was addressed by the authors in former papers. In this paper we have presented an improved algorithm to reconstruct mass imbalance and pitch angle deviation of the rotor of a WT at the same time. Including pitch angle deviations into the reconstruction leads to a nonlinear forward operator. The regularization of the inverse problem requires the minimization of the Tikhonov functional. Straight forward direct search method proved to be to slow and not stable enough for our purpose. Hence we employed gradient based minimization methods. To that end, we have derived the Frechet derivative of the forward operator that occurs in the gradient of the Tikhonov functional. The accelerated algorithm was successfully applied to several test examples with artificial vibration data. In case of real vibration data we expect equally good reconstruction results as in the test computations, provided that the necessary aerodynamic data of the blades are available.

Competing interests

The authors declare that they have no competing interests.

Authors' contributions

Both authors contributed equally to the writing of this paper. All authors read and approved the final manuscript.

Author details

[1]Austrian Academy of Sciences, RICAM, Altenbergerstr. 69, Linz, 4040, Austria. [2]Johannes Kepler University Linz, Altenbergerstr. 69, Linz, 4040, Austria.

Acknowledgements

The authors work supported by the Austrian Research Promotion Agency (FFG).

References

1. Donth A, Grunwald A, Heilmann C, Melsheimer M: **Improving performance of wind turbines through blade angle optimization and rotor balancing.** EWEA O/M strategies 350; 2011.
2. Caselitz P, Giebhardt J: **Rotor condition monitoring for improved operational safety of offshore wind energy converters.** *ASME J Sol Energy Eng* 2005, **127**:253-261.
3. Hameed Z, Hong YS, Cho YM, Ahn SH, Song CK: **Condition monitoring and fault detection of wind turbines and related algorithms: a review.** *Renew Sustain Energy Rev* 2009, **13**:1-39.

4. Ramlau R, Niebsch J: **Imbalance estimation without test masses for wind turbines.** *ASME J Sol Energy Eng* 2009, **131**(1):011010.
5. Niebsch J, Ramlau R, Nguyen TT: **Mass and aerodynamic imbalance estimates of wind turbines.** *Energies* 2010, **3**:696-710.
6. Engl HW, Hanke M, Neubauer A: *Regularization of Inverse Problems*. Dordrecht: Kluwer Academic; 1996.
7. Gasch R, Knothe K: *Strukturdynamik 2*. Berlin: Springer; 1989.
8. Ingram G: **Wind turbine blade analysis using the blade element momentum method.** Durham University, Durham; 2005.
9. Ahlström A: **Aeroelastic simulations of wind turbine dynamics.** *PhD thesis*. Royal Institute of Technology, Department of Mechanics, Stockholm; 2005.
10. Louis AK: *Inverse und Schlecht Gestellte Probleme*. Stuttgart: Teubner; 1989.
11. Kaltenbacher B, Neubauer A, Scherzer O: *Iterative Regularization Methods for Nonlinear Ill-Posed Problems*. Berlin: de Gruyter; 2008.
12. Ramlau R, Teschke G: **Tikhonov replacement functionals for iteratively solving nonlinear operator equations.** *Inverse Probl* 2005, **21**(5):1571-1592.
13. Ramlau R: **TIGRA – an iterative algorithm for regularizing nonlinear ill-posed problems.** *Inverse Probl* 2003, **19**(2):433-4670.
14. Ramlau R: **Morozov's discrepancy principle for Tikhonov regularization of nonlinear operators.** *Numer Funct Anal Optim* 2002, **23**(1,2):147-172.
15. Scherzer O: **The use of Morozov's discrepancy principle for Tikhonov regularization for solving nonlinear ill-posed problems.** *Computing* 1993, **51**:45-60.
16. Ramlau R: **On the use of fixed point iterations for the regularization of nonlinear ill-posed problems.** *J Inverse Ill-Posed Probl* 2005, **13**(2):175-200.

11

Mathematics in paper

Jari Hämäläinen*

*Correspondence:
jari.hamalainen@lut.fi
Centre of Computational Engineering and Integrated Design (CEID) and Department of Mathematics and Physics, Lappeenranta University of Technology, P.O. Box 20, Lappeenranta, FI-53851, Finland

Abstract

This paper aims to give a general overview of industrial mathematics applied to papermaking. Modelling challenges vary from computational fluid dynamics (CFD) to finite-element analyses (FEA) when the paper web transforms from a multiphase flow to a solid fibre network - to a ready paper. Also, different length scales are present from fibre level to machine level problems, *i.e.* from millimetres up to one hundred metres. Mathematical modelling of papermaking is aiming at optimizing the process and the end-product. Thus, computational tools for optimal shape design and optimal control purposes have been developed.

1 Introduction

We write and read mathematics on paper but we seldom think about what mathematics there is *in* paper, that is, what mathematics is behind papermaking. Paper is made from wood fibres mixed in water with some chemical and filler ingredients called as fibre suspension. A modern paper machine is producing a two kilometres long and ten metres wide paper web in a minute with extremely evenly distributed solid materials (see Figure 1). A papermaking production line begins from a so-called headbox which distributes a dilute (1% solids content) fibre suspension coming from a pump evenly across the width of a paper machine. A ten metre wide and one centimetre thick free jet flies from a headbox to a forming section between porous wires and most of water are removed in fractions of a second. After mechanical dewatering the solid contents of a wet paper web increases up to 50%. The rest of water is removed by evaporation in a drying section. Paper machines vary depending on paper grades they produce, that is, a paper machine for tissue paper is different from a machine for copy paper or newspaper, but the basic concept is the same from a headbox to a dryer.

2 Mathematical modelling of fibre suspension flows in multiple scales

Modelling approaches vary a lot depending on a length scale we are dealing with [1] (see Figure 2). Here, the small scale means the fibre length which is 1-2 millimetres when considering fibre suspension flows. Or, when the structure of a single fibre is studied, then we are talking about orders of magnitude smaller scales. Indeed, one trend today in pulp and paper industry is to refine fibres to micrometre particles. The other extreme is the mill-level scale when the whole paper mill is considered. A fibre suspension flow can be modelled by using either a Lagrangian or an Eulerian multi-phase approach. A flow of tens of thousands of fibres can be simulated with Lagrangian methods but only in a small volume, 1 cm^3, say. When dealing with unit processes like the headbox, this approach is not practical anymore. Thus, averaging is needed resulting in Eulerian multi-phase flow

Figure 1 The dry-end view of a modern paper machine.

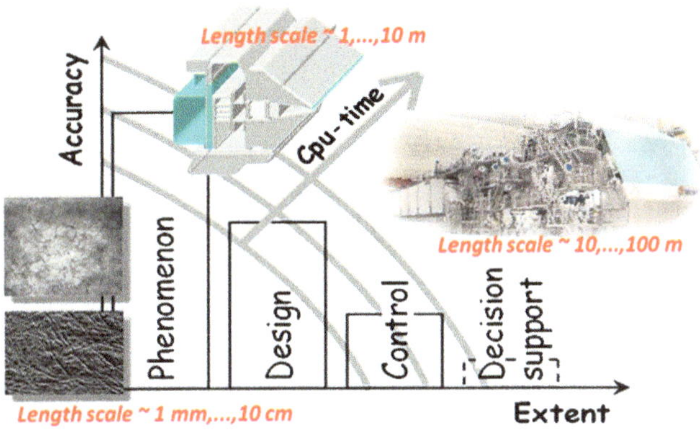

Figure 2 Mathematical modelling of fibre suspension flows and paper machine processes in different scales.

models or rheological models for the mixture. Specific mathematical models have also been developed for certain phenomena, namely to fibre orientation and fibre flocculation. All this is further complicated with the presence of turbulence interacting with the fibre phase.

3 Optimized fluid flows

Whenever a process is modelled, it is also tried to be optimized to some extent in every-day R&D actions. The first immediate optimization application in papermaking is the optimal shape design of the headbox such that it distributes the flow evenly on a ten metre wide forming section resulting in even solid material and weight profile of a paper web. Furthermore, a wrong fibre alignment in thickness-directional layers in a web causes undesired end-properties of the paper sheet produced. Problems in copy machines or printing processes are typically caused by cockling and curling of paper, which in turn, are caused by non-optimal fibre orientations in a paper sheet (see Figure 3).

Figure 3 **Curling and cockling of a paper sheet due to moisture expansion of non-optimally oriented fibres in paper.**

From a mathematical point of view, the optimal shape design and optimal control of a headbox leads to CFD-based optimization [2]. Optimization software tools have been developed for the headbox, both for shape and control purposes. The optimal shape design tool can be computationally more expensive because the optimal design needs to be found only once for a certain industrial application and there are not any specific time limitations. But, computational time is limited when using CFD-based optimization for control purposes. Otherwise, the control actions take place too late. One such optimal control software called HOCS Fibre has been developed by world-leading machine supplier Metso Paper Inc. (www.metso.com) and computational technology enterprise Numerola (www.numerola.fi). The software tool is not used for on-line control system at the moment, but off-line for trouble-shooting purposes [3]. It is based on depth-averaged two-dimensional Navier-Stokes equations and adjoint state technique in gradient-based optimization. One cost function evaluation, *i.e.*, one flow field simulation, takes a few seconds in HOCS Fibre when full three-dimensional Navier-Stokes solution takes hours. HOCS Fibre has been used to tens of paper mills and the results are illustrated in Figure 4. The red markers illustrate the fibre orientation misalignment before optimization and the green markers after optimization. These are measured from a ready paper before and after HOCS Fibre optimizations.

There are also aims to model the whole paper mill and to optimize it. Now, it is not anymore so clear what optimal means, that is, there are plenty of objectives to be considered simultaneously. This leads to multi-objective optimization which is a basis for decision-support systems for papermaking [4]. Mathematical models of the whole mill are further simplified from the model of the headbox fluid flow optimizers. In addition, paper quality parameters are also prediction based on statistical correlations derived from measured data. So, there are tens of algebraic correlations in addition to first-principle process models.

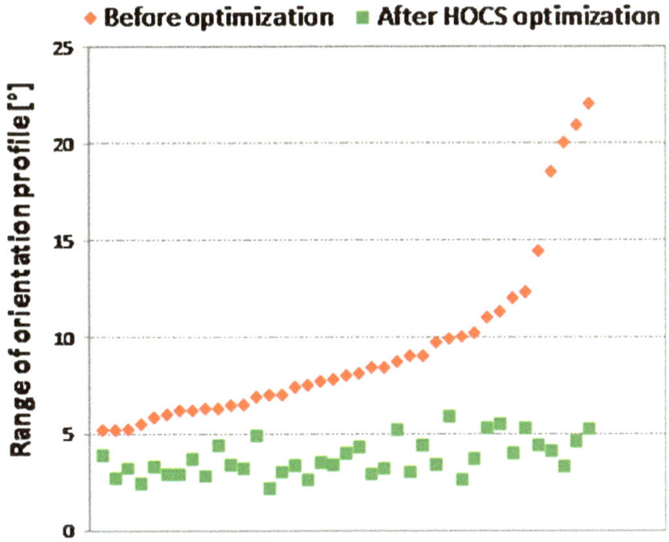

Figure 4 Min-max ranges of fibre orientation angles before and after optimization. Both are measured values in mill conditions.

These optimization examples illustrate that there are not 'good' or 'bad' models, but different levels of complexity can be used in different applications.

4 From rheology of fluids to rheology of solids

High-concentration fibre suspension can be considered as a generalized-Newtonian fluid obeying a shear-thinning viscosity model. It may also include memory effects. Rheological behaviour can also be detected for a ready paper [5]. Several studies have shown that a paper sheet is an elasto-visco-plastic material. And, mechanical properties and corresponding material parameters of the paper depend strongly on the fibrous structure of the paper sheet, which in turn, has been determined by fluid dynamics in the beginning of the papermaking process, in the headbox.

5 Concluding remarks

Papermaking is full of mathematics. Some examples of fluid dynamical and solid mechanical modelling were given in this article. Model-based optimization tools are also natural applications for mathematical models. But, plenty of mathematical modelling of papermaking was left out, *e.g.* the whole papermaking chemistry as well as free-surface flows, fluid-structure interactions (FSI) and phase-change problems (*e.g.* evaporation). And, totally different area of mathematics is the economy of papermaking. Pulp and paper industry is facing outstanding challenges in reducing investment and operating costs while aiming at top-quality products and increased production. Today this is further challenged by competition of raw materials with bio-energy and bio-fuel productions and other new wood-based products.

When you are writing mathematical formulae on a paper sheet next time, you may remember that there is already a lot of mathematics in the paper sheet.

Competing interests
The author declares that he has no competing interests.

References

1. Hämäläinen J, Lindström S, Hämäläinen T, Niskanen H: **Papermaking fibre suspension flow simulations at multiple scales.** *J Eng Math* 2011, **71**(1):55-79.
2. Hämäläinen J, Hämäläinen T, Madetoja E, Ruotsalainen H: **CFD-based optimization for complete industrial process: papermaking.** In *Optimization and Computational Fluid Dynamics*. Edited by Thevenin, D, Janiga, G. Berlin: Springer; 2008.
3. Avikainen M, Hämäläinen J, Tarvainen P: **HOCS Fibre: CFD-based software for fibre orientation profile optimization for conventional and dilution headboxes.** *Nord Pulp Pap Res J* 2010, **25**(4):456-462.
4. Hämäläinen J, Madetoja E, Ruotsalainen H: **Simulation-based optimization and decision support for conflicting objectives in papermaking.** *Nord Pulp Pap Res J* 2010, **25**(3):405-410.
5. Hämäläinen J, Eskola R, Erkkilä A-L, Leppänen T: **Rheology in papermaking - from fibre suspension flows to mechanics of solid paper.** *Korea-Australia Rheol J* 2011, **23**(4):211-217.

12

Application of a three-dimensional fiber lay-down model to non-woven production processes

Martin Grothaus[1], Axel Klar[1,2], Johannes Maringer[1*], Patrik Stilgenbauer[1] and Raimund Wegener[2]

*Correspondence:
johannes.maringer@itwm.fraunhofer.de
[1]Technische Universität Kaiserslautern, Gottlieb-Daimler Str., Kaiserslautern, 67663, Germany
Full list of author information is available at the end of the article

Abstract

In this work we present the industrial application of fiber lay-down models that enable an efficient simulation of non-woven structures. The models describe the deposition of fibers on a moving conveyor belt with the help of stochastic differential equations on manifolds. The model parameters have to be estimated from more complex models in combination with measurements of the resulting non-woven. In the application we discuss especially a three-dimensional fiber model for a typical industrial problem from non-woven production processes.

Keywords: fiber lay-down; stochastic differential equations on manifolds; parameter identification

1 Introduction

Technical textiles have attracted great attention to diverse branches of industry over the last decades due to their comparatively cheap manufacturing. By overlapping thousands of individual slender fibers, random fiber webs emerge yielding non-woven materials that find applications e.g. in textile, building and hygiene industry as integral components of baby diapers, closing textiles, filters and medical devices, to name but a few. A typical method of production is given by melt-spinning processes, see Figure 1 for a corresponding plant. There melted polymers are extruded through narrow nozzles leading to plenty single fibers. Then these fibers are stretched and spun until they solidify due to cooling air streams. The latter also account for swirling of the fibers before they are laid down on a moving conveyor belt. The resulting fiber web eventually passes through several processing steps of reworking and reinforcement before the finished product is obtained. The quality of the non-woven can be measured, for example, in terms of homogeneity, basis weight, or permeability and is already determined to a great extend by the fiber deposition. Fluctuations in these properties due to stochastic influences during the production process cause difficulties for the manufacturer, such as unintended holes in filter materials. An objective in industry is the simulation of the deposited fiber web and its optimization with respect to the desired characteristics. Since in this process, raw material costs play a more decisive role as compared to the costs of procedure, the optimization aims at raw material saving that corresponds to a small value of basis weight of the non-woven, while quality assurance. The primary objective is to generate the microstructure of the technical textile given the relevant production parameters. Then, on the one hand, the

Figure 1 Plant for melt-spinning process (Photo by Oerlikon Neumag).

microstructure will be further investigated with respect to the above quality parameters. On the other hand, a microstructure simulation can be used to predict material properties of the textile. The material properties can be optimized solving the associated inverse and optimization problems. Finally, this leads to a complete virtual material design of technical textiles.

Since the mathematical treatment of the whole process at a stroke is not possible due to its complexity, a hierarchy of models that adequately describe partial aspects of the process chain has been developed in research during the last years. In [1] the conjunction of those models has been elaborated using asymptotic analysis, similarity estimates and parameter identification. More precisely, the description of the fiber dynamics in turbulent air flows has been realized by a stochastic generalized string model that has been deduced from a special Cosserat rod under Kirchhoff constraint, consult [2, 3] for its derivation. This provides the basis for the software tool FIDYST[a] that enables the full simulation of fiber motion under the influence of surrounding turbulent air flows up to the fiber lay-down on the transport belt. In this manner, the microstructure of virtual fiber webs can be in principle generated. However, this approach is computationally expensive and the complexity is strongly affected by the number of fibers. For this reason, a class of stochastic surrogate models has been proposed in order to compute only the image of the deposited fiber web, instead of describing the full fiber dynamics that lead to this web. These surrogate models have in common the structure of a system of ordinary stochastic differential equations that enables the fast computation of a considerable number of fibers. They contain parameters that are related to the physical production process, as for example turbulence influence and fiber coiling, and that have to be identified from the full simulation of a few fibers with help of the complex string model. Based on the original two-dimensional version in [4], the surrogate models have been frequently improved within the last years. This includes the addition of a moving conveyor belt [5] or, more general, the consideration of more specific types of production processes as for instance rotational spinning processes [1]. Further enhancements are given by a smooth version in [6] where the curvatures of the fibers are taken into account, and by modelling the lay-down in three dimensions [7, 8]. The latter has been intended as first step into a realistic generation of three-dimensional microstructures that can be used for study of flow resistance and elasticity of the nonwoven. The mathematical analysis of these models is sophisticated due to degeneracies of the associated Fokker-Planck equations. The trend to equilibrium for the original two-dimensional model has been investigated using Dirichlet forms and operator semi-group techniques in [9], with a hypocoercivity method in [10] and with probabilistic methods in [11–13]. The higher-dimensional models have been accordingly analyzed in [13, 14]. With the help of various techniques from asymptotic analysis connections between these models have been shown and model reduction in the situation of large or small turbulence have been performed in [4–8]. A possible strategy for estimating the parameters of the two-dimensional models has been provided as a heuristic approach in [1] which has been successfully tested by FIDYST simulations of transversal as well as rotational spinning processes in the stated source. In [7] an extension of this identification strategy to the three-dimensional case has been proposed on condition that information about the fiber orientation is available. In this work we realize this approach in a real industrial problem for the first time. At the example of a pilot plant of the company Oerlikon Neumag, we discuss the calibration of the smooth three-dimensional surrogate model and illustrate the computation of corresponding fiber webs. For this we need both the full simulation of a few representative fibers with regard to the problem setting and image processing data from CT-scans of the resulting non-woven.

This paper is structured as follows. At first, we recapitulate the class of surrogate fiber lay-down models developed in the above mentioned papers. This is done in Section 2 where we use a description that can be embedded in the context of geometric Langevin equations on regular submanifolds as investigated in [15]. Via this approach, the highly geometric nature of the class of fiber lay-down models is illustrated.

In Section 3 we describe a strategy for estimating the parameters of the surrogate models. The application of the models to the production process in the pilot plant is done in Section 4 where we are able to simulate virtual fiber webs that are numerically investigated with regard to their quality characteristics.

2 The fiber lay-down models

As mentioned in the Introduction, in this section we recapitulate the class of surrogate models developed in [1, 7, 8, 13, 15]. In the surrogate fiber lay-down models the paths of the deposited fibers (as images of arc-length parametrized curves) are described by simplified stochastic differential equations. For this purpose, distinctive process parameters that influence the form of the fiber web are incorporated. These are the typical throwing ranges of the fibers and their coiling behavior, whereas the deposition itself is perturbed by some random force which is affected by the entanglement of the fibers above the lay-down region. Other characteristics involve the fiber stiffness and the fiber orientation in space. Additionally, depending on the production process, different specifications can be handled using appropriate reference curves. In this work we restrict ourselves on the situation of a transversal spinning process, that means the fiber lay-down starting from fixed spinning positions onto a straight moving conveyor belt. Therein the belt velocity as well as the spinning speed are predetermined parameters given by the production process. The other parameters have to be appropriately estimated, see Section 3.

Conceptually, we distinguish between two major types of fiber lay-down models, a basic and a smooth version. For better clarity we hide the real environment of the production process for the moment and focus on the situation of a non-moving conveyor belt. The motion of the belt as well as anisotropy effects due to physical constraints can be easily included later on, when we consider the practically relevant 2D and 3D models once more

Figure 2 Scheme ($d = 2$) of fiber curve ξ with normalized tangent ω.

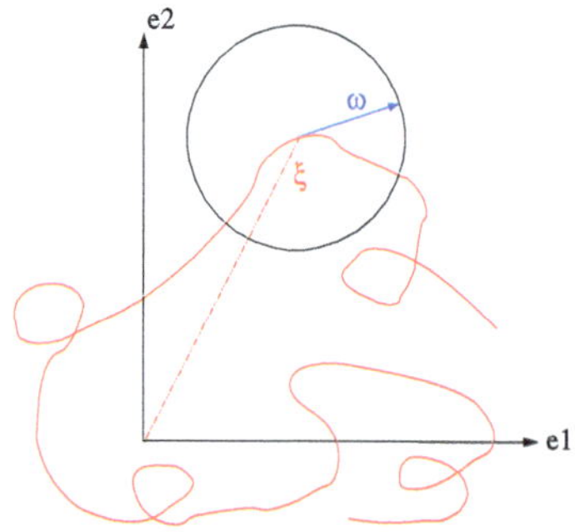

separately. In general, the basic model is formulated as a Stratonovich stochastic differential equation with state space $\mathbb{R}^d \times \mathbb{S}^{d-1}$, $d \in \mathbb{N}$, $d \geq 2$, of the form

$$\begin{aligned} d\boldsymbol{\xi}_t &= \boldsymbol{\omega}_t \, dt, \\ d\boldsymbol{\omega}_t &= -\Pi[\boldsymbol{\omega}_t]\nabla V(\boldsymbol{\xi}_t)\, dt + A\Pi[\boldsymbol{\omega}_t] \circ d\mathbf{W}_t, \end{aligned} \tag{1}$$

see [8, 13]. $\mathbf{W}$ denotes a d-dimensional standard Brownian motion. For more details on manifold-valued stochastic differential equations, see e.g. [16]. Here the arc-length parametrized curve $\boldsymbol{\xi} : \mathbb{R}_0^+ \to \mathbb{R}^d$ represents one deposited fiber. $\boldsymbol{\xi}$ is really arc-length parametrized since $\boldsymbol{\omega}$ lives on $\mathbb{S}^{d-1}$, compare Figure 2. The drift term in the second equation models the coiling behavior of the fiber, where $\Pi[\boldsymbol{\omega}] := I - \boldsymbol{\omega} \otimes \boldsymbol{\omega}$ denotes the projection of the tangent onto the sphere $\mathbb{S}^{d-1}$ with I the identity matrix and $x \otimes y = xy^T$. The throwing ranges of the fiber can be controlled with the help of a suitable potential V. The second term in the equation for $\boldsymbol{\omega}$ describes a Brownian motion on $\mathbb{S}^{d-1}$ with constant noise amplitude $A \in \mathbb{R}_0^+$ and expresses the stochastic forces, i.e. the effect of the turbulent air flows that perturb the deposition of the fibers as desired. We remark that the basic model (1) can also be viewed as a geometric Langevin equation having spherical velocities, see [15].

Taking the curvature of the fibers into consideration, which is unavoidable for characterizing bending effects, the smooth fiber lay-down model is developed: The non-differentiable Brownian motion in (1) is replaced by a spherical Ornstein-Uhlenbeck process, see [7] and [15]. The state space of the new model is then $\mathbb{R}^d \times T\mathbb{S}^{d-1}$, $d \in \mathbb{N}$, $d \geq 2$ where $T\mathbb{S}^{d-1}$ indicates the tangent bundle of the sphere, see [15]. This gives the following Stratonovich stochastic differential equation for the smooth model

$$\begin{aligned} d\boldsymbol{\xi}_t &= \boldsymbol{\omega}_t \, dt, \\ d\boldsymbol{\omega}_t &= -\Pi[\boldsymbol{\omega}_t]\nabla V(\boldsymbol{\xi}_t)\, dt + \boldsymbol{v}_t \, dt, \\ d\boldsymbol{v}_t &= (\boldsymbol{\omega}_t \otimes \boldsymbol{v}_t)\nabla V(\boldsymbol{\xi}_t)\, dt - \lambda \boldsymbol{v}_t \, dt - |\boldsymbol{v}_t|^2 \boldsymbol{\omega}_t \, dt + \mu \Pi[\boldsymbol{\omega}_t] \circ d\mathbf{W}_t. \end{aligned} \tag{2}$$

Therein, the parameter $\mu \in \mathbb{R}_0^+$ takes on the role of the noise amplitude, whereas the stiffness of the fibers can be related to $\lambda \in \mathbb{R}_0^+$. The connection to (1) is given by the white noise limit (compare [7]), i.e. the basic model can be viewed as a model for non-stiff fibers.

For the efficient simulation of fiber lay-down, however, these coordinate-free formulations (1) and (2) are rather inadequate. Therefore, we consider their description in local coordinates. We introduce a parametrization of $\mathbb{S}^{d-1}$ or $T\mathbb{S}^{d-1}$ by

$$\boldsymbol{\omega} = \boldsymbol{\tau}(\boldsymbol{\theta}) := \boldsymbol{\tau}_{d-1}(\theta_1, \dots, \theta_{d-1}),$$
$$\boldsymbol{\nu}(\boldsymbol{\theta}, \boldsymbol{\kappa}) = \sum_{i=1}^{d-1} \kappa_i \mathbf{n}_i(\boldsymbol{\theta}).$$

Here $\boldsymbol{\tau}_i$ is inductively defined as

$$\boldsymbol{\tau}_i(\theta_1, \dots, \theta_i) := \begin{pmatrix} \boldsymbol{\tau}_{i-1}(\theta_1, \dots, \theta_{i-1}) \sin\theta_i \\ \cos\theta_i \end{pmatrix}, \quad \theta_i \in (0, \pi), i \geq 2,$$
$$\boldsymbol{\tau}_1(\theta_1) := \begin{pmatrix} \cos\theta_1 \\ \sin\theta_1 \end{pmatrix}, \quad \theta_1 \in \mathbb{R}/2\pi\mathbb{Z},$$

and the spherical unit vectors $\mathbf{n}_i$ are given by $\mathbf{n}_i := \frac{1}{|\partial_{\theta_i}\boldsymbol{\tau}|}\partial_{\theta_i}\boldsymbol{\tau}$. So note that the parameters take values in $\boldsymbol{\theta} \in \mathbb{R}/2\pi\mathbb{Z} \times (0, \pi)^{d-2}$ (or in $\mathbb{R}/2\pi\mathbb{Z}$ in case $d = 2$) and $\boldsymbol{\kappa} \in \mathbb{R}^{d-1}$ respectively.

Then the basic model in spherical polar coordinates (see [13]) reads

$$\begin{aligned} \mathrm{d}\boldsymbol{\xi}_t &= \boldsymbol{\tau}(\boldsymbol{\theta}_t)\,\mathrm{d}t, \\ \mathrm{d}\theta_{j,t} &= \left(-\mathcal{G}_j(\boldsymbol{\theta}_t)\mathbf{n}_j(\boldsymbol{\theta}_t) \cdot \nabla V(\boldsymbol{\xi}_t) + \frac{A^2}{2}\mathcal{G}_j^2(\boldsymbol{\theta}_t)(j-1)\cot\theta_{j,t}\right)\mathrm{d}t + A\mathcal{G}_j(\boldsymbol{\theta}_t)\,\mathrm{d}W_t^{(j)} \end{aligned} \tag{3}$$

with $j = 1, \dots, d-1$ and $\mathcal{G}_j := |\partial_{\theta_j}\boldsymbol{\tau}|^{-1} = \prod_{i=j+1}^{d-1} \frac{1}{\sin(\theta_i)}$ (where the empty product in case $j = d - 1$ is defined to be equal to 1). The respective smooth model in local coordinates, compare with [15], is now given by

$$\begin{aligned} \mathrm{d}\boldsymbol{\xi}_t &= \boldsymbol{\tau}(\boldsymbol{\theta}_t)\,\mathrm{d}t, \\ \mathrm{d}\theta_{j,t} &= -\mathcal{G}_j(\boldsymbol{\theta}_t)\mathbf{n}_j(\boldsymbol{\theta}_t) \cdot \nabla V(\boldsymbol{\xi}_t)\,\mathrm{d}t + \mathcal{G}_j(\boldsymbol{\theta}_t)\kappa_{j,t}\,\mathrm{d}t, \\ \mathrm{d}\kappa_{j,t} &= \sum_{i,n=1}^{d-1} \Delta_{inj}(\boldsymbol{\theta}_t)\kappa_{n,t}\big(\mathbf{n}_i(\boldsymbol{\theta}_t) \cdot \nabla V(\boldsymbol{\xi}_t) - \kappa_{i,t}\big)\,\mathrm{d}t - \lambda\kappa_{j,t}\,\mathrm{d}t + \mu\,\mathrm{d}W_t^{(j)}, \end{aligned} \tag{4}$$

where $j = 1, \dots, d-1$ and $\Delta_{inj} = \mathcal{G}_i\,\partial_{\theta_i}\mathbf{n}_n \cdot \mathbf{n}_j$. In all these local coordinate representations $\mathbf{W} = (W^{(1)}, \dots, W^{(d-1)})$ denotes a standard $(d-1)$-dimensional Brownian motion. It is worth mentioning that stationary states can explicitly be computed from the respective associated Fokker-Planck equations. The equilibrium for the smooth case reads

$$p_s(\boldsymbol{\xi}, \boldsymbol{\theta}, \boldsymbol{\kappa}) = Ce^{-(d-1)V(\boldsymbol{\xi})} \prod_{i=2}^{d-1} (\sin\theta_i)^{i-1} \exp\left(-\frac{\lambda}{\mu^2}\sum_{i=1}^{d-1}\kappa_i^2\right) \tag{5}$$

with some finite constant $C > 0$. Here the empty product is again equal to 1 in case $d = 2$. In the basic case we obtain the stationary distribution by integrating (5) over $\boldsymbol{\kappa}$.

Now that we have the abstract framework available, we recur to the application point of view and look at the essential cases $d = 2$ and $d = 3$ more closely.

2.1 The 2D model

Setting $d = 2$ and $\theta_1 = \alpha$ in (3) leads to the original version of the fiber lay-down models as it has been introduced in [4] as

$$\begin{aligned} \mathrm{d}\boldsymbol{\xi}_t &= \boldsymbol{\tau}(\alpha_t)\,\mathrm{d}t, \\ \mathrm{d}\alpha_t &= -\mathbf{n}_1(\alpha_t)\cdot\nabla V(\boldsymbol{\xi}_t)\,\mathrm{d}t + A\,\mathrm{d}W_t. \end{aligned} \tag{6}$$

A virtual fiber web can now be simulated by simultaneous use of this model for a large number of fibers, neglecting the influence of fiber-fiber-contact. The reference points indicating different spinning nozzles are included by adding an appropriate constant to the $\boldsymbol{\xi}_t$-process. Using this 2D model, for example, the basis weight distribution of the non-woven can straightforwardly be determined assuming a uniform thickness of the fibers.

To obtain more realistic and smoother fiber paths the Brownian motion is replaced by an Ornstein-Uhlenbeck process. Setting $d = 2$, $\theta_1 = \alpha$ and $\kappa_1 = \kappa$ in (4) the smooth model reads

$$\begin{aligned} \mathrm{d}\boldsymbol{\xi}_t &= \boldsymbol{\tau}(\alpha_t)\,\mathrm{d}t, \\ \mathrm{d}\alpha_t &= -\mathbf{n}_1(\alpha_t)\cdot\nabla V(\boldsymbol{\xi}_t)\,\mathrm{d}t + \kappa_t\,\mathrm{d}t, \\ \mathrm{d}\kappa_t &= -\lambda\kappa_t\,\mathrm{d}t + \mu\,\mathrm{d}W_t. \end{aligned} \tag{7}$$

This is a natural smoother version of (6). See also [6] for a similar model.

The moving conveyor belt can be included in (6) or (7) as an additional reference curve, i.e.

$$\mathrm{d}\boldsymbol{\xi}_t = \boldsymbol{\tau}(\alpha_t)\,\mathrm{d}t + \nu\mathbf{e}_1\,\mathrm{d}t$$

gives the deviation of the fiber from a reference point determined by the position of the corresponding spinning nozzle, where $\nu = \frac{v_{\text{belt}}}{v_{\text{in}}} \geq 0$ defines the ratio between belt speed and spinning speed of the fiber, see [1]. The image of the fiber on the belt, denoted by $(\boldsymbol{\eta}_t)_{t\geq 0}$, is then obtained by $\boldsymbol{\eta}_t = \boldsymbol{\xi}_t - \nu t\mathbf{e}_1$. Unfortunately, there is no explicit computable stationary state available in case $\nu > 0$. For a non-moving conveyor belt ($\nu = 0$) the equilibrium is given by (5).

2.2 The 3D model

In [7, 8, 15] three-dimensional fiber lay-down models have been developed. Due to physical constraints, such as the inpenetrable conveyor belt or gravitation, the possibility of anisotropic fiber orientations in the resulting fiber web has to be taken into account. Therefore, it is necessary to amend the isotropic 3D models, i.e. (3) and (4) with $d = 3$, $\boldsymbol{\theta} = (\alpha, \theta)$ and $\boldsymbol{\kappa} = (\kappa_1, \kappa_2)$ in order to enable this anisotropy. This has been realized via some weighting parameter $B \in [0,1]$. With the abbreviations

$$\begin{aligned} p(\boldsymbol{\xi},\alpha,\theta) &:= \frac{1}{\sin\theta}\mathbf{n}_1(\alpha)\cdot\nabla V(\boldsymbol{\xi}), \\ q(\boldsymbol{\xi},\alpha,\theta) &:= \mathbf{n}_2(\alpha,\theta)\cdot\nabla V(\boldsymbol{\xi}) \end{aligned}$$

one obtains the following (anisotropic) basic 3D model

$$\begin{aligned} \mathrm{d}\boldsymbol{\xi}_t &= \boldsymbol{\tau}(\alpha_t,\theta_t)\,\mathrm{d}t, \\ \mathrm{d}\alpha_t &= -p(\boldsymbol{\xi}_t,\alpha_t,\theta_t)\,\mathrm{d}t + \frac{1}{\sin\theta_t}A\,\mathrm{d}W_t^{(1)}, \\ \mathrm{d}\theta_t &= -Bq(\boldsymbol{\xi}_t,\alpha_t,\theta_t)\,\mathrm{d}t + \frac{1}{2}A^2\cot\theta_t\,\mathrm{d}t + A\sqrt{B}\,\mathrm{d}W_t^{(2)} \end{aligned} \tag{8}$$

as well as the (anisotropic) smooth 3D model as

$$\begin{aligned} \mathrm{d}\boldsymbol{\xi}_t &= \boldsymbol{\tau}(\alpha_t,\theta_t)\,\mathrm{d}t, \\ \mathrm{d}\alpha_t &= -p(\boldsymbol{\xi}_t,\alpha_t,\theta_t)\,\mathrm{d}t + \frac{1}{\sin\theta_t}\kappa_{1,t}\,\mathrm{d}t, \\ \mathrm{d}\theta_t &= -Bq(\boldsymbol{\xi}_t,\alpha_t,\theta_t)\,\mathrm{d}t + B\kappa_{2,t}\,\mathrm{d}t, \\ \mathrm{d}\kappa_{1,t} &= p(\boldsymbol{\xi}_t,\alpha_t,\theta_t)\kappa_{2,t}\cos\theta_t\,\mathrm{d}t - \kappa_{1,t}\kappa_{2,t}\cot\theta_t\,\mathrm{d}t - \lambda\kappa_{1,t}\,\mathrm{d}t + \mu\,\mathrm{d}W_t^{(1)}, \\ \mathrm{d}\kappa_{2,t} &= -p(\boldsymbol{\xi}_t,\alpha_t,\theta_t)\kappa_{1,t}\cos\theta_t\,\mathrm{d}t + \kappa_{1,t}^2\cot\theta_t\,\mathrm{d}t - B\lambda\kappa_{2,t}\,\mathrm{d}t + \mu\sqrt{B}\,\mathrm{d}W_t^{(2)}. \end{aligned} \tag{9}$$

Here $\mathbf{W} = (W^{(1)}, W^{(2)})$ denotes a standard two-dimensional Brownian motion. We note that for $B = 0$ the respective 2D models (6), (7) are obtained. Basically, the weighting parameter allows for manipulation of the marginal θ-distribution. The stationary solution of (9) is given by

$$p_s(\boldsymbol{\xi},\theta,\kappa_1,\kappa_2) = Ce^{-V_B(\boldsymbol{\xi})}(\sin\theta)^{\frac{1}{B}}e^{-\frac{\lambda}{\mu^2}(\kappa_1^2+\kappa_2^2)}, \tag{10}$$

compare also with Figure 3. Here $V_B := (B+1)V$ denotes a rescaled potential. In the following we deal with this rescaled formulation, so we drop off the B to facilitate the reading. The equilibrium state for the basic model (8) is obtained by integrating (10) over the curvatures (κ_1,κ_2). Again the moving conveyor belt can be incorporated via

$$\mathrm{d}\boldsymbol{\xi}_t = \boldsymbol{\tau}(\alpha_t,\theta_t)\,\mathrm{d}t + v\mathbf{e}_1\,\mathrm{d}t.$$

Exemplary fiber trajectories computed with (9) for varying parameter B with fixed parameters $\lambda = 1$, $\mu = 1$, $v = 0$ and chosen potential $V(\boldsymbol{\xi}) = \frac{1}{2}\|\boldsymbol{\xi}\|^2$ are illustrated in Figures 4-6.

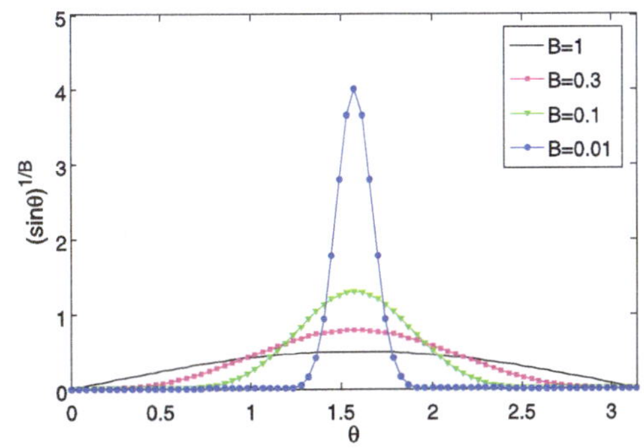

Figure 3 Behavior of the stationary θ-density for decreasing B.

Figure 4 Behavior of fiber trajectories in dependence on the weighting parameter. Here: $B = 0.1$.

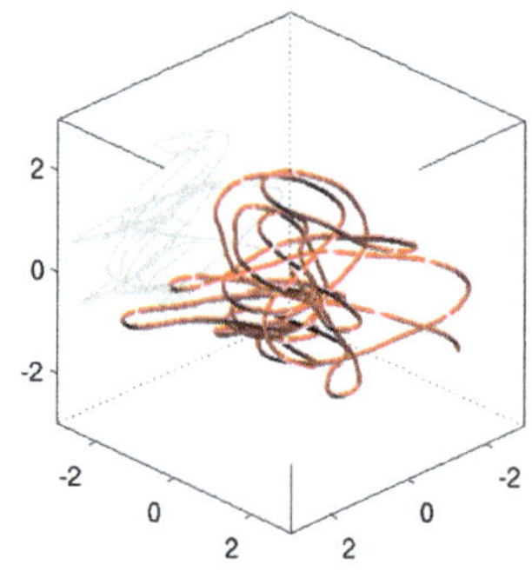

Figure 5 Behavior of fiber trajectories in dependence on the weighting parameter. Here: $B = 0.01$.

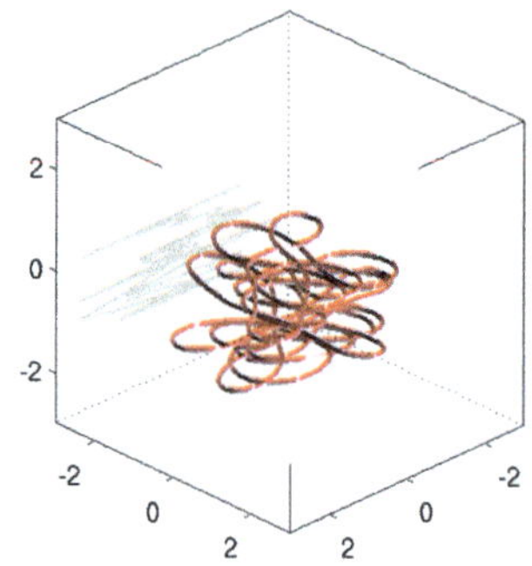

Figure 6 Behavior of fiber trajectories in dependence on the weighting parameter. Here: $B = 0.001$.

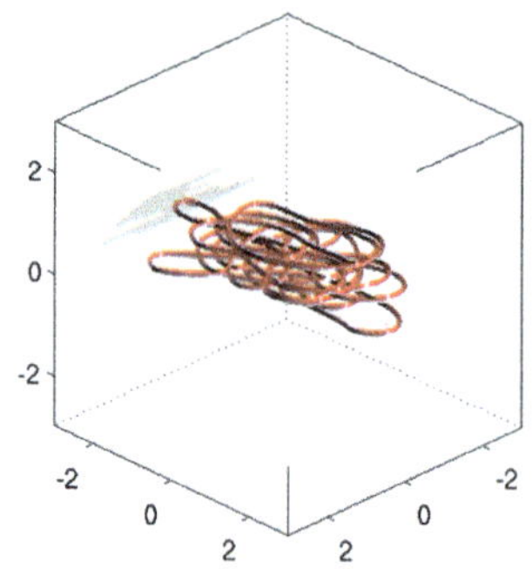

3 Parameter estimation

In this chapter, we specify our strategy to identify the parameters of the surrogate models. Here we focus on the calibration of the smooth 3D model (9) in the situation of a moving conveyor belt. More precisely, we follow the proposed approach from [8] that combine the heuristic identification method for the 2D models (compare [1]) with information about fiber orientations in the resulting non-woven gained from CT-scan measurements. The motivation for this approach is associated with the utilization of FIDYST as simulation tool of the physical fiber lay-down process. Its computations involve the motion of the fibers in turbulent air flows until they reach the conveyor belt where they are deposited. In this way FIDYST provides a two-dimensional texture that represents the fiber positions on the belt, see [17]. The entanglement of the fibers during their deposition is responsible for their typical forms, i.e. looping, buckling and coiling, see also [18, 19]. As already mentioned these characteristics are recovered by some parameters of the surrogate models. Apart from the supposedly known speed ratio ν, we need estimations of the parameters μ, λ and

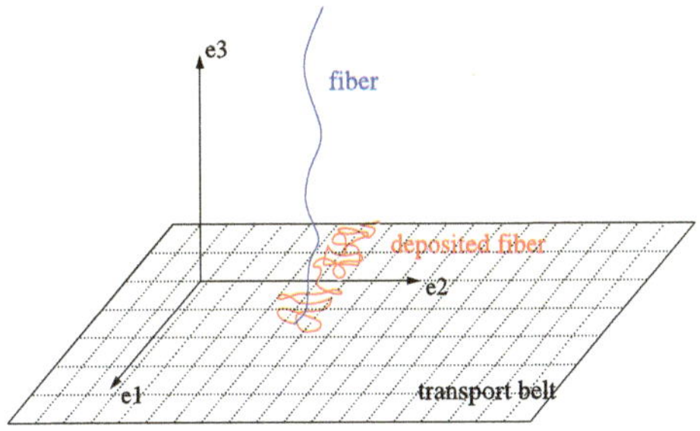

Figure 7 Sketch of fiber lay-down and fixing of the coordinate system.

B as well as the shape of the potential V. We use a potential of the following form

$$V(\boldsymbol{\xi}) = \mathcal{V}(\xi_1, \xi_2) + \Phi(\xi_3), \quad \text{with } \mathcal{V}(\xi_1, \xi_2) = \frac{\xi_1^2}{2\sigma_1^2} + \frac{\xi_2^2}{2\sigma_2^2},$$

where $\sigma_1, \sigma_2 > 0$ represents the standard deviations of the fiber throwing on the belt, compare Figure 7 for a specification of the coordinate system. Furthermore, Φ denotes a confining potential taking account of the location of the belt, such that ξ_3 is restricted to $[0, d_f]$, where d_f denotes the thickness of the non-woven.

In summary, the parameters $(\sigma_1, \sigma_2, \mu, \lambda)$ are identified from a data sample of two-dimensional fiber locations, denoted by $\boldsymbol{\eta}$ and provided by FIDYST, whereas the anisotropy parameter B is obtained from CT-scans. For the following considerations, we suppose that the reference curve $\boldsymbol{\gamma} = -v\mathbf{e}_1 t$ is known and that the deviation process $\boldsymbol{\xi} = \boldsymbol{\eta} - \boldsymbol{\gamma}$ is centered in the origin. At first, we dedicate ourselves to the afore-mentioned heuristic approach from [1] that enables the estimation of the 2D parameters. The advantage of this method lies in the use of the characteristic parameters that are actually observable in the process. Since the subsequently defined functional is closely related to the parameter space, the identification algorithm is very robust. An alternative method in the context of fiber lay-down models using occupation times can be find in [20]. For more systematic approaches we refer to [21, 22]. Let $\mathcal{D} = (\mathbf{D}_1, \ldots, \mathbf{D}_N) \in (\mathbb{R}^2 \times \mathbb{R} \times \mathbb{R})^N$ with $\mathbf{D}_i = (\boldsymbol{\eta}_{t_i}, \alpha_{t_i}, \kappa_{t_i}), 1 \le i \le N$, be a set of data points obtained from a FIDYST simulation. Here an equidistant time grid with $\Delta t = t_{i+1} - t_i$ is used. The angles α_{t_i} and curvatures κ_{t_i} are reconstructed from the fiber points $\boldsymbol{\eta}_{t_i}$ by finite difference approximations. We note that in [1] the emphasis is on the calibration of the model from [6], whereas we want to work with the smooth model (7) here. Therefore, with $\boldsymbol{\xi}_t = (\xi_{t,1}, \xi_{t,2})$, we consider a slightly different functional of characteristic properties than stated in [1]:

$$\mathcal{F}(\mathcal{D}) = \big(\mathcal{S}(\xi_{t,1}), \mathcal{S}(\xi_{t,2}), \mathcal{S}(\kappa_t), \mathcal{K}(\kappa_t)\big),$$

where we define for a real valued stochastic process $(X_t)_{t \in \{t_1, \ldots, t_N\}}$,

$$\mathcal{S}(X_t) := \sqrt{\frac{1}{N}\sum_{i=1}^{N}\left(X_{t_i} - \frac{1}{N}\sum_{j=1}^{N} X_{t_j}\right)^2},$$

$$\mathcal{K}(X_t) := \max_{k \in \{1, \ldots, \bar{k}\}} \sqrt{\frac{\sum_{i=1}^{N-k}(X_{t_{i+k}} - X_{t_i})^2}{k(N-k)\Delta t}}, \quad \bar{k} \ll N,$$

Table 1 Predetermined parameters by the pilot plant and size of data samples

	Predetermined			FIDYST		IPD
	v_{belt} [m/s]	v_{spin} [m/s]	f_{diam} [m]	Δt [m]	# points	# points
Sample *I*	0.633	79.400	$12 \cdot 10^{-6}$	0.001	13×19,850	4,352,392
Sample *II*	4.167	79.400	$12 \cdot 10^{-6}$	0.001	13×17,850	500,583

which approximate the standard deviations of the processes and of their increments. In order to find the optimal calibration $\mathbf{P} = (\sigma_1, \sigma_2, \frac{\mu}{\sqrt{2\lambda}}, \mu)$ of the surrogate model (7) with moving belt and potential $\mathcal{V}$, denoted by $\mathcal{D}_{\text{sur}}(\mathbf{P})$, we fix the minimization problem

$$\mathbf{P}^* = \operatorname{argmin}_{\mathbf{P}} \left\| \mathcal{F}\big(\mathcal{D}_{\text{sur}}(\mathbf{P})\big) - \mathcal{F}\big(\mathcal{D}_{\text{fid}}\big) \right\|_2,$$

where $\mathcal{D}_{\text{fid}}$ indicates the data sample obtained from a FIDYST simulation. This can be solved by a relaxated quasi Newton method with unit Jacobian, i.e. $\mathbf{P}^{(n+1)} = \mathbf{P}^{(n)} + \omega[\mathcal{F}(\mathcal{D}_{\text{sur}}(\mathbf{P}^{(n)})) - \mathcal{F}(\mathcal{D}_{\text{fid}})]$ and starting point $\mathbf{P}^{(0)} = \mathcal{F}(\mathcal{D}_{\text{fid}})$, compare [1]. We note that $\mathcal{F}$ is a very good estimator for $\mathbf{P}$, if the fiber process is close to its stationary state (5) with $d = 2$, i.e. for adequately large data sample and small speed ratio ν, two conditions which are usually fulfilled in the production processes considered here. It is worth mentioning that this choice of $\mathcal{F}$ is completely detached from the white noise limit situation and can therefore be used for larger ranges of applications compared to [1].

We next describe how the data from CT-scan measurements of a non-woven can be handled to disclose the orientation of the fibers therein. This method of measurement is suitable for the examination, since the resolution of the CT-scan is around 1 μm, which is about an order of magnitude smaller than a typical fiber diameter, see Table 1. The CT-scan provides a real-valued three-dimensional matrix with so-called gray values. These raw image data are edited by image processing, which translates them into local fiber directions at each pixel with the help of an eigenvalue analysis of the Hessian matrix of the second partial derivatives of the gray values, for more details consult [23]. In other words, the image processing data (IPD) yield spherical polar angles that determine the orientation of the tangents at the fiber points. With these we generate the corresponding density distribution $\breve{p}^{(CT)}(\alpha,\theta)$ with $(\alpha,\theta) \in \breve{U} = \mathbb{R}/2\pi\mathbb{Z} \times (0, \frac{\pi}{2}]$. The restriction to the hemisphere is obligatory, since the data of the CT-scan do not reveal the temporal course of the (undistinguishable) fiber paths. Thus the tangents contain no more than an unsigned directional information. Therefore, the density $\breve{p}^{(CT)}(\alpha,\theta)$ has to be distinguished from the (unknown) angular density distribution obtained by our model (9) with moving conveyor belt, in the following denoted by $p^{(M)}(\alpha,\theta)$ with $(\alpha,\theta) \in U = \mathbb{R}/2\pi\mathbb{Z} \times (0,\pi)$. The latter can be convert to the hemisphere via

$$\breve{p}^{(M)}(\alpha,\theta) = \frac{1}{2}\big(p^{(M)}(\alpha,\theta) + p^{(M)}(\alpha+\pi, \pi-\theta)\big), \quad (\alpha,\theta) \in \breve{U} \tag{11}$$

which allows us to compare the densities.

Assuming that the fiber lay-down process is close to equilibrium, or in other words, if ν is sufficiently small, we can identify the parameter B without great effort. In that case, we demand the equality of the standard deviation σ_B of the θ-marginal of the distribution function obtained from the CT-scan, denoted by $\breve{p}_\theta^{(CT)}$, and the explicitly computed standard deviation $\bar{\sigma}_B$ of the θ-marginal of the stationary distribution (10) of our model (with $\nu = 0$). The latter reads $\bar{\sigma}_B = (\int_0^\pi C_B(\sin\theta)^{\frac{1}{B}}(\theta - \frac{\pi}{2})^2 \, d\theta)^{\frac{1}{2}}$ with normalization constant

$C_B^{-1} = \int_0^{\pi} (\sin\theta)^{\frac{1}{B}}\, d\theta$. Besides, the influence of the belt speed on the size of B is expected to be small in contrast to the other parameters, since it involves the deviation from the planar structure. In the case of very large ν, the estimation of B can be straightforwardly embedded in the above heuristic identification approach with a little more effort. For our purposes, however, the described identification strategy will turn out to be adequate anyway.

4 Application

In the following we want to demonstrate the adaptation of the surrogate models on the basis of a industrial test case describing a real non-woven production process. From a pilot plant of the company Oerlikon Neumag process data have been available that have been used to initialize FIDYST simulations of the full process. Plant specific parameters, such as machine geometry and prevailing air flows have been included in these computations. In addition, an associated non-woven product has been produced by the pilot plant with the same configuration data. Pieces of this non-woven have been cut and have been analyzed in CT-scans. The statistics of the fiber orientation in the non-woven given as image processing data (IPD) complement the FIDYST informations as described in the previous chapter.

We discuss two different data samples, denoted by sample I and sample II. Associated characteristic values and data sizes are summarized in Table 1. The resolution of FIDYST is something to be viewed critically. A more accurate resolution increases the effort and leads to unreasonable computational costs, compare Chapter 1. At least the surrogate models use the same grid sizes and they can be viewed as optimal discrete substitutes for FIDYST, see also below. The different numbers of IPD points indicate a different usable sample size of the pieces of nonwoven. A larger number of points leads to smoother α-marginals $\breve{p}_{\alpha}^{(CT)}(\alpha,\theta)$, compare Figures 8, 9, otherwise, it is not relevant. The essential difference be-

Figure 8 Simulated fibers (red) with one highlighted filament (blue). Here: FIDYST, sample I.

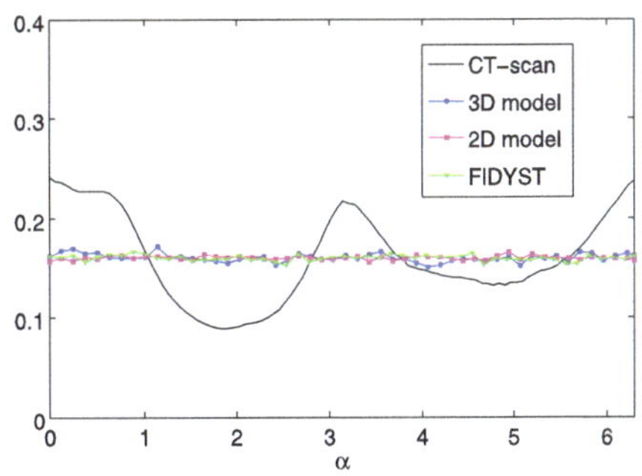

Figure 9 Simulated fibers (red) with one highlighted filament (blue). Here: surrogate model, sample I.

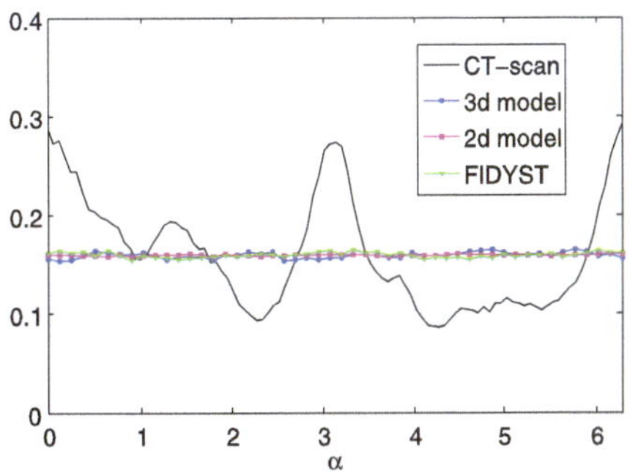

Table 2 Identified parameters to be used in the surrogate model

	σ_1 [m]	σ_2 [m]	μ [$m^{-3/2}$]	λ [m^{-1}]	B	v
Sample *I*	0.0050	0.0049	49,096	1,039	0.398	0.0080
Sample *II*	0.0055	0.0045	49,885	997	0.375	0.0525

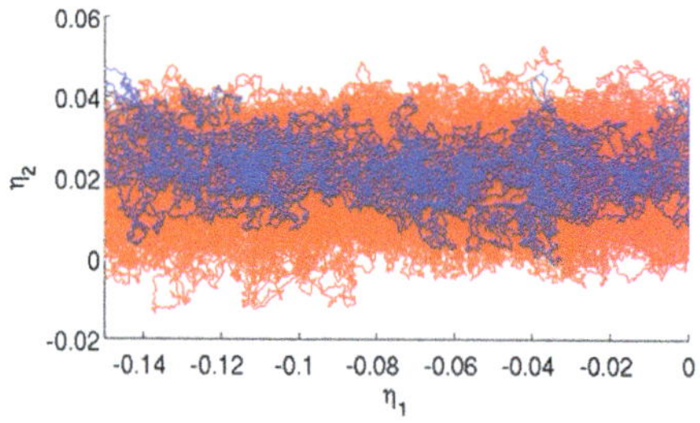

Figure 10 Simulated fibers (red) with one highlighted filament (blue). Here: FIDYST, sample *II*.

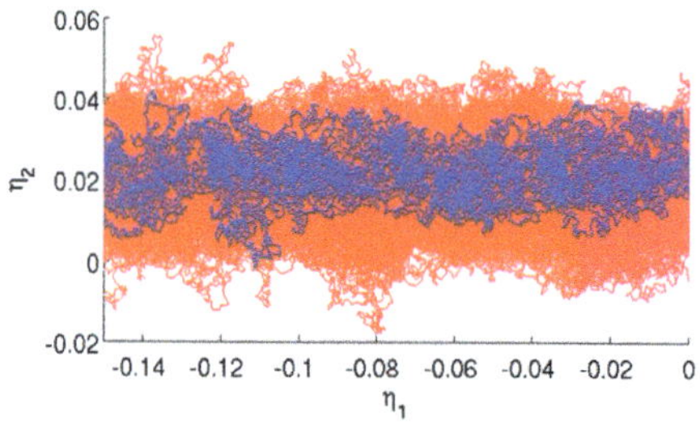

Figure 11 Simulated fibers (red) with one highlighted filament (blue). Here: surrogate model, sample *II*.

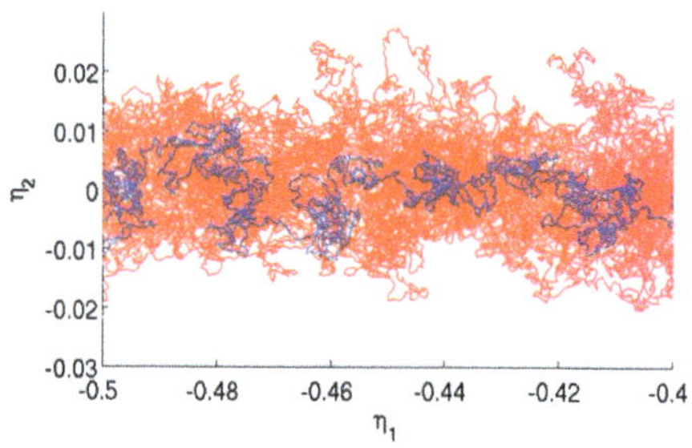

Figure 12 Marginal angular distribution $\breve{p}_\theta^{(CT)}(\alpha, \theta)$, $\breve{p}_\theta^{(M)}(\alpha, \theta)$. Here: sample *I*.

tween the two samples is the belt speed. Nevertheless, the speed ratio v is sufficiently small in both cases, which can be ascertained by the comparison between the standard deviation $\bar{\sigma}_B$ and the one of our model with given v using Monte-Carlo simulations. Thus the calibration of the surrogate model (9) is carried out as described above. The estimated parameters are summarized in Table 2. In Figures 10-13 we illustrate a comparison of fibers computed by FIDYST and the calibrated 2D model (7) with moving transport belt. Qualitatively the same fiber lay-down structures are observed. The parameter B is identified from the CT-scans data as outlined above. This leads to θ-marginal distributions as in Figures 14, 15.

Furthermore, in Figures 8, 9 we compare the α-marginal $\breve{p}_\alpha^{(CT)}(\alpha, \theta)$ of the CT-scan data with the corresponding distributions obtained from FIDYST simulations $\breve{p}_\alpha^{(M_F)}(\alpha, \theta)$ and from the surrogate models $\breve{p}_\alpha^{(M_{2d})}(\alpha, \theta)$ and $\breve{p}_\alpha^{(M_{3d})}(\alpha, \theta)$, which can be computed with the help of (11), (where $\theta = \pi/2$ is fixed for $p^{(M_F)}$ and $p^{(M_{2d})}$). It seems conspicuous, that the large amplitude in the CT-scans is not reached by the models. On the contrary, both FIDYST

Figure 13 Marginal angular distribution $\breve{p}_{\theta}^{(CT)}(\alpha, \theta)$, $\breve{p}_{\theta}^{(M)}(\alpha, \theta)$. Here: sample *II*.

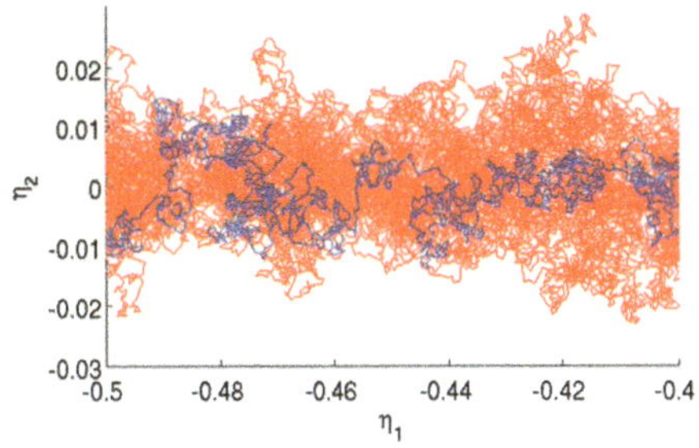

Figure 14 Marginal angular distribution $\breve{p}_{\alpha}^{(CT)}(\alpha, \theta)$, $\breve{p}_{\alpha}^{(M)}(\alpha, \theta)$. Here: sample *I*.

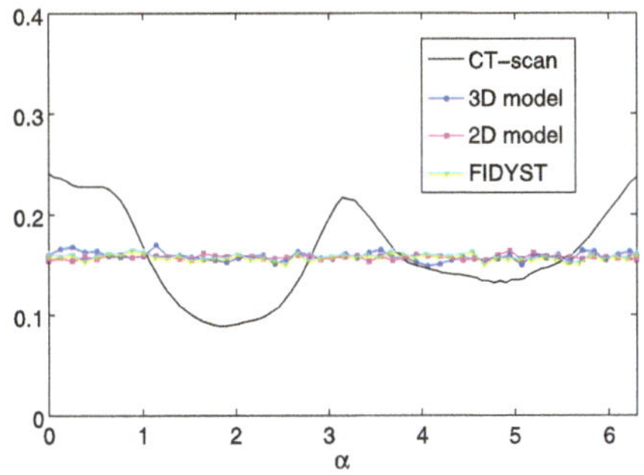

Figure 15 Marginal angular distribution $\breve{p}_{\alpha}^{(CT)}(\alpha, \theta)$, $\breve{p}_{\alpha}^{(M)}(\alpha, \theta)$. Here: sample *II*.

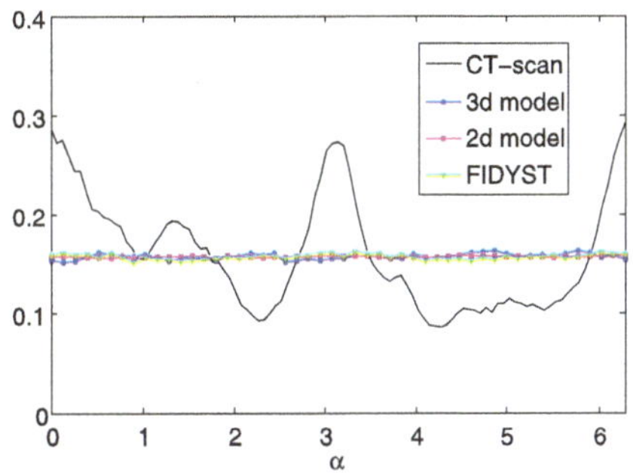

and the surrogate models (7) and (9) show an almost uniform distribution, which is not surprising because of the very small speed ratio. Indeed, the surrogate models show the same quantitative behavior as the FIDYST simulation.

This discrepancy between the CT-scan measurements and the models with respect to the α-distribution might be explained by the fact that the CT-scan has analyzed pieces of non-woven that had already passed through several process steps of reworking and reinforcement which had influenced the angular distribution. On the contrary, both, FIDYST and the surrogate fiber lay-down models have described the texture before post-processing. To investigate this issue more closely, it would be of great interest to have CT-scans of the deposited fibers without effects due to post-processing.

Finally, we use the calibrated 3D model to simulate virtual fiber webs by superposing hundred fibers. The distance of neighboring spinning positions $a \approx 0.0015$ m can be determined from FIDYST simulations, whereas the nonwoven thickness d_f has to be appropriately chosen in the potential V. Unfortunately, it is technically impossible to get at this parameter. We can only estimate in terms of the fiber diameter f_{diam}. Exemplary simulations for the two samples with a presumed non-woven thickness $d_f = 200 f_{\text{diam}}$ are shown

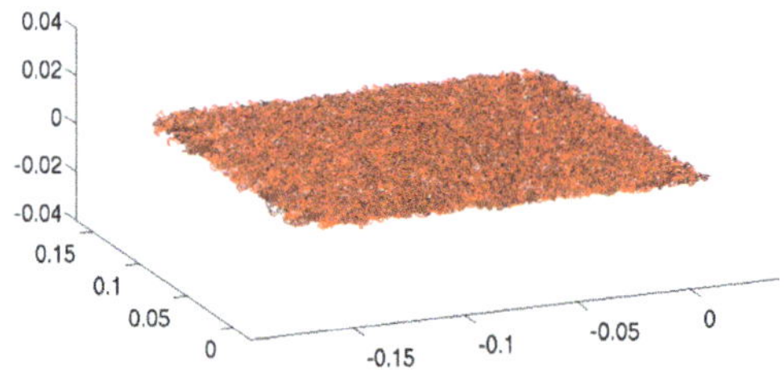

Figure 16 Simulation of a fiber web corresponding to sample *I*. Here: side view.

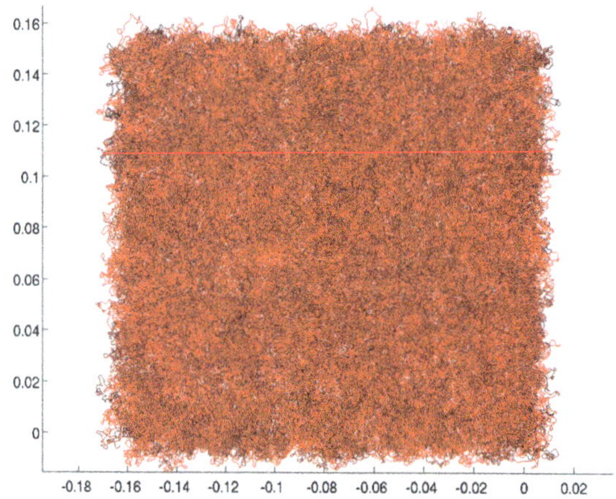

Figure 17 Simulation of a fiber web corresponding to sample *I*. Here: top view.

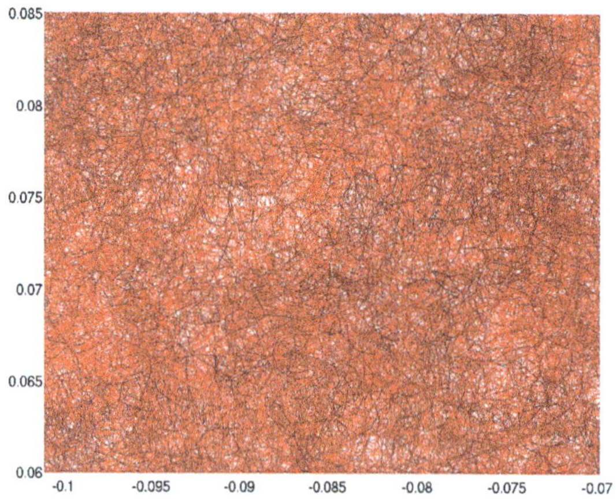

Figure 18 Simulation of a fiber web corresponding to sample *I*. Here: top view (zoomed in).

in Figures 16-21. To compare the quality of the resulting fiber webs, we investigate the homogeneity more closely. As is common in the practical application, we look at the basis weight distribution. For this purpose, a 2D grid is placed over a fixed area of the simulated fiber web (projected to 2D) and the basis weight of each cell is numerically determined. This weight M of a given cell is proportional to the time the fiber process is present in that cell. In Figures 22-27 the relative deviations $\frac{|E[M]-M|}{E[M]}$ between the (expected) averaged and the actual basis weight in the cells are illustrated for different grid sizes. The standard de-

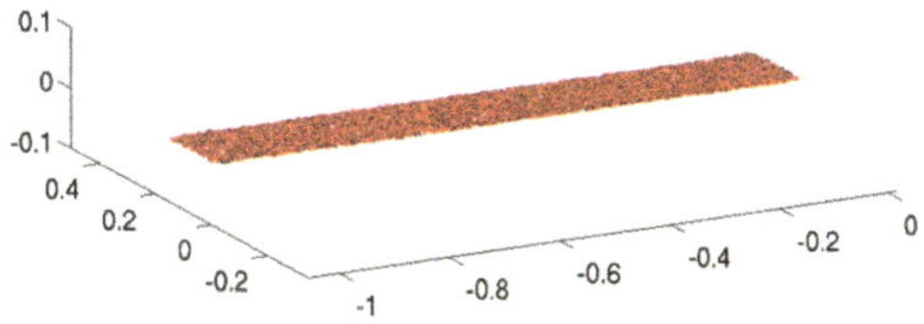

Figure 19 Simulation of a fiber web corresponding to sample *II*. Here: side view.

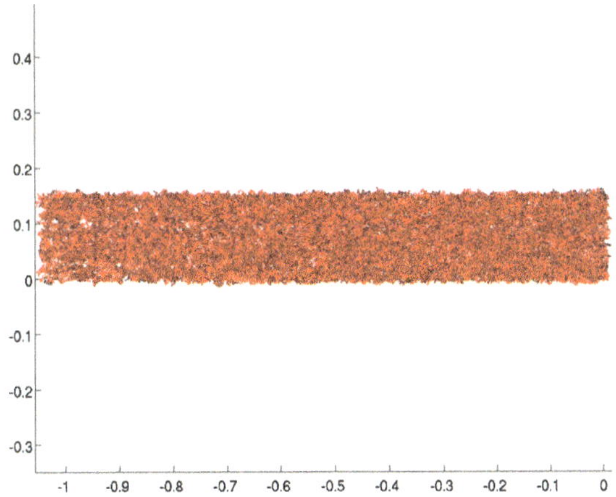

Figure 20 Simulation of a fiber web corresponding to sample *II*. Here: top view.

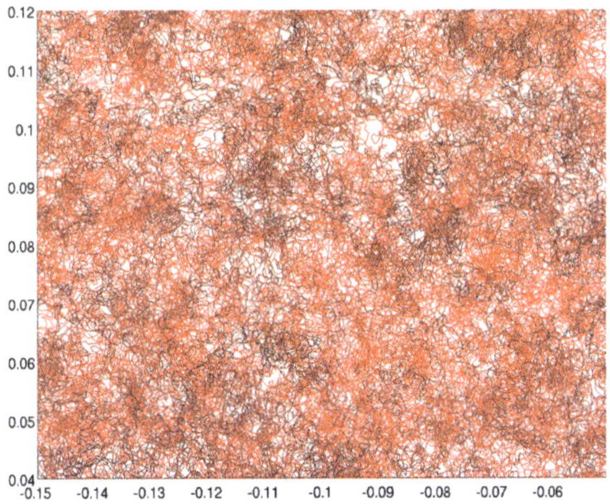

Figure 21 Simulation of a fiber web corresponding to sample *II*. Here: top view (zoomed in).

viations of these values give the corresponding coefficients of variation (CV-values) which are summarized in Table 3. As specimen we choose a square of size 100 cm^2 for each sample. Due to the respective smaller CV-values we can state that the simulated fiber web associated to sample I is more homogeneous than the one given by sample II and hence better quality characteristics of nonwovens are expected for the machine configuration of sample I, i.e. for a slower conveyor belt. Due to the different belt speed, however, the basis weight with respect to sample II is six times smaller compared to sample I.

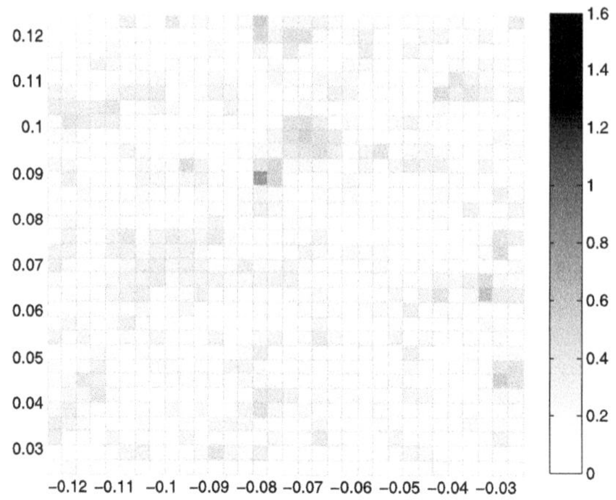

Figure 22 Relative deviation of basis weight per grid cell. Here: sample *I*, 32 × 32-grid.

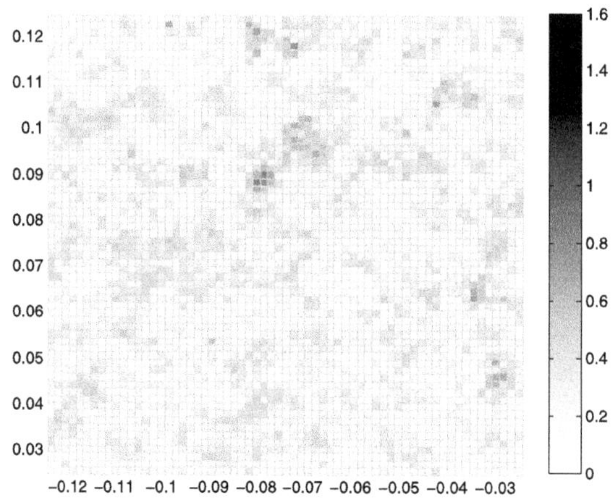

Figure 23 Relative deviation of basis weight per grid cell. Here: sample *I*, 64 × 64-grid.

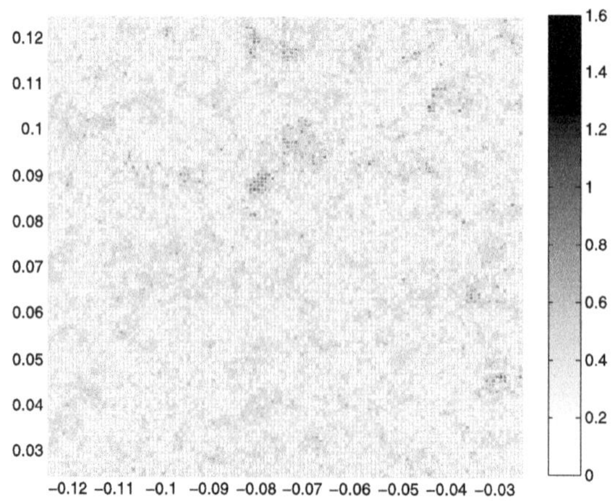

Figure 24 Relative deviation of basis weight per grid cell. Here: sample *I*, 128 × 128-grid.

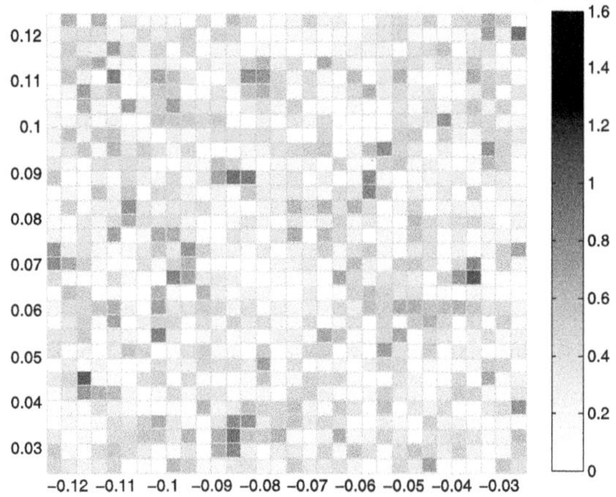

Figure 25 Relative deviation of basis weight per grid cell. Here: sample *II*, 32 × 32-grid.

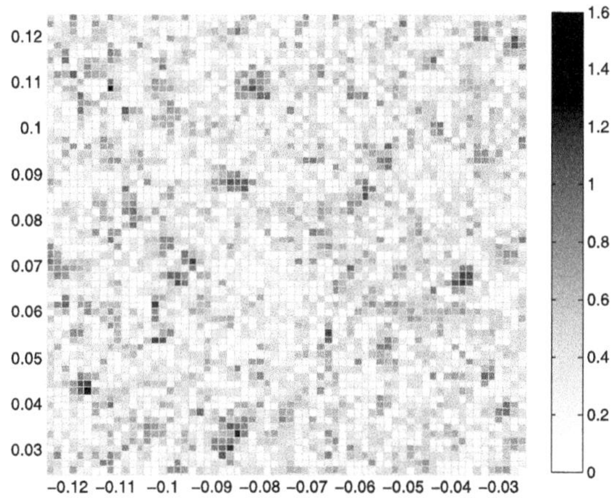

Figure 26 Relative deviation of basis weight per grid cell. Here: sample *II*, 64 × 64-grid.

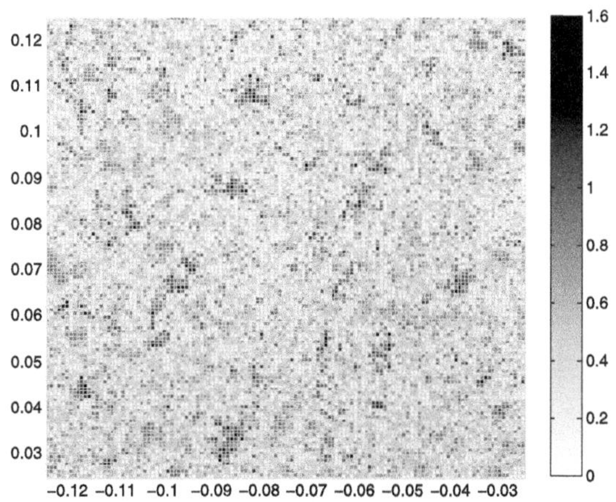

Figure 27 Relative deviation of basis weight per grid cell. Here: sample *II*, 128 × 128-grid.

Table 3 Basis weight distribution in terms of CV-values (in %) of the simulated fiber webs

	Grid		
	32 x 32	**64 x 64**	**128 x 128**
Sample *I*	16.4	18.9	22.6
Sample *II*	31.8	38.9	49.1

5 Conclusion and outlook

We presented the application of a 3D surrogate fiber lay-down model to an industrial problem. The parameters are identified on the basis of experimental data. The calibrated model enables the efficient simulation of a whole virtual fiber web. Further modifications of the model will include, for example, the impenetrability of the fibers. This is examined in further studies. To predict material properties like permeability of the textile, the geometric model presented here has to be used as a basic model for a complex flow simulation to determine macroscopic properties of the textile.

Competing interests

The authors declare that they have no competing interests.

Authors' contributions

All authors contributed equally to the writing of this paper. All authors read and approved the final manuscript.

Author details

[1]Technische Universität Kaiserslautern, Gottlieb-Daimler Str., Kaiserslautern, 67663, Germany. [2]Fraunhofer-Institut für Techno- und Wirtschaftsmathematik, Fraunhofer Platz 1, Kaiserslautern, 67663, Germany.

Acknowledgements

This work has been supported by Bundesministerium für Bildung und Forschung (BMBF), Verbundprojekt ProFil, 03MS606. Moreover, we thank the colleagues from the departments Transport Processes and Image Processing at Fraunhofer ITWM.

Endnote

[a] FIDYST: **Fi**ber **Dy**namics **S**imulation **T**ool developed at Fraunhofer ITWM, Kaiserslautern.

References

1. Klar A, Marheineke N, Wegener R: **Hierarchy of mathematical models for production processes of technical textiles.** *Z Angew Math Mech* 2009, **89**(12):941-961.
2. Marheineke N, Wegener R: **Fiber dynamics in turbulent flows: general modeling framework.** *SIAM J Appl Math* 2006, **66**(5):1703-1726.
3. Marheineke N, Wegener R: **Modeling and application of a stochastic drag for fibers in turbulent flows.** *Int J Multiph Flow* 2011, **37**:136-148.
4. Götz T, Klar A, Marheineke N, Wegener R: **A stochastic model and associated Fokker-Planck equation for the fiber lay-down process in nonwoven production processes.** *SIAM J Appl Math* 2007, **67**(6):1704-1717.
5. Bonilla LL, Götz T, Klar A, Marheineke N, Wegener R: **Hydrodynamic limit of a Fokker-Planck equation describing fiber lay-down processes.** *SIAM J Appl Math* 2007/08, **68**(3):648-665.
6. Herty M, Klar A, Motsch S, Olawsky F: **A smooth model for fiber lay-down processes and its diffusion approximations.** *Kinet Relat Models* 2009, **2**(3):489-502.
7. Klar A, Maringer J, Wegener R: **A smooth 3D model for fiber lay-down in nonwoven production processes.** *Kinet Relat Models* 2012, **5**:97-112.
8. Klar A, Maringer J, Wegener R: **A 3D model for fiber lay-down in nonwoven production processes.** *Math Models Methods Appl Sci* 2012, **22**(9):1250020.
9. Grothaus M, Klar A: **Ergodicity and rate of convergence for a nonsectorial fiber lay-down process.** *SIAM J Math Anal* 2008, **40**(3):968-983.
10. Dolbeault J, Klar A, Mouhot C, Schmeiser C: **Exponential rate of convergence to equilibrium for a model describing fiber lay-down processes.** *Appl Math Res Express* 2012. doi:10.1093/amrx/abs015.
11. Kolb M, Savov M, Wübker A: **(Non-)ergodicity of a degenerate diffusion modeling the fiber lay down process.** *SIAM J Math Anal* 2013, **45**:1-13.
12. Kolb M, Savov M, Wübker A: **Geometric ergodicity of a hypoelliptic diffusion modelling the melt-spinning process of nonwoven materials**; 2012 [arXiv:1112.6159]
13. Grothaus M, Klar A, Maringer J, Stilgenbauer P: **Geometry, mixing properties and hypocoercivity of a degenerate diffusion arising in technical textile industry**; 2012 [arXiv:1203.4502]
14. Grothaus M, Stilgenbauer P: **Hypocoercivity for Kolmogorov backward evolution equations and applications**; 2012 [arXiv:1207.5447]

15. Grothaus M, Stilgenbauer P: **Geometric Langevin equations on submanifolds and applications to the stochastic melt-spinning process of nonwovens and biology**. *Stoch Dyn* 2013, **13**(4):1350001.
16. Hsu EP: *Stochastic Analysis on Manifolds*. Providence: Am. Math. Soc.; 2002. [*Graduate Studies in Mathematics*]
17. Hietel D, Wegener R: **Simulation of spinning and lay-down processes**. *Tech Text* 2005, **3**:145-148.
18. Hearle J, Sultan M, Govender S: **The form taken by threads laid on a moving belt**. *J Text Inst* 1976, **67**:373-386.
19. Mahadevan L, Keller J: **Coiling of flexible ropes**. *Proc R Soc Lond A* 1996, **452**(1950):1679-1694.
20. Bock W, Götz T, Liyanage P: **Parameter estimation of fiber lay-down in nonwoven production - an occupation time approach**; 2011 [arXiv:1112.3550]
21. Kutoyants YA: *Statistical Inference for Ergodic Diffusion Processes*. London: Springer; 2004. [*Springer Series in Statistics*]
22. Gobet E, Hoffmann M, Reiß M: **Nonparametric estimation of scalar diffusions based on low frequency data**. *Ann Stat* 2004, **32**(5):2223-2253.
23. Redenbach C, Rack A, Schladitz K, Wirjadi O, Godehardt M: **Beyond imaging: on the quantitative analysis of tomographic volume data**. *Int J Mater Res* 2012, **2**:217-227.

13

Counteracting ring formation in rotary kilns

M Pisaroni[1], R Sadi[2] and D Lahaye[1*]

*Correspondence:
d.j.p.lahaye@tudelft.nl
[1] Scientific Computing Group, Delft Institute of Applied Mathematics, Faculty of Electrical Engineering, Mathematics and Computer Science, Delft University of Technology, Delft, The Netherlands
Full list of author information is available at the end of the article

Abstract

Avoiding the formation of rings in rotary kilns is an issue of primary concern to the cement production industry. We developed a numerical combustion model that revealed that in our case study rings are typically formed in zones of maximal radiative heat transfer. This local overheating causes an overproduction of the liquid phase of the granular material, which tends to stick to the oven's wall and to form rings. To counteract for this phenomenon, we propose to increase the amount of secondary air injected to cool the oven. Experimental validation at the plant has confirmed that our solution is indeed effective. For the first time in years, the kiln has been operating without unscheduled shut-downs, resulting in hugely important cost savings.

Keywords: rotary kiln; computational fluid dynamics; ring formation

1 Introduction

1.1 Rotary kilns

Rotary kilns [1] are long cylindrical industrial furnaces that for their operation are slightly tilted as shown in Figure 1. They are used in a wide range of material processing industries. The material to be processed is fed into the upper end of the cylinder, leaving a considerable amount of freeboard or empty space. As the kiln rotates about its axis, the material bed gradually moves down towards the lower end, and undergoes a certain amount of stirring and mixing. The material leaves the oven at the lower end to be further processed. The heating of the kiln serves to drive the specific bed reactions, which, for either kinetic and thermodynamic reasons, require high temperatures. The CFD simulation of these devices has been considered in *e.g.* [2–4].

We will look into direct fired kilns in which the energy necessary to heat the material to the level required for the intended reactions is generated by the combustion of hydrocarbon fuels continuously fed to a burner placed in the freeboard. This energy is subsequently transferred by heat exchange between the gas phase and the material bed. The lateral surface of the kiln is covered by an isolating lining of refractory material. Hot gases flow in a direction opposite to that of the material bed through the kiln. The heat transfer between the freeboard and the bed is a rather complex phenomenon as it occurs along various paths determined by the physics of radiation. Various transport processes therefore constitute the working principle of the ovens considered.

We study a rotary kiln used by Almatis B.V. in Rotterdam for a production of calcium-aluminate cement, a white, high purity hydraulic bonding agent providing controlled setting times and strength required in today's high performance refractories operating at temperatures up to 1,800 degrees Celsius. This cement is made by fusing a mixture of a

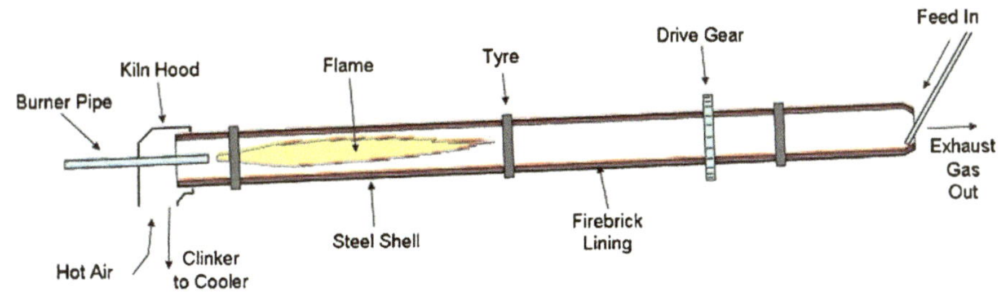

Figure 1 General layout of a direct fired, counter-current fed rotary kiln.

calcium-bearing limestone and an aluminum-bearing material. The following two considerations will be important in our approach to the modeling of the kiln. The first is that the material bed occupies no more than five percent of the total volume of the kiln. The second is that the combustion of fuel by the burner is aided by inflow of preheated air through a secondary air inlet.

1.2 Ring formation

As material slides and tumbles slowly through a heated rotary kiln, a thin layer of dust invariably forms on its inner lateral surface. Some zones of the kiln may be particularly prone to particle accumulation in such a way that the combination of particular thermal and flow conditions results in the formation of cylindrical deposits of solid crystalline structures referred to as *rings*. As such rings grow thicker, they form a dam in the freeboard, hindering the flow of material and flue gasses through the kiln.

In the kiln we consider, we observed in particular so-called *front-end/mid-kiln* rings [5]. They are formed close to the burner and are presumably caused by the direct impingement of the burner flame on the lateral surface causing local overheating. These are the most common and most troublesome type of rings as they cannot be reached from the kiln's exterior and are therefore impossible to remove while the kiln is in operation. In severe cases, ring dams grow rapidly and cause the unscheduled shutdown of the kiln. In the last three years we registered on average at least one ring formation per month, of which seventy percent caused the unforeseen shutdown of the kiln for on average three days. Each of these kiln outages causes very important production and turnover losses.

This paper is structured as follows. In Section 2 we motivate the development of a computational fluid dynamics model for the kiln. In Section 3 we give details of this model's operational setup and geometry, the polyhedral mesh discretizing the spatial coordinates and the partial differential equations governing the gas flow in the oven, the combustion of hydrocarbon fuel in the burner, the reactions of the chemical species and the radiative heat transfer in the freeboard. In Section 4 we present numerical results that convincingly show that our model allows to localize and counteract the ring formation by increasing the amount of air through the secondary air inlet. In Section 5 finally we report on an experimental validation at the plant that confirms the effectiveness of the change in operating conditions that we propose.

2 Localizing and counteracting ring formation

Experience at the plant has shown that neither trial and error nor simplified mathematical modeling approaches provide sufficient insight to eliminate the problem of ring formation.

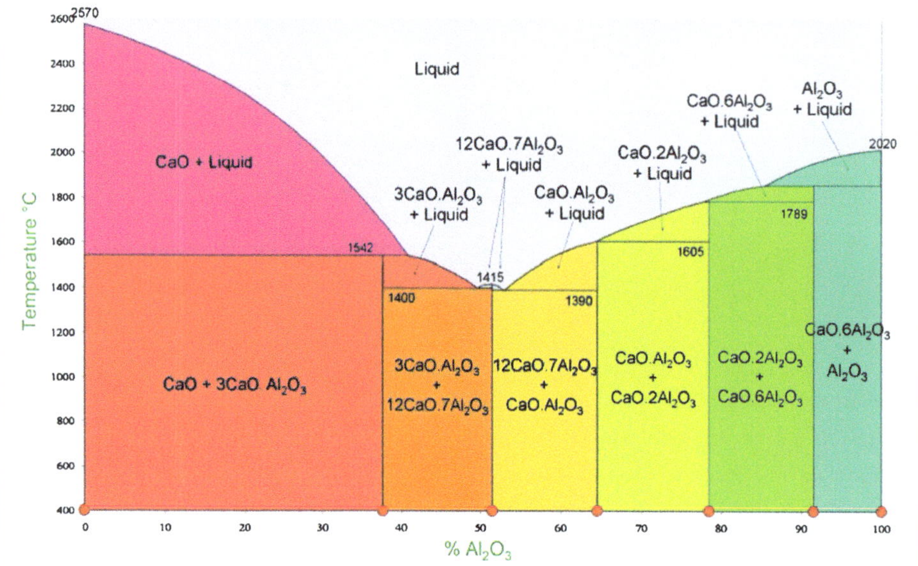

Figure 2 Phase diagram of one of the materials processed in the kiln.

Measuring the temperature inside the kiln, using thermocouples for instance, has proven to be difficult due to the harsh operating conditions. We therefore decided to develop a sufficiently detailed mathematical model enabling to link the ring formation to the temperature, radiative heat and gas flow profiles in the kiln. This model will be described in more details in the next section and has revealed that rings are typically formed in zones of maximal radiative heat transfer. In a phase-diagram this local overheating can be linked to the overproduction of the amount of liquid phase of the material. Such a phase diagram is shown in Figure 2 for one of the raw materials to be processed.

The insight linking the ring formation to local overheating proved to be instrumental in altering the operating conditions of the kiln in such a way to avoid ring formation. Our solution avoids drastic changes in the geometry of the kiln, in the configuration of the burner and in the properties of the fuel or the lining. We instead alter the fuel/air composition by altering the amount of preheated air injected through the secondary air inlet shown in Figure 3. This allows us to eliminate peaks in temperature and radiative heat transfer profiles, to limit the amount of liquid phase production and therefore avoid ring formation. The change in operating conditions we propose does maintain the temperature distribution required by the stringent quality specifications on the end product.

3 Computational fluid dynamics model

The detailed mathematical model we developed is a multi-physics model that takes the following phenomena into account: the reactive gas flow and temperature, chemical species and radiative heat transfer distribution in the kiln, the turbulent non-premixed combustion of hydro-carbon gasses in the burner, the insulating properties of the lining, the rotary motion of the kiln and the forced convection on the outside surface.

The material bed only occupies a small fraction of the volume of the kiln and has a negligible limited impact of temperature distribution. We therefore do not take the material bed into account in our model and simulate an empty kiln.

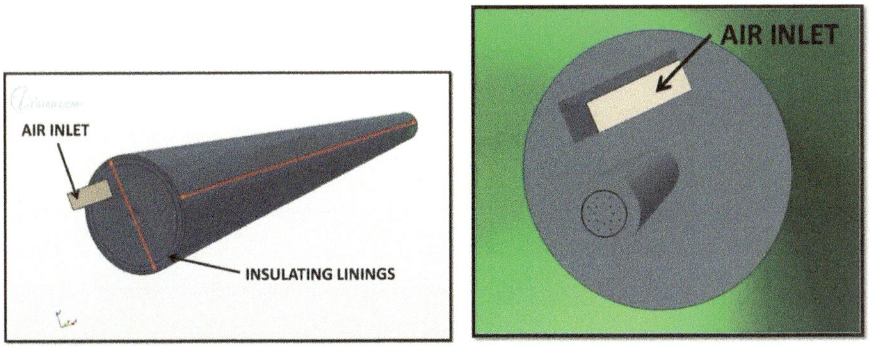

(a) Exterior view on the complete model showing air inlet and insulating lining.

(b) Interior view on the burner and air inlet region.

Figure 3 Geometry model of the kiln.

In the kiln hot gases are generated by a flame projected from a burner-pipe placed inside the freeboard. The setup of the burner is shown in Figure 3 and Figure 4. The burner injects fuel in axial and radial direction and is cooled by an amount of air forced through the cooling slots. The flow of cooling air through the rectangular air inlet breaks the circular symmetry of the configuration and requires a model in three spatial dimensions to be resolved.

The most important physical phenomenon that takes place in this burner region is the turbulent non-premixed combustion of the fuel injected from the burner with the secondary air. Combustion, even without turbulence, is an intrinsically complex process involving a large range of chemical time and length scales. Some of the chemical phenomena controlling flames take place in short times over thin layers and are associated with very large mass fractions, temperature and density gradients. The full description of chemical mechanisms in laminar flames may require hundreds of species and thousands of reactions leading to considerable numerical difficulties. Turbulence itself is probably the most complex phenomenon in non-reacting fluid mechanics. Various time and length scales are involved and the description of turbulence remains to date an open questions. The modeling of the kiln therefore requires resorting to a set of assumptions that are described in the remainder of this section.

3.1 The geometry

The geometry model of the kiln is shown in Figure 3. Figure 3(a) and Figure 3(b) give an exterior view on the complete kiln and a more detailed interior view of the burner region, respectively. The challenging aspect of this geometry is that the inlets of the burner are a factor thousand times smaller than the axial length of the kiln, imposing challenges in the mesh generation process.

3.2 Grid generation

The mesh model is shown in Figure 4. Figure 4(a), Figure 4(b) and Figure 4(c) give an exterior view of the mesh in the air inlet region, an interior view of the mesh and a detailed view of the mesh on the burner, respectively. We employed a *polyhedral mesh* of 2.8 million cells with local refinement in the critical inlets and burner regions. The main difficulties in meshing this geometry were found in balancing the accuracy required capture

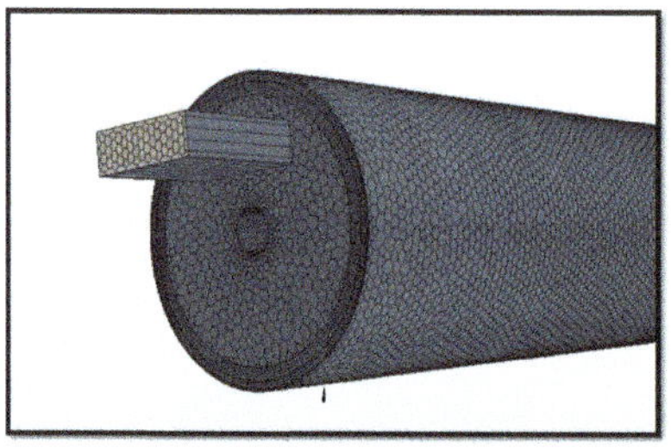

(a) Exterior view on the air inlet region.

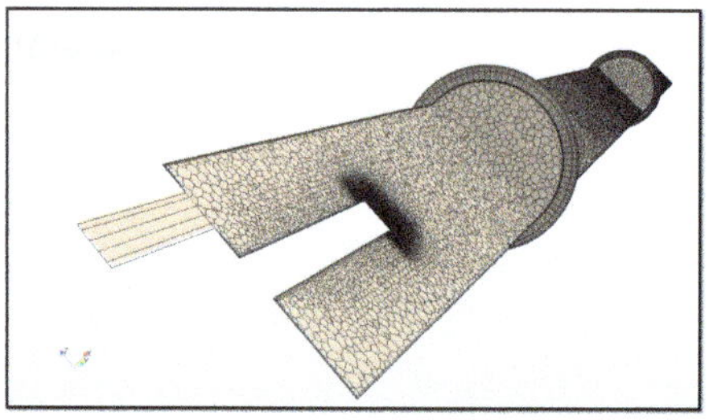

(b) Interior view showing local refinement in burner region.

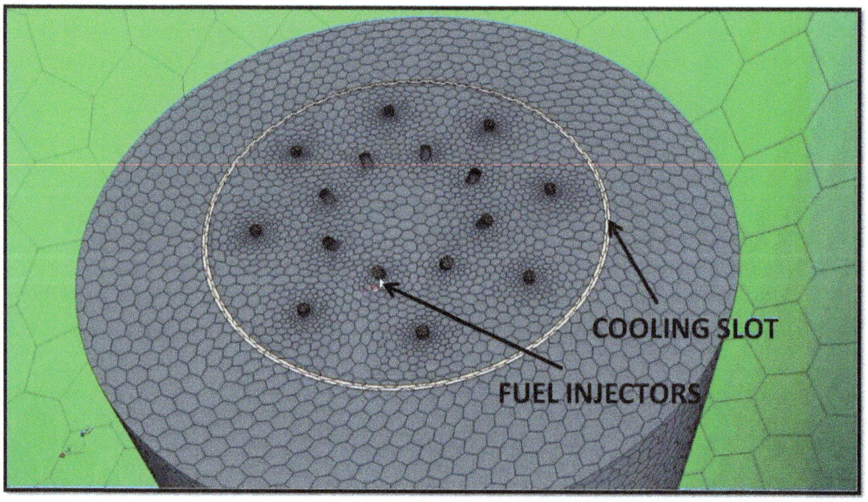

(c) Detailed view on the burner region showing fuel injectors and cooling slots.

Figure 4 Mesh model of the kiln.

the flow around small features in the burner with the overall computational cost. Polyhedral meshes provide a balanced solution in complex mesh generation problems of this kind.

Tetrahedra are the simplest type of volume elements. As their faces are plane segments, both face and volume centroid locations are well defined. A disadvantage is that tetrahedra cannot be stretched too much. To achieve a reasonable accuracy a much larger number of control volumes is needed than if structured meshes are used. Furthermore, as tetrahedral control volumes have only four neighbors, and computing gradients at cell centers using standard approximations can be problematic.

Polyhedra offer the same automatic meshing benefits as tetrahedra while overcoming their disadvantages. A major advantage of polyhedral cells is that they have many neighbors (typically of order 10) allowing gradients to be much better approximated. Obviously more neighbors implies more storage and computational operations per cell, but this is more than compensated by a higher accuracy. Polyhedral cells are also less sensitive to stretching than tetrahedra. A polyhedron with 12 faces for instance has six optimal directions which, together with the larger number of neighbors, leads to a more accurate solution with a lower cell count. Comparisons in many practical tests have verified that with polyhedral meshes, one needs about four times fewer cells, half the amount of memory and a tenth to a fifth of computing time compared to tetrahedral meshes to reach solutions of the same accuracy. In addition, solvers on polyhedral meshes were found to converge more robustly with respect to change in their parameters. A more detailed analysis of polyhedral meshes can be found in [6].

3.3 Governing reacting flow equations

In this section we present the conservation equations for reacting flows we used. The equations are derived from the Navier-Stokes (NS) equations by adding terms that account for reacting flows. The reacting gas is a non-isothermal mixture of multiple species which must be tracked individually. As heat capacities change significantly with temperature and composition, the transport coefficients require specific attention. In this subsection we will describe the Navier-Stokes and Reynolds-Averaged Navier-Stokes flow model, the non-realizable K-Epsilon turbulence model and the Standard Eddy Break Up combustion model. A more detailed derivation of these equations can be found in *e.g.* [7, 8].

Species are defined by their mass fraction defined as

$$Y_\ell = \frac{m_\ell}{m} \quad \text{for } \ell = 1:N, \tag{1}$$

where N is the number of species in the reacting mixture, m_ℓ the mass of species ℓ in a volume V and m the mass of gas in the volume, respectively. The conservation of mass can then be written as

NS: Conservation of mass

$$\frac{D\rho}{Dt} = \frac{\partial \rho}{\partial t} + \frac{\partial \rho u_i}{\partial x_i} = 0, \tag{2}$$

where $\rho = m/V$ is the density of the gas and u_i its three dimensional velocity field, respectively. The conservation of species ℓ for $\ell = 1:N$ can then be written as

NS: Conservation of species

$$\frac{\partial \rho Y_\ell}{\partial t} + \frac{\partial}{\partial x_i}\big(\rho(u_i + V_{\ell,i})Y_\ell\big) = \dot{\omega}_\ell, \tag{3}$$

where $V_{\ell,i}$ the ith component of the diffusion velocity V_ℓ of species ℓ and $\dot{\omega}_\ell$ the chemical reaction rate of species ℓ. The conservation of momentum for the gas can for $j = 1:3$ be expressed as:

NS: Conservation of momentum

$$\begin{aligned} \frac{\partial}{\partial t}(\rho u_j) + \frac{\partial}{\partial x_i}(\rho u_i u_j) &= \frac{\partial \sigma_{ij}}{\partial x_i} + \rho \sum_{\ell=1}^{N} Y_\ell f_{\ell,j} \\ &= -\frac{\partial p}{\partial x_j} + \frac{\partial \tau_{ij}}{\partial x_i} + \rho \sum_{\ell=1}^{N} Y_\ell f_{\ell,j}, \end{aligned} \tag{4}$$

where p denotes pressure and τ_{ij} and $f_{\ell,j}$ are the components of the Reynolds stress tensor and of the volume force acting on species ℓ, respectively. We will express the conservation of energy using the sensible enthalpy of the mixture h_s as independent variable. To introduce this quantify, we will first denote the enthalpy of species ℓ as h_ℓ. This quantify is the sum of the of a sensible and chemical part, *i.e.*,

$$h_\ell = h_{\ell,s} + \Delta h^0_{f,\ell}, \tag{5}$$

where the last term represents the enthalpy of formation of the species at a particular reference temperature T_0. The enthalpy of the mixture is then defined as a weighted average that again can be decomposed into a sensible and chemical part, as follows

$$h = \sum_{\ell=1}^{N} Y_\ell h_\ell = \sum_{\ell=1}^{N} Y_\ell h_{\ell,s} + \sum_{\ell=1}^{N} Y_\ell \Delta h_{f,\ell}^0$$
$$= h_s + \sum_{\ell=1}^{N} Y_\ell \Delta h_{f,\ell}^0. \tag{6}$$

We will denote the heat diffusion coefficient and the temperature by λ and T, respectively. The energy flux q_i is the sum a heat diffusion term derived from Fourier's Law and a term associated with the diffusion of species with different enthalpies, *i.e.*,

$$q_i = -\lambda \frac{\partial T}{\partial x_i} + \rho \sum_{\ell=1}^{N} h_\ell Y_\ell V_{\ell,i}. \tag{7}$$

In the energy conservation equation the following three source terms will play a role: the heat source due to radiative heat flux denoted as Q, the viscous heating term denoted as Φ, where $\Phi = \tau_{ij} \frac{\partial u_i}{\partial x_j}$ and heat release due to combustion denoted as $\dot{\omega}_T$, where

$$\dot{\omega}_T = -\sum_{\ell=1}^{N} \Delta h_{f,\ell}^0 \dot{\omega}_\ell. \tag{8}$$

The work done by the gas on the species can be expressed as $\rho \sum_{\ell=1}^{N} Y_\ell f_{\ell,i} V_{\ell,i}$. With all these quantities introduced, the conservation of energy in terms of h_s can be expressed as:

NS: Conservation of energy

$$\rho \frac{Dh_s}{Dt} = \dot{\omega}_T + \frac{Dp}{Dt} - \frac{\partial}{\partial x_i}\left(\lambda \frac{\partial T}{\partial x_i}\right) + \Phi + \dot{Q}$$
$$- \frac{\partial}{\partial x_i}\left(\rho \sum_{\ell=1}^{N} h_{s,\ell} Y_\ell V_{\ell,i}\right) + \rho \sum_{\ell=1}^{N} Y_\ell f_{\ell,i} V_{\ell,i}. \tag{9}$$

Turbulent combustion results from the two-way interaction between chemistry and turbulence. When a flame interacts with a turbulent flow, the combustion modifies the turbulence in two ways. The heat released induces high flow accelerations through the flame front and the temperature changes generate large changes in kinematic viscosity. These phenomena may either generate or damp turbulence and are referred to as flame-generated turbulence and relaminarization due to combustion, respectively. The turbulence conversely modifies the flame structure. This may either enhance the chemical reactions or completely inhibit it, leading to flame quenching. Compared to premixed flames, turbulent non-premixed flames exhibit some specific features that have to be taken into account. Non-premixed flames do not propagate as they localized on the fuel-oxidizer interface. This property is useful for safety purposes but it also has consequences on the chemistry-turbulence interaction. Without propagation speed, a non-premixed flame is unable to impose its own dynamics on the flow field and is therefore more sensitive to turbulence.

The description of the turbulent non-premixed combustion processes in a computational fluid dynamics model may be achieved using three levels of accuracy in the computations. Either a Reynolds Averaged Navier Stokes (RANS), a Large Eddy Simulations (LES) or a Direct Numerical Simulations (DNS) model can be adopted. In current engineering practice, the RANS model is extensively used because it is less demanding in terms of resources. Its validity however is limited by the closure models describing turbulence and combustion and the need for some form of callibration. Considering the complexities and the dimensions of our kiln, using the RANS model is the only feasible choice.

3.3.1 RANS model

In constant density flows, Reynolds averaging consist in splitting any quantity ξ in mean and fluctuating component ($\xi = \overline{\xi} + \xi'$). In variable density flow Favre [9] mass-weighted averages are usually preferred, *i.e.*, $\widetilde{f} = \frac{\overline{\rho f}}{\overline{\rho}}$. Any quantity f can therefore be split into:

$$f = \widetilde{f} + f'', \quad \text{where } \widetilde{f''} = 0.$$

The RANS equations derived from the reacting Navier-Stokes equation given above are then given by the equation for conservation of mass

RANS: Conservation of mass

$$\frac{\partial \overline{\rho}}{\partial t} + \frac{\partial \overline{\rho}\widetilde{u}_i}{\partial x_i} = 0, \tag{10}$$

the equation for conservation of species ℓ for $\ell = 1 : N$

RANS: Conservation of species

$$\frac{\partial}{\partial t}(\overline{\rho}\widetilde{Y}_\ell) + \frac{\partial}{\partial x_i}(\overline{\rho}\widetilde{u}_i\widetilde{Y}_\ell) = \overline{\dot{\omega}}_\ell - \frac{\partial}{\partial x_i}\left(\overline{V_{\ell,i}Y_\ell} + \overline{\rho}\widetilde{u_i''Y_\ell''}\right), \tag{11}$$

the equation for conservation of momentum for $j = 1 : 3$

RANS: Conservation of momentum

$$\frac{\partial}{\partial t}(\overline{\rho}\widetilde{u}_j) + \frac{\partial}{\partial x_i}(\overline{\rho}\widetilde{u}_i\widetilde{u}_j) + \frac{\partial \overline{p}}{\partial x_j} = \frac{\partial}{\partial x_i}\left(\overline{\tau}_{ij} - \overline{\rho}\widetilde{u_i''u_j''}\right), \tag{12}$$

and finally the equation for conservation of momentum

RANS: Conservation of energy

$$\begin{aligned}\frac{\partial}{\partial t}(\overline{\rho}\widetilde{h}_s) + \frac{\partial}{\partial x_i}(\overline{\rho}\widetilde{u}_i\widetilde{h}_s) = \overline{\dot{\omega}}_T + \overline{\frac{Dp}{Dt}} \\ + \frac{\partial}{\partial x_i}\left(\overline{\lambda\frac{\partial T}{\partial x_i}} - \overline{\rho}\widetilde{u_i''h_s''}\right) \\ + \overline{\Phi} - \frac{\partial}{\partial x_i}\left(\overline{\rho\sum_{k=1}^{N} h_{s,\ell}Y_\ell V_{\ell,i}}\right).\end{aligned} \tag{13}$$

The averaging procedure introduces unclosed quantities that have to be modeled. Without entering in the details we list here the two main unclosed terms that will be described in the next sections:

- Reynolds stresses: $\overline{\rho}\widetilde{u_i''u_j''}$.
- Species chemical reaction rates: $\overline{\dot{\omega}}_\ell$.

3.3.2 *Turbulence model*

Using the turbulence viscosity assumption by Boussinesq [10], the Reynolds stresses can be represented as

$$\begin{aligned}\overline{\rho u_i''u_j''} &= \overline{\rho}\widetilde{u_i''u_j''} \\ &= -\mu_t\left(\frac{\partial \widetilde{u}_i}{\partial x_j} + \frac{\partial \widetilde{u}_j}{\partial x_i} - \frac{2}{3}\delta_{ij}\frac{\partial \widetilde{u}_K}{\partial x_K}\right) + \frac{2}{3}\overline{\rho}k,\end{aligned} \tag{14}$$

where $\mu_t = \overline{\rho}\nu_t$ is the turbulent dynamic viscosity and δ_{ij} the Kronecker delta. The turbulent kinetic energy k in turn can be expressed as

$$k = \frac{1}{2}\sum_{j=1}^{3}\widetilde{u_j''u_j''}. \tag{15}$$

Modeling the turbulent viscosity μ_t is the central problem in turbulence computations. Many approaches exist. In this work we use a classical turbulence model developed for non-reacting flows, namely the *Realizable K-Epsilon model* [11]. Heat release effects on the Reynolds stresses are not explicitly taken into account in this approach and the turbulent viscosity is modeled as

$$\mu_t = \overline{\rho}C_\mu\frac{k^2}{\varepsilon}, \tag{16}$$

where ε is the rate of energy dissipation. In this model the critical coefficient C_μ is a function of mean flow and turbulence properties, rather than assumed to be constant as in the standard model. This allows to satisfy certain mathematical constraints on the normal stresses consistent with the physics of turbulence and is referred to as *realizability*.

From the Boussinesq relationship in Equation (14) and the eddy viscosity definition in Equation (16) it is possible to obtain the following expression for the normal Reynolds stress $\overline{u^2}$ in an incompressible strained mean flow U

$$\overline{u^2} = \frac{2}{3}k - 2\nu_t\frac{\partial U}{\partial x}, \tag{17}$$

where $\nu_t = \frac{\mu_t}{\rho}$. It can be shown that $\overline{u^2}$, which by definition is a positive quantity, become negative (*non-realizable*), when the strain is large enough to satisfy

$$\frac{k}{\varepsilon}\frac{\partial U}{\partial x} > \frac{1}{3C_\mu} \approx 3.7. \tag{18}$$

The easiest way to ensure the realizability is to make C_μ in (16) variable [12].

The critical coefficient C_μ can be expressed as a function of the mean strain and rotation rates, the angular velocity of the system rotation and the turbulence fields as follows

$$C_\mu = \frac{1}{A_0 + A_s \frac{kU^*}{\varepsilon}}, \tag{19}$$

where

$$U^* = \sqrt{S_{ij}S_{ij} + \tilde{\Omega}_{ij}\tilde{\Omega}_{ij}}, \tag{20}$$

$$\tilde{\Omega}_{ij} = \Omega_{ij} - 2\varepsilon_{ijk}\omega_k \quad \text{and} \quad \Omega_{ij} = \overline{\Omega}_{ij} - \varepsilon_{ijk}\omega_k, \tag{21}$$

where $\overline{\Omega}_{ij}$ is the mean rate of rotation tensor viewed in a rotating reference frame with the angular velocity ω_k. The parameters A_0 and A_s in (19) can be computed as

$$A_0 = 4.04, \qquad A_s = \sqrt{6}\cos\phi, \tag{22}$$

$$\phi = \frac{1}{3}\cos^{-1}(\sqrt{6}W), \tag{23}$$

$$W = \frac{S_{ij}S_{jk}S_{ki}}{\tilde{S}}, \qquad \tilde{S} = \sqrt{S_{ij}S_{ij}}, \tag{24}$$

$$S_{ij} = \frac{1}{2}\left(\frac{\partial u_j}{\partial x_i} + \frac{\partial u_i}{\partial x_j}\right). \tag{25}$$

The turbulent kinetic energy k and its dissipation rate ε in Equation (16) are described by the following two balance equations

$$\frac{\partial}{\partial t}(\overline{\rho}k) + \frac{\partial}{\partial x_i}(\overline{\rho}k\tilde{u}_i) = \frac{\partial}{\partial x_j}\left[\left(\mu + \frac{\mu_t}{\sigma_k}\right)\frac{\partial k}{\partial x_j}\right] + P_k + P_b - \rho\varepsilon - Y_M + S_k, \tag{26}$$

$$\frac{\partial}{\partial t}(\overline{\rho}\varepsilon) + \frac{\partial}{\partial x_i}(\overline{\rho}\varepsilon\tilde{u}_i) = \frac{\partial}{\partial x_j}\left[\left(\mu + \frac{\mu_t}{\sigma_\varepsilon}\right)\frac{\partial \varepsilon}{\partial x_j}\right] + C_{1\varepsilon}\frac{\varepsilon}{k}(P_k + C_{3\varepsilon}P_b) - C_{2\varepsilon}\rho\frac{\varepsilon^2}{k} + S_\varepsilon, \tag{27}$$

where P_k is the production term of turbulent kinetic energy due to the mean velocity gradients, P_b the production of turbulent kinetic energy due to buoyancy, Y_M the dilatation dissipation term that accounts for the contribution of the fluctuating dilatation in compressible turbulence to the overall dissipation rate, S_k and S_ε user defined source terms for turbulent kinetic energy and dissipation, and σ_k and σ_ε the turbulent Prandtl numbers for k and ε, respectively. $C_{1\varepsilon}$, $C_{2\varepsilon}$ and $C_{3\varepsilon}$ are model constants.

Another weakness of traditional K-Epsilon turbulence models is their modeling of the dissipation rate ε. Indeed, the well-known spreading (or dispersion) rate anomaly refers to the fact that traditional models do reasonably well in predicting the spreading rate of a planar jet but perform unexpectedly poor for rounds jets. This weakness can be traced back to a deficiency in traditional ε-equations. The realizable model proposed by Shih [11] was developed to repair this deficiency and addresses as such an issue that is of primary importance in our study.

3.3.3 Combustion model

The averaged equation for conservation of species (11) can be rewritten in compact form for $\ell = 1 : N$ as

$$\frac{\partial}{\partial t}(\overline{\rho}\widetilde{Y}_\ell) + \frac{\partial}{\partial x_i}(\overline{\rho}\widetilde{u}_i\widetilde{Y}_\ell) = -\nabla \cdot J_\ell + \overline{\dot{\omega}}_\ell, \tag{28}$$

where J_ℓ is the mass diffusion flux of species ℓ. The previous equation is solved in a CFD code for $N - 1$ species where N is the total number of fluid phase chemical species present in the system. Since the mass fraction of the species must sum to unity, the Nth mass fraction is determined as one minus the sum of the $N - 1$ solved mass fractions. To minimize numerical error, the Nth species should be selected as that species with the overall largest mass fraction.

In turbulent flows the mass diffusion flux is computed as

$$J_\ell = -\left(\overline{\rho}D_\ell + \frac{\mu_t}{Sc_t}\right), \tag{29}$$

where Sc_t is the turbulent Schmidt number and D_ℓ is the molecular diffusivity of species ℓ.

The species chemical reaction rate unclosed term $\dot{\omega}_\ell$ must be modeled with a combustion model. A combustion model describes the two-way interaction between properties of the turbulent flow produced by the flame and the chemical reactions. It serves to compute the reaction state space and the quantities it influences, namely the fluid density, viscosity, and temperature. It accounts for the processes that occur at length and time scales that we cannot resolve on a grid either in space or time due to limitations in computational resources. The choice of combustion model is decided by knowing the Damkohler number, defined as $Da = \frac{t_{\text{mix}}}{t_{\text{rxn}}}$, where t_{mix} is the mixing time scale and t_{rxn} is the reaction time scale. When the Damkohler number is very large, as in the case of the kiln, the reaction rate is controlled by the turbulent mixing that brings reactants together at the molecular scale. In this limit, the Standard Eddy Break Up (EBU) [13] model is fairly accurate because it assumes that the reaction occurs instantaneously upon micromixing.

The EBU combustion model tracks individual mean species concentrations on the grid through transport equations. The reaction rates used in these equations are calculated as functions of the mean species concentrations, turbulence characteristics and, depending on the specific model used, temperature. A mean enthalpy equation is solved in addition to the species transport equations. The mean temperature, density and viscosity are then calculated knowing the mean enthalpy and species concentrations. In the EBU used, the individual species in the global reaction are assumed to be transported at different rates according to their own governing equations.

The reaction rate is modeled through an expression that takes the turbulent micromixing process into account. This is done through dimensional arguments. Thus, for a reaction of the form

$$\nu_F F + \nu_O O \longrightarrow \nu_{P1}P_1 + \nu_{P2}P_2 + \cdots + \nu_{Pj}P_j, \tag{30}$$

where F stands for fuel, O oxidiser and P products of the reaction, the reaction rate for reaction ℓ is assumed to be

$$\overline{\omega}_\ell = \frac{\overline{\rho}}{M}\left(\frac{1}{\tau_{\mathrm{mix}}}\right)A_{\mathrm{ebu}}\min\left\{\overline{Y}_F, \frac{\overline{Y}_O}{s_O}, B_{\mathrm{ebu}}\left(\frac{\overline{Y}_{P1}}{s_{P1}} + \cdots + \frac{\overline{Y}_{Pj}}{s_{Pj}}\right)\right\}, \tag{31}$$

where $s_O = \frac{v_O M_O}{v_F M_F}$, $s_{Pj} = \frac{|v_{Pj}|M_{Pj}}{v_F M_F}$, v is the molar stoichiometric coefficient for species j in reaction ℓ, M is molecular weight of species. Equation (30) essentially states that the integrated micromixing rate is proportional to the mean (macroscopic) concentration of the limiting reactant divided by the time scale of the large eddies ($\frac{k}{\varepsilon} = \tau_{\mathrm{mix}}$). $\overline{Y}_F$, $\overline{Y}_O$, $\overline{Y}_P$ are respectively the mean concentrations of fuel, oxidizer, and products. A_{ebu} and B_{ebu} are model constants with typical values of 0.5 and 4.0 respectively. The values of these constants are fitted according to experimental results and they are suitable for most cases of general interest.

In our simulations we used a reduced combustion mechanism with 6 species and 4 reactions to account for a fuel that is a mixture of different alkanes. This mixture consists for 95% of CH_4 and for 5% of C_2H_6, C_3H_8 and C_4H_{10}.

The above models are discretized by a finite volume technique using second order upwinding for the convective terms [14–16]. The flow equations are solved in a segregated approach in which the SIMPLE algorithm realizes the velocity-pressure coupling. The energy equation is solved for the chemical thermal enthalpy using again a segregated approach. The temperature is computed from the enthalpy according to the equation of state. At each outer non-linear iteration the resulting linear systems are solved using an algebraic multigrid preconditioner for a suitable Krylov subspace acceleration [17].

3.4 Additional information

In situations in which the media separating hot walls is transparent for thermal radiation as in the case of dry air, radiation can only occur as a surface phenomenon. In the our case however, the gas in the freeboard of the kiln will absorb, emit and scatter the thermal radiation intensity emitted from the hot walls of the kiln. This process is governed by the radiative transfer equation (RTE) that is implemented in a Participating Media Radiation Model. The model is discretized in solid angle by the discrete ordinate method described in detail in [18, 19].

3.5 Software implementation

Simulations were performed using the STAR-CCM+ software suite [20] on a ten-nodes Linux cluster having Intel Duo and Quad Core processors at a clock speed between 2.20 GHz and 3.16 GHz running a Slackware 13 64-bit distribution. Iterating the three-dimensional combustion model to equilibrium state required between 3,500 and 4,000 non-linear iterations and between three and three and a half days of computation time.

4 Computational results

In this section we discuss how our combustion model validates the difference between two operating conditions of the oven. These operating conditions are listed in Table 1 and differ in the amount of preheated air that flows through the secondary air inlet shown in Figure 3. In the first and second line in Table 1 the ratio of volume air to volume of fuel

Table 1 Standard and new configuration of operating conditions of the kiln

Operating conditions	Air to fuel ratio
Standard	9
New	12

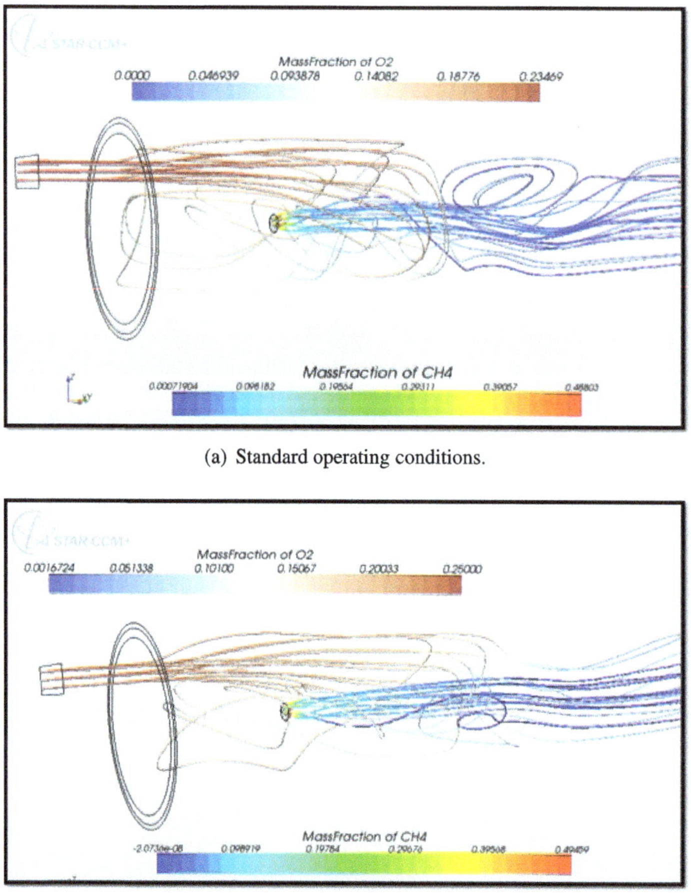

(a) Standard operating conditions.

(b) New operating conditions.

Figure 5 Effect of change in operating conditions on the streamlines of methane and oxigen in the kiln.

equals nine and twelve, respectively. The ratio of nine corresponds to the kiln's operation conditions that were commonly used before the start of this work and will be referred to as the *standard* operating conditions. The ratio of twelve correspond to the *new* operation conditions that we propose. In switching from the standard to the new configuration, the flux of fuel injected in axial and radial direction and the flux of air through the cooling slot is kept the same. The effect of this switch in operation conditions on the mass of oxygen and methane is shown in Figure 5. This figure shows that with in the new operating conditions the oxigen is transported further down in the kiln.

The objective of this section is to show how according to the combustion model described in the previous section the new configuration is less prone to the development of rings than the standard configuration. The two key arguments in this demonstration are

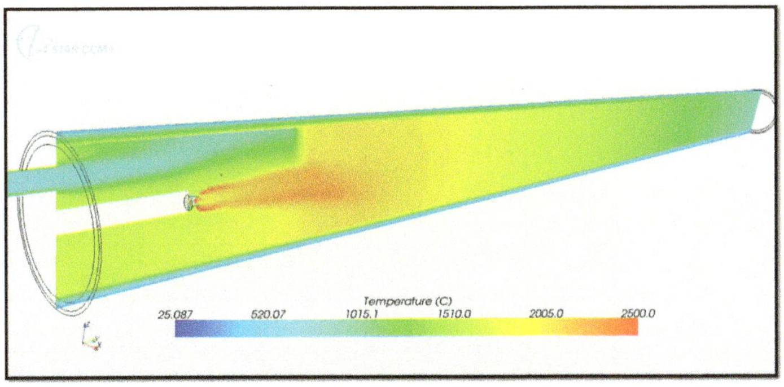

(a) Standard operating conditions.

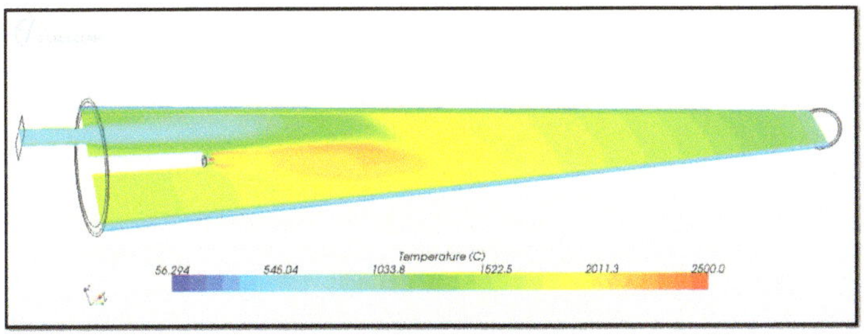

(b) New operating conditions.

Figure 6 Effect of change in operating conditions on the temperature on an axial section of the kiln through the air inlet.

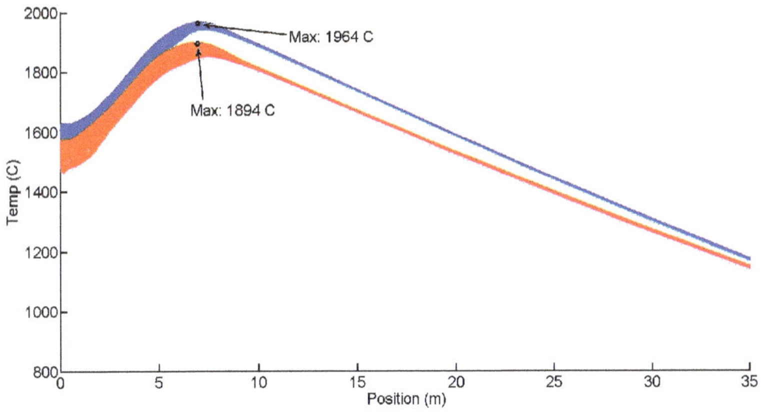

Figure 7 Temperature *vs.* position (blue standard operating conditions, red new operating conditions).

the temperature and radiative heat distribution on the inside wall of the kiln. These will be elaborated separately in the two subsections.

4.1 Effect on temperature

In Figure 6, Figure 7 and Figure 8 we show how the change in operating conditions affects the computed temperature profile. Figure 6 shows the temperature distribution on an axial section through the air inlet. Figure 7 shows the temperature evaluated at the centroid of each cell of the surface mesh on the inside wall versus the centroid's axial distance.

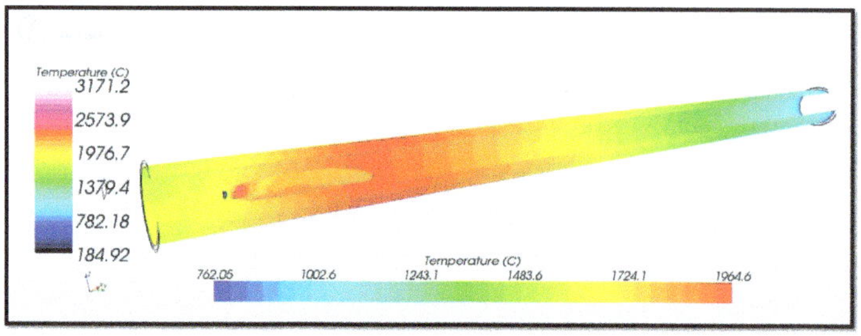

(a) Standard operating conditions.

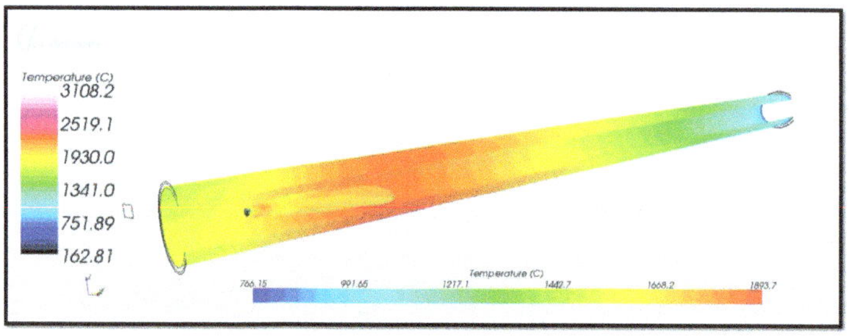

(b) New operating conditions.

Figure 8 Effect of change in operating conditions on the temperature on the inside wall of the kiln.

This wall forms the interface between the hot gas and the innermost part of the insulating lining. For positions closer to low end of the kiln the temperature values shown vary more with the circumferential angle due to the higher temperature gradients caused by the injected air. The lines in Figure 7 are therefore thicker for small values of the abscis. Figure 8 finally shows the temperature distribution on the inside wall of the kiln. In this figure the flame propagating from the burner is represented by an iso-surface that encloses a region in which the methane concentration is larger than or equal to 0.01. The vertical and horizontal color bars indicate the temperature of the flame and the wall, respectively.

Figure 6 clearly illustrates the effect of switching to the new operating conditions as it shows in the new conditions, the relatively cold preheated air is transported further into the kiln acting as a coolant on the wall. Figure 7 shows that this coolant reduces the wall peak temperature drops by 3.5% from 1,964 degrees Celcius to 1,894 degrees Celcius. It also shows that switching to the new operating conditions does not significantly alter the global distribution characteristics of the temperature which is of paramount importance in the material production process. Figure 8 is particularly interesting as it shows that in both configurations the peak in temperature is situated in a zone at four and halve to seven meters from the burner. This zone coincides with the zone in which rings are typically formed. It is therefore plausible that a reduction in peak temperature by switching to the new operating conditions will result in a reduction of the amount of liquid phase of the material being formed, and therefore impede the formation of rings. This hypothesis will be confirmed by looking into the computed radiative heat transfer distribution as we will do in the next subsection.

4.2 Effect on radiative heat transfer

In Figure 9, Figure 10 and Figure 11 and we show how the change in operating conditions affects the computed radiative heat transfer on the inside wall of the kiln. Figure 9

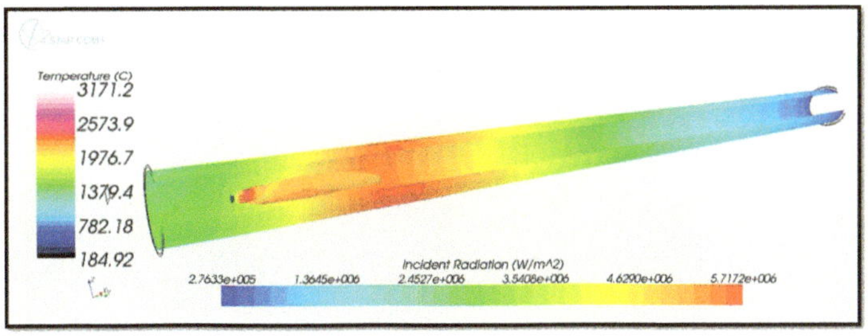

(a) Standard operating conditions.

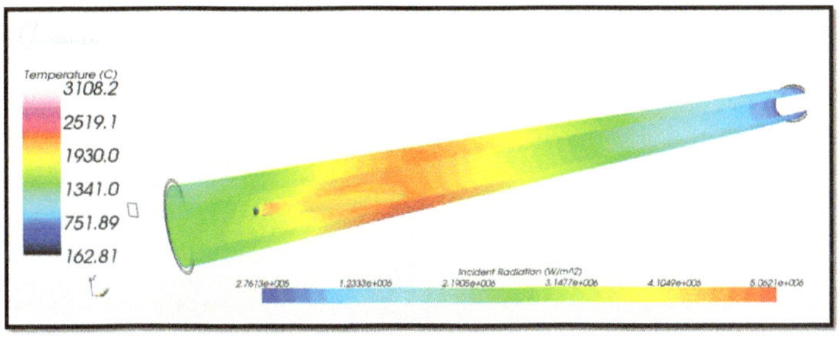

(b) New operating conditions.

Figure 9 Outside view on the effect of change in operating conditions on radiative heat transfer on the inside wall.

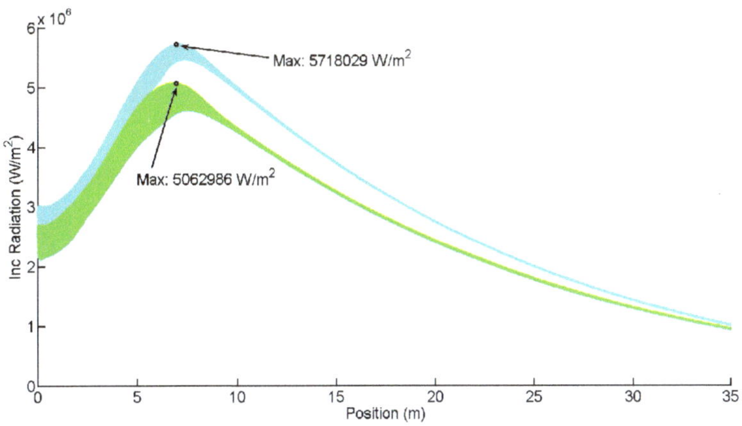

Figure 10 Incident radiation *vs.* position (blue standard operating conditions, red new conditions).

and Figure 10 are the equivalent for radiative heat transfer of Figure 7 and Figure 8 for the temperature, respectively. Figure 11 finally gives an view on the radiative heat transfer distribution from the inside of the kiln.

Figure 9 confirms that as expected changing to the new operating conditions results in a lowering of the peak in radiative heat transfer. Figure 10 shows that peak in the radiative heat transfer drops by 13.5% from 5,718,029 W/m^2 to 5,062,986 W/m^2 and that the region of high values is reduced in size. Figure 11 gives another representation of this fact. By switching to the new operating conditions, the material in the kiln is less likely to absorb an excessive amount of heat, effectively limiting the amount of liquid phase of the material and therefore the formation of rings.

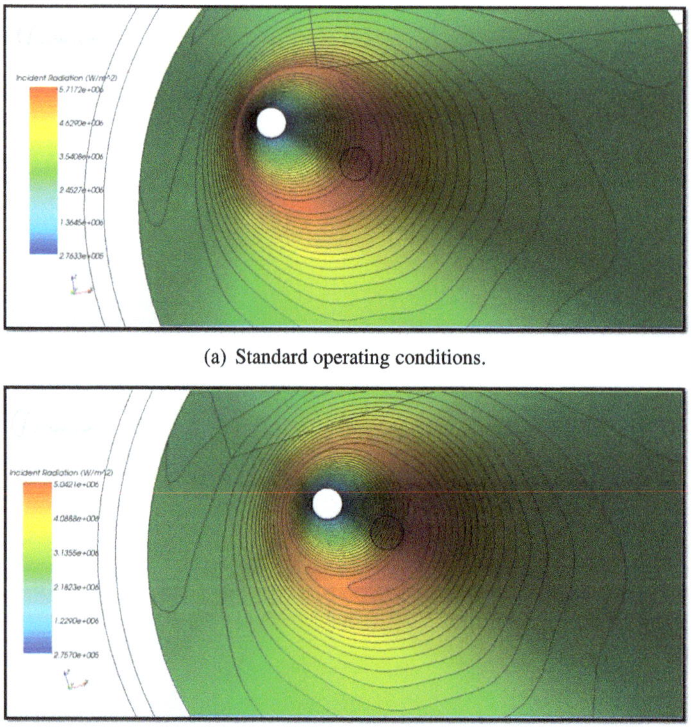

(a) Standard operating conditions.

(b) New operating conditions.

Figure 11 Inside view on the effect of change in operating conditions on radiative heat transfer on the inside wall.

5 Experimental validation at the plant

The change in operating conditions we propose was experimentally validated at the plant for the first time on August 28th, 2011. That day the formation of a massive ring was reported. After assessing the situation, it was decided to change to the new operating conditions. The effect of this decision over time is shown in Figure 12. Four hours after the decision was taken, it was observed that the ring stopped growing in size as shown in Figure 12(a). After twenty-four hours, the ring started to break into lumps due to the vibration of the drive gears as shown in Figure 12(b). These lumps are transported through the kiln and leave it at the lower end by the rotary motion. After forty hours, the kiln is clean and in stable operating conditions as shown in Figure 12(c). Careful analysis has revealed that with the new operating conditions, the final cement product still meets the stringent quality specifications. After August 28th, 2011 the benefits of the changing to the new operating conditions has been repeatedly observed when maintenance of other circumstances required switching back to the standard operating conditions.

6 Conclusions

We developed a numerical model allowing to access the effectiveness of measures implemented to counteract the formation of rings in a rotary cement kiln in use by Almatis B.V. in Rotterdam. In this three-dimensional combustion model, the gas flow, the temperature profile, radiative heat distribution and the concentration of hydro-carbon species in the kiln is taken into account. Simulations show that deluting the air-fuel mixture with air reduces peaks in radiative heat transfer in zones critical to ring formation. This reduc-

(a) After four hours.

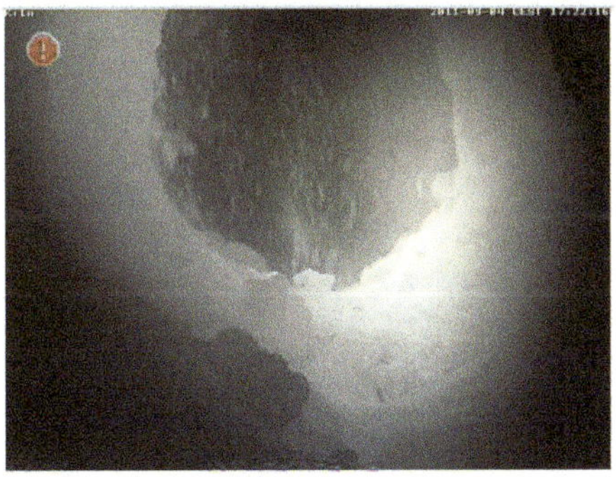

(b) After twenty-four hours.

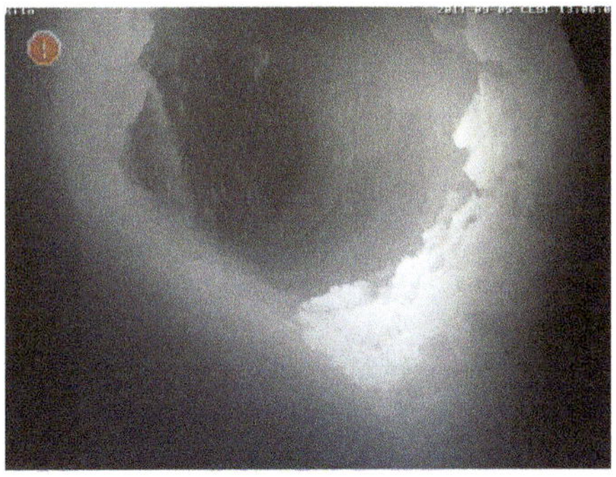

(c) After forty hours.

Figure 12 Experimentally observed time evolution of the ring after the new operating conditions have been imposed.

tion results in turn in less heat being absorbed by the granular material bed, effectively reducing the amount of material liquid phase prone to sticking to the kiln's surface and to forming rings. The validity of our model has been experimental observed at the Almatis plant in Rotterdam. Since August 28th, 2011, the kiln has been operation without unscheduled shut-downs, resulting in hugely important cost savings.

Competing interests

The authors declare that they have no competing interests.

Authors' contributions

MP developed and validated the CFD model. RS provided supervision from within the industrial partner. DL provided supervision from within the academic partner.

Author details

[1]Scientific Computing Group, Delft Institute of Applied Mathematics, Faculty of Electrical Engineering, Mathematics and Computer Science, Delft University of Technology, Delft, The Netherlands. [2]Almatis B.V., Theemsweg 30, Botlek, Rotterdam, 3197 KM, The Netherlands.

Acknowledgements

We thank Marco Talice for his advice worth more than platina and the CD-ADAPCO London office for the support in using their software.

References

1. Boateng AA: *Rotary Kilns: Transport Phenomena and Transport Processes*. Stoneham: Butterworth-Heinemann; 2006.
2. Mastorakos E, Massias A, Tsakiroglou CD, Goussis DA, Burganos VN, Payatakes AC: **CFD predictions for cement kilns including flame modelling, heat transfer and clinker chemistry**. *Appl Math Model* 1999, **23**(1):55-76.
3. Marias F, Roustan H, Pichat A: **Modelling of a rotary kiln for the pyrolysis of aluminium waste**. *Chem Eng Sci* 2005, **60**(16):4609-4622.
4. Giannopoulos D, Kolaitis DI, Togkalidou A, Skevis G, Founti MA: **Quantification of emissions from the co-incineration of cutting oil emulsions in cement plants: part i: NO_x, CO and VOC**. *Fuel* 2007, **86**(7-8):1144-1152.
5. Tran HN, Barham D: **An overview of ring formation in lime kilns**. *Tappi J* 1991, **74**(1):131-136.
6. Peric M: Flow simulation using control volumes of arbitrary polyhedral shape. *ERCOFTAC Bulletin* September 2004, **62**.
7. Williams FA: *Combustion Theory*. 2nd edition. New York: Westview Press; 1994.
8. Kuo KK: *Principles of Combustion*. 2nd edition. New York: Wiley-Interscience; 2005.
9. Poinsot T, Veynante D: *Theoretical and Numerical Combustion*. 2nd edition. Ann Arbor: Edwards; 2005.
10. Hinze JO: *Turbulence*. New York: McGraw-Hill; 1975.
11. Shih TH, Liou WW, Shabbir A, Zhu J: **A new k-ϵ Eddy-viscosity model for high Reynolds number turbulent flows - model development and validation**. *Comput Fluids* 1995, **24**(3):227-238.
12. Reynolds WC: *Fundamentals of Turbulence for Turbulence Modeling and Simulation*; 1987. [*Lecture Notes for Von Karman Institute*.] (Agard Report No. 755.)
13. Magnussen BF, Mjertager BH: **On mathematical modeling of turbulent combustion**. In *16th Symp. (Int.) on Combustion*. Pittsburgh: The Combustion Institute; 1976: 719-727.
14. Demirdzic I, Peric M: **Finite volume method for prediction of fluid flow in arbitrarily shaped domains with moving boundaries**. *Int J Numer Methods Fluids* 1990, **10**(7):771-790.
15. Mathur SR, Murthy JY: **A pressure-based method for ustructured meshes**. *Numer Heat Transf, Part B, Fundam* 1997, **31**(2):195-215.
16. Murthy JY, Mathur SR: **Radiative heat transfer in axisymmetric geometries using an unstructured finite-volume method**. *Numer Heat Transf, Part B, Fundam* 1998, **33**(4):397-416.
17. Trottenberg U, Oosterlee C, Schüller A: *Multigrid*. San Diego: Academic Press; 2001.
18. Modest MF: *Radiative Heat Transfer*. New York: McGraw-Hill; 1993.
19. Siegel R, Howell JR: *Thermal Radiation Heat Transfer*. 3rd edition. Washington: Hemisphere; 1992.
20. **Star-CCM+ Version 6.06.011 user guide**; 2011.

Permissions

All chapters in this book were first published in JMI, by Springer; hereby published with permission under the Creative Commons Attribution License or equivalent. Every chapter published in this book has been scrutinized by our experts. Their significance has been extensively debated. The topics covered herein carry significant findings which will fuel the growth of the discipline. They may even be implemented as practical applications or may be referred to as a beginning point for another development.

The contributors of this book come from diverse backgrounds, making this book a truly international effort. This book will bring forth new frontiers with its revolutionizing research information and detailed analysis of the nascent developments around the world.

We would like to thank all the contributing authors for lending their expertise to make the book truly unique. They have played a crucial role in the development of this book. Without their invaluable contributions this book wouldn't have been possible. They have made vital efforts to compile up to date information on the varied aspects of this subject to make this book a valuable addition to the collection of many professionals and students.

This book was conceptualized with the vision of imparting up-to-date information and advanced data in this field. To ensure the same, a matchless editorial board was set up. Every individual on the board went through rigorous rounds of assessment to prove their worth. After which they invested a large part of their time researching and compiling the most relevant data for our readers.

The editorial board has been involved in producing this book since its inception. They have spent rigorous hours researching and exploring the diverse topics which have resulted in the successful publishing of this book. They have passed on their knowledge of decades through this book. To expedite this challenging task, the publisher supported the team at every step. A small team of assistant editors was also appointed to further simplify the editing procedure and attain best results for the readers.

Apart from the editorial board, the designing team has also invested a significant amount of their time in understanding the subject and creating the most relevant covers. They scrutinized every image to scout for the most suitable representation of the subject and create an appropriate cover for the book.

The publishing team has been an ardent support to the editorial, designing and production team. Their endless efforts to recruit the best for this project, has resulted in the accomplishment of this book. They are a veteran in the field of academics and their pool of knowledge is as vast as their experience in printing. Their expertise and guidance has proved useful at every step. Their uncompromising quality standards have made this book an exceptional effort. Their encouragement from time to time has been an inspiration for everyone.

The publisher and the editorial board hope that this book will prove to be a valuable piece of knowledge for researchers, students, practitioners and scholars across the globe.

List of Contributors

Bruno Figliuzzi
Centre de Morphologie Mathématique, École des Mines ParisTech, 77300, Fontainebleau, France

Dominique Jeulin
Centre de Morphologie Mathématique, École des Mines ParisTech, 77300, Fontainebleau, France

Anaël Lemaître
UMR Navier, 2 allée Kepler, 77420, Champs-sur-Marne, France

Gabriel Fricout
ArcelorMittal Global R&D, 57283, Maizières-lès-Metz Cedex, France

Jean-Jacques Piezanowski
ArcelorMittal Global R&D, 57283, Maizières-lès-Metz Cedex, France

Paul Manneville
LadHyX, École Polytechnique, 91128, Palaiseau, France

Nicodemus Banagaaya
Department of Mathematics and Computer Science, Eindhoven University of Technology, Eindhoven, The Netherlands

Wil Schilders
Department of Mathematics and Computer Science, Eindhoven University of Technology, Eindhoven, The Netherlands

Corien Prins
Centre for Analysis, Scientific computing and Applications (CASA), Technische Universiteit Eindhoven, P.O. Box 513, 5600 MB Eindhoven, The Netherlands

Jan ten Thije Boonkkamp
Centre for Analysis, Scientific computing and Applications (CASA), Technische Universiteit Eindhoven, P.O. Box 513, 5600 MB Eindhoven, The Netherlands

Teus Tukker
Philips Research, High Tech Campus 44, 5656 AE Eindhoven, The Netherlands

Wilbert IJzerman
Philips Lighting, High Tech Campus 44, 5656 AE Eindhoven, The Netherlands

Tor Gillberg
Simula Research Laboratory, Martin Linges vei, Oslo, Norway
Kalkulo AS, Martin Linges vei, Oslo, Norway

Are Magnus Bruaset
Simula Research Laboratory, Martin Linges vei, Oslo, Norway

Øyvind Hjelle
Kalkulo AS, Martin Linges vei, Oslo, Norway

Mohammed Sourouri
Simula Research Laboratory, Martin Linges vei, Oslo, Norway

Florian Augustin
Department of Mathematics, Technische Universität München, Boltzmannstr. 3, Garching, 85748, Germany

Albert Gilg
Corporate Technology, Siemens AG, Otto-Hahn-Ring 6, Munich, 81730, Germany

Meinhard Paffrath
Corporate Technology, Siemens AG, Otto-Hahn-Ring 6, Munich, 81730, Germany

Peter Rentrop
Department of Mathematics, Technische Universität München, Boltzmannstr. 3, Garching, 85748, Germany

Manuel Villegas
Department of Mathematics, Technische Universität München, Boltzmannstr. 3, Garching, 85748, Germany

Utz Wever
Corporate Technology, Siemens AG, Otto-Hahn-Ring 6, Munich, 81730, Germany

Aurelio Mauri
Process R & D, Micron Technology, via Olivetti 2, Agrate Brianza, MB 20864, Italy

Andrea Bortolossi
IBM Italia, Via Circonvallazione Idroscalo, Segrate, MI 20090, Italy

Giovanni Novielli
STMicroelectronics, via Olivetti 2, Agrate Brianza, MB 20864, Italy

Riccardo Sacco
Dipartimento di Matematica, Politecnico di Milano, Piazza L. da Vinci 32, Milano, 20133, Italy

Elisabetta Felaco
Department Mathematik, Universität Hamburg, Bundesstraße 55, Hamburg, 20146, Germany

Ingenuin Gasser
Department Mathematik, Universität Hamburg, Bundesstraße 55, Hamburg, 20146, Germany

Dominic Amann
Institut für Numerische Mathematik, Technische Universität Graz, Steyrergasse 30, Graz, 8010, Austria

Andreas Blaszczyk
Corporate Research, ABB Switzerland Ltd., Baden-Dättwil, 5405, Switzerland

Günther Of
Institut für Numerische Mathematik, Technische Universität Graz, Steyrergasse 30, Graz, 8010, Austria

Olaf Steinbach
Institut für Numerische Mathematik, Technische Universität Graz, Steyrergasse 30, Graz, 8010, Austria

Carsten Othmer
Vehicle Technology, Group Research, Volkswagen AG, 38436 Wolfsburg, Germany

Jenny Niebsch
Austrian Academy of Sciences, RICAM, Altenbergerstr. 69, Linz, 4040, Austria Johannes Kepler University Linz, Altenbergerstr. 69, Linz, 4040, Austria

Ronny Ramlau
Austrian Academy of Sciences, RICAM, Altenbergerstr. 69, Linz, 4040, Austria Johannes Kepler University Linz, Altenbergerstr. 69, Linz, 4040, Austria

Jari Hämäläinen
Centre of Computational Engineering and Integrated Design (CEID) and Department of Mathematics and Physics, Lappeenranta University of Technology, P.O. Box 20, Lappeenranta, FI-53851, Finland

Martin Grothaus
Technische Universität Kaiserslautern, Gottlieb-Daimler Str., Kaiserslautern, 67663, Germany

Axel Klar
Technische Universität Kaiserslautern, Gottlieb-Daimler Str., Kaiserslautern, 67663, Germany Fraunhofer-Institut für Techno- und Wirtschaftsmathematik, Fraunhofer Platz 1, Kaiserslautern, 67663, Germany

Johannes Maringer
Technische Universität Kaiserslautern, Gottlieb-Daimler Str., Kaiserslautern, 67663, Germany

Patrik Stilgenbauer
Technische Universität Kaiserslautern, Gottlieb-Daimler Str., Kaiserslautern, 67663, Germany

Raimund Wegener
Fraunhofer-Institut für Techno- und Wirtschaftsmathematik, Fraunhofer Platz 1, Kaiserslautern, 67663, Germany

M Pisaroni
Scientific Computing Group, Delft Institute of Applied Mathematics, Faculty of Electrical Engineering, Mathematics and Computer Science, Delft University of Technology, Delft, The Netherlands

R Sadi
Almatis B.V., Theemsweg 30, Botlek, Rotterdam, 3197 KM, The Netherlands

D Lahaye
Scientific Computing Group, Delft Institute of Applied Mathematics, Faculty of Electrical Engineering, Mathematics and Computer Science, Delft University of Technology, Delft, The Netherlands